Hypertext 3.0

Parallax  Re-visions of
Culture and Society

Stephen G. Nichols, Gerald Prince, and
Wendy Steiner, Series Editors

Hypertext 3.0

Critical Theory and
New Media in an Era of
Globalization

George P. Landow

The Johns Hopkins University Press
Baltimore

© 1992, 1997, 2006 The Johns Hopkins University Press
All rights reserved. Published 2006
Printed in the United States of America
9 8 7 6 5 4 3 2 1

The Johns Hopkins University Press
2715 North Charles Street
Baltimore, Maryland 21218-4363
www.press.jhu.edu

Library of Congress Cataloging-in-Publication Data

Landow, George P.
    Hypertext 3.0 : critical theory and new media in an era of globalization / George P.
Landow.— [3rd ed.]
        p. cm. — (Parallax)
    Rev. ed. of: Hypertext 2.0. 1997.
    Includes bibliographical references and index.
    ISBN 0-8018-8256-7 (hardcover : alk. paper) — ISBN 0-8018-8257-5 (pbk. : alk. paper)
    1. Criticism. 2. Literature and technology. 3. Hypertext systems. I. Title: Hypertext
three point zero. II. Landow, George P. Hypertext 2.0 III. Title. IV. Parallax (Baltimore, Md.)
    PN81.L28 2006
    801'.95—dc22   2005007788

A catalog record for this book is available from the British Library.

For Ruth—always 1.0

Contents

Preface

Why Hypertext 3.0? When I wrote *Hypertext 2.0* in 1997, the need was obvious: developments in hardware and software since the appearance of the first version led me to remove most references to Intermedia, replacing them with discussions of the World Wide Web and other hypermedia systems (Storyspace, Microcosm, CD-ROM proprietary environments). In addition, I added a chapter on writing for e-space, included examples from new hypertext fiction, and so on. Since the appearance of *Hypertext 2.0*, several developments have occurred that again led to the need for a new version. These changes include (1) the enormous growth of the Web and its use in literary, business, and political applications; (2) the development of Weblogs, or blogs, as a widely available form of read-write hypertext—the first widely available Web mode that begins to approach the vision of the first hypertext theorists; (3) the rapid growth of interest in animated text, using Flash, now that enough Web users have broadband access to make such large files practicable in Internet applications; (4) the increasing importance of our understanding of postcoloniality and globalization; and (5) some first steps toward a theory of digital cinema (*Hypertext 2.0* briefly discussed this topic with emphasis on examples rather than on their theoretical implications.) To the earlier discussions of the convergence of hypertext, critical theory, and editorial theory, I also propose to consider two additional possible points of convergence, postcolonial culture and interactive cinema.

Perhaps most important, given the optimistic and even celebratory tone taken by most writers on hypermedia, has been the notorious dotcom bust, which Vincent Mosco has so effectively described in terms of its relations to cyberspace as a cultural myth. In *The Digital Sublime: Myth, Power, and*

Cyberspace (2004), he explains "the extraordinary boom-and-bust cycle" (6) of the 1990s by placing it in the context of mythic notions of cyberspace promulgated by those like Nicholas Negroponte and other proponents of "technomania" (21). According to Mosco, cyberspace functioned as one of those cultural myths that provide "stories that animate individuals and societies by providing paths to transcendence that lift people out of the banality of everyday life. They offer another reality, a reality once characterized by the promise of the sublime" (3).

Convinced by the demise of the Cold War and the magic of a new technology, people accepted the view that history as we once knew it was ending and that, along with the end of politics as we knew it, there would be an end to the laws promulgated by that most dismal of sciences, economics. Constraints once imposed by scarcities of resources, labor, and capital would end, or at least loosen significantly, and a new economics of cyberspace (a "network economics") would make it easier for societies to grow and, especially, to grow rich . . . What made the dotcom boom a myth was not that it was false but that it was alive, sustained by the collective belief that cyberspace was opening a new world by transcending what we once knew about space, time, and economics. (4)

These "myths," Mosco argues, point "to an intense longing for a promised community, a public democracy" (15). Like all myths, they make "socially and intellectually tolerable what would otherwise be experienced as incoherence" (29) and otherwise shield people from political and economic realities (31) because they "mask the continuities that make the power we observe today, for example in the global market and in globe-spanning companies like Microsoft and IBM." Cyberspace myths, which purport to lead us to a golden future in which geography and history end, create "amnesia about old politics and older myths" (83). Mosco's solution involves valuably reminding us that during the past two centuries almost every major development of technology—electricity, telegraph, telephone, radio, television, cable television, and so on—brought with it similar mythic claims.

One of the few weaknesses in his convincing, if limited, analysis lies in the fact that it so emphasizes myth as a social construction born from a community's need that it never inquires if any of these myths about cyberspace proved to have roots in fact. Thus, although he several times assures the reader that "myths are not true or false, but living or dead" (3), in practice he always acts as if all statements about cyberspace are false. Unlike William J. Mitchell in *Me++: The Cyborg Self and the Networked City* (2003), Mosco never inquires if *some* of the claims about location-independent work, business applications of the Internet, or hypermedia in education proved correct. After all, a great many computer-related enterprises—educational, artistic, and commercial—continue to thrive.

In fact, since I wrote the first version of *Hypertext* the situation of computing in humanities, arts, and culture has changed dramatically. When I first tried to explain the nature and possibilities of hypermedia, most of my readers had little contact with computing, but that had changed by *Hypertext 2.0.* The situation has now changed dramatically once more, and a book like this one now finds itself situated very differently within our culture, particularly the humanities, than was the case only a short time ago. For example, when I first explained the characteristics of a document within a hypertext environment, contrasting it to a page of print, I had to describe and explain three things: (1) how one used a computer—even how one used a mouse and drop-down menus; (2) the basic effects of digital information technology; and (3) the characteristic qualities and experience of hypertext itself. Such is no longer necessary, and such is no longer adequate. It is not simply a matter that many of you have become skillful users of e-mail, discussion lists, Google, and the World Wide Web. Equally important, you have experienced numerous digital applications, genres, and media that do not take the specific form of hypertext. Some of these, such as Weblogs, show a important relation to hypermedia, but others, like computer games, have only a few points of convergence with it. Still others of increasing economic, educational, and cultural importance, such as animated text, text presented in PDF (portable document format) format, and streaming sound and video, go in very different directions, often producing effects that fundamentally differ from hypermedia.

Let me emphasize here that I do not propose to evaluate nonhypertextual developments of digital information technology according to the degree to which they resemble hypertext and hypermedia. I am also not interested in presenting hypermedia as an overarching umbrella concept under which to gather all other digital forms. I shall, however, compare these other kinds of digitality to hypermedia on the assumption that doing so will help us better understand characteristic effects and applications of all these new media.

The situation—in particular, the academic standing and fashionability—of poststructuralism has also changed markedly since the first version of *Hypertext,* though in a way perhaps opposite to that of hypermedia. Whereas hypertext and other forms of digital media have experienced enormous growth, poststructuralism and other forms of critical theory have lost their centrality for almost everyone, it seems, but theorists of new media. One might claim to see a parallel between the dotcom bust and the general loss of academic standing by critical theory, but websites, blogs, discussion lists, and new media arts flourish despite the bankruptcy of many ill-conceived computer-related businesses, some of which never managed to produce anything more than vaporware.

I don't believe this change in situation lessens the value of one of the main approaches of this book, its use of hypertext and late-twentieth-century critical theory to illuminate each other. As I stated repeatedly in the earlier versions of this book, the writings of Roland Barthes, Jacques Derrida, and other critical theorists neither caused the development of hypermedia nor coincided exactly with it. Nonetheless, their approach to textuality remains very helpful in understanding our experience of hypermedia. And vice versa. I have had many students in my hypertext and literary theory class who have told me that they found the writings of Barthes, Derrida, Michel Foucault, and Gilles Deleuze and Félix Guattari easier to understand after the experience of reading and writing hypertexts. Others have agreed that these theorists, particularly Derrida and Barthes, provide useful ways to think about hypertext.

Perhaps the single most important development in the world of hypermedia has been the steady development of read-write systems—of the kind of systems, in other words, that the pioneering theorists Vannevar Bush and Theodor H. Nelson envisioned. Blogs, wikis, and the Portal Maximizer by Active Navigation all represent attempts to bring to the Web the features found in hypertext software of the 1980s that made readers into authors.

Acknowledgments

Because my first acquaintance with the idea of hypertext goes back to 1986 or 1987, when members of Brown University's long-vanished Institute for Research in Information and Scholarship (IRIS) recruited me to join the Intermedia project, I owe special thanks to its founding director, William G. Shipp, to its later co-directors, Norman K. Meyrowitz and Marty J. Michel, and to my friend and colleague Paul Kahn, who was project coordinator during the creation of *The Dickens Web* and later Intermedia projects and who served as the institute's final director. Nicole Yankelovich, IRIS project coordinator during the initial development and application stages of Intermedia, always proved enormously resourceful, helpful, and good humored even in periods of crisis, as did Julie Launhardt, assistant project coordinator. In the final years of the project, the late James H. Coombs, who created many of the key parts of the second stage of Intermedia, provided invaluable assistance.

Jay Bolter enticed me into using Storyspace, and I am most grateful to him, Michael Joyce, and Mark Bernstein of Eastgate Systems for their continuing assistance.

I owe an especial debt to my enthusiastic and talented graduate and undergraduate research assistants between 1987 and 1992, particularly Randall Bass, David C. Cody, Shoshana M. Landow, Jan Lanestedt, Ho Lin, David Stevenson, Kathryn Stockton, Gary Weissman, Gene Yu, and Marc Zbyszynski. My students at Brown University, the University Scholars Program at the National University of Singapore (NUS), and the Faculty of Computer Science at NUS have provided a continual source of inspiration and delight.

The development of Intermedia was funded in part by grants and contracts from International Business Machines, Apple Computer, and the

ACKNOWLEDGMENTS Annenberg/Corporation for Public Broadcasting Project, and I am grateful to them for their support. A Mellon Foundation grant and one from Dr. Frank Rothman, the provost of Brown University, enabled me to transfer the Intermedia materials created for English and creative writing courses into Storyspace. The generosity of Daniel Russell, then of Apple Computers, made it much easier for me to carry out my research in the 1990s after the closing of IRIS, when my university found itself able to offer little assistance or encouragement.

Since 2000 NUS has funded the web servers in New York and Southeast Asia on which reside the most recent descendants of materials originally created in Intermedia and Storyspace—The *Victorian, Postcolonial Literature and Culture,* and *Cyberspace, Hypertext, and Critical Theory* sites—and in 2001–2 NUS funded postdoctoral fellows and senior research fellows, who created materials for the sites, including Philip V. Allingham, Marjorie Bloy, Leong Yew, Tamara S. Wagner, and John van Whye. I also have to thank the hundreds of international contributors, particularly Philip V. Allingham, contributing editor of the *Victorian Web,* who have shared so many thousands of documents with readers of these sites. I would like to thank Peyton Skipwith of the Fine Art Society, London, and Peter Nahum for generously granting permission to include the images and text from their catalogues, thus permitting me to create the *Victorian Web*'s sections on painting and the decorative arts. I am especially grateful to the authors of two dozen out-of-print scholarly books and contributors of many other Victorian texts who have generously shared their work with *Victorian Web,* thus making possible the *Victorian Web* Books section that explores what is happening to the forms of humanistic scholarship in a digital age. Thanks, too, to the readers of my websites who were responsible for their receiving 17 million hits/page views in March 2002 (95% of them for the *Victorian Web*). Aloysius Tay Wee Kok, head of information technology at the University Scholars Program, and his crew of technicians have set up and maintained the servers in both the United States and Singapore with the assistance of Joseph Aulisi of Macktez.com.

I also owe a debt of gratitude to many colleagues and students who shared their work with me: Mark Amerika, J. David Bolter, Alberto Cecchi, Robert Coover, Daniela Danielle, Cicero da Silva, Jay Dillemuth, Carolyn Guyer, Terence Harpold, Paul Kahn, Robert Kendall, David Kolb, Deena Larson, Gary Marchionini, Stuart Moulthrop, and Marc Nanard kindly provided me with draft, prepublication, or prerelease versions of their work; and Cambridge University Press, Dynamic Diagrams, Eastgate Systems, MetaDesign West,

ACKNOWLEDGMENTS PWS Publishing, Oxford University Press, Routledge, and Voyager have provided published versions of their electronic publications.

I would also like to thank for their advice, assistance, and encouragement Irina Aristarkhova, David Balcom, Bruno Bassi, Gui Bonsiepe, George Bornstein, Katell Briatte, Leslie Carr, Laura Borràs Castanyer, Hugh Davis, Marilyn Deegan, Emanuela del Monaco, Jacques Derrida, Umberto Eco, Markku Eskelinen, Susan Farrell, Niels Ole Finnemann, Patrizia Ghislandi, Antoni J. Gomez-Bosquet, Diane Greco, Robert Grudin, Anna Gunder, Wendy Hall, E. W. B. Hess-Littich, Elaine Yee Lin Ho, Raine Koskimaa, Jean-Louis Lebrave, José Lebrero, Michael Ledgerwood, Gunnar Liestøl, Peter Lunenfeld, Cathy Marshall, Graham McCulloch, Bernard Mcguirk, Tom Meyer, J. Hillis Miller, Andrew Morrrison, Elli Mylonas, Patrizzia Nerozzi, Geoffrey Nunberg, Sutayut Osornprasop, Allesandro Pamini, Paolo Petta, Allen Renear, Massimo Riva, Peter Robinson, Lothar Roisteck, Luisella Romeo, James Rosenheim, Daniel Russell, Marco Santoro, Valentina Sestini, Ture Schwebs, Shih Choon Fong, Rosemary Michelle Simpson, Christine Tamblyn, Jeff Taylor, Robert Trappl, Paul Tucker, Frank Turner, Gregory Ulmer, Andy van Dam, Karin Wenz, Rob Wittig, and the members of CHUG.

Among the many students and others who have shared their hypermedia projects with me since the late 1980s I have to thank Mark Amerika, Diego Bonilla, Don Bosco, Sarah Eron, Ian Flitman, Nicholas Friesner, Amanda Griscom, Jeremy Hight, Taro Ikai, Shelley Jackson, Ian M. Lyons, Abigail Newman, Nitin Sawhney, David Balcom, Jeff Pack, Ian Smith, Owen Strain, Noah Wardrip-Fruin, David Yun, and Leni Zumas,

When I presented the idea for the first version of this work to the Johns Hopkins University Press, Eric Halpern, then editor in chief, was open-minded enough to have enthusiasm for a project that editors at other presses thought too strange or too unintelligible to consider. I greatly appreciate the encouragement I received from him and the support for the second version by Douglas Armato and Willis Regier, then director of the Press. Michael Lonegro, my editor for 3.0, has added to my experience of Johns Hopkins University Press assistance with his valuable encouragement and suggestions. Jim Johnston, design and production manager when the first version was produced, and Glen Burris, the book's designer, deserve thanks for tackling something new in a new way. Thanks, too, to Maria denBoer, who copyedited this version, for contributing much to whatever grace, clarity, and accuracy this book may possess.

Finally, I would like to thank my children, Shoshana and Noah, who have

ACKNOWLEDGMENTS listened for years to my effusions about links, webs, lexias, web views, and local tracking maps. Noah's technical expertise about information architecture, blogging, and countless arcane details of hardware and software made many of my projects possible, and he keeps introducing me to new areas of digital culture. My most important debt, of course, is to my wife, Ruth, to whom this book is dedicated. It was she who coined the titles *Hypertext 2.0* and 3.0 and who taught me everything I know about Internet shopping. In the course of encouraging my explorations of hypermedia, she has become a true member of the digerati—someone who has worn off the characters on several keyboards while editing a magazine on the other side of the world via the Internet and who sends me a stream of e-mail even when we are in the same room. Of all the debts I have incurred while writing this book, I enjoy most acknowledging the one to her.

Hypertext 3.0

1

Hypertext:
An Introduction

Hypertextual Derrida,

Poststructuralist Nelson?

When designers of computer software examine the pages of *Glas* or *Of Grammatology*, they encounter a digitalized, hypertextual Derrida; and when literary theorists examine *Literary Machines*, they encounter a deconstructionist or poststructuralist Nelson. These shocks of recognition can occur because over the past several decades literary theory and computer hypertext, apparently unconnected areas of inquiry, have increasingly converged. Statements by theorists concerned with literature, like those by theorists concerned with computing, show a remarkable convergence. Working often, but not always, in ignorance of each other, writers in these areas offer evidence that provides us with a way into the contemporary *episteme* in the midst of major changes. A paradigm shift, I suggest, has begun to take place in the writings of Jacques Derrida and Theodor Nelson, Roland Barthes and Andries van Dam. I expect that one name in each pair will be unknown to most of my readers. Those working in computing will know well the ideas of Nelson and van Dam; those working in literary and cultural theory will know equally well the ideas of Derrida and Barthes.[1]

All four, like many others who write on hypertext and literary theory, argue that we must abandon conceptual systems founded on ideas of center, margin, hierarchy, and linearity and replace them by ones of multilinearity, nodes, links, and networks. Almost all parties to this paradigm shift, which marks a revolution in human thought, see electronic writing as a direct response to the strengths and weaknesses of the printed book, one of the major landmarks in the history of human thought. This response has profound implications for literature, education, and politics.

The many parallels between computer hypertext and critical theory have many points of interest, the most important of which, perhaps, lies in the fact that critical theory promises to theorize hypertext and hypertext promises to embody and thereby test aspects of theory, particularly those concerning textuality, narrative, and the roles or functions of reader and writer. Using hypertext, digital textuality, and the Internet, students of critical theory now have a laboratory with which to test its ideas.[2] Most important, perhaps, an experience of reading hypertext or reading with hypertext greatly clarifies many of the most significant ideas of critical theory. As J. David Bolter points out in the course of explaining that hypertextuality embodies poststructuralist conceptions of the open text, "what is unnatural in print becomes natural in the electronic medium and will soon no longer need saying at all, because it can be shown" (*Writing Space*, 143).

The Definition of Hypertext and Its History as a Concept

In *S/Z*, Roland Barthes describes an ideal textuality that precisely matches that which has come to be called *computer hypertext*—text composed of blocks of words (or images) linked electronically by multiple paths, chains, or trails in an open-ended, perpetually unfinished textuality described by the terms *link, node, network, web,* and *path.* "In this ideal text," says Barthes, "the networks [*réseaux*] are many and interact, without any one of them being able to surpass the rest; this text is a galaxy of signifiers, not a structure of signifieds; it has no beginning; it is reversible; we gain access to it by several entrances, none of which can be authoritatively declared to be the main one; the codes it mobilizes extend *as far as the eye can reach,* they are indeterminable . . . ; the systems of meaning can take over this absolutely plural text, but their number is never closed, based as it is on the infinity of language" (5–6 [English translation]; 11–12 [French]).

Like Barthes, Michel Foucault conceives of text in terms of network and links. In *The Archaeology of Knowledge,* he points out that the "frontiers of a book are never clear-cut," because "it is caught up in a system of references to other books, other texts, other sentences: it is a node within a network . . . [a] network of references" (23).

Like almost all structuralists and poststructuralists, Barthes and Foucault describe text, the world of letters, and the power and status relations they involve in terms shared by the field of computer hypertext. *Hypertext,* a term coined by Theodor H. Nelson in the 1960s, refers also to a form of electronic text, a radically new information technology, and a mode of publication.[3] "By 'hypertext,'" Nelson explains, "I mean non-sequential writing—text that branches and allows choices to the reader, best read at an interactive screen. As popularly conceived, this is a series of text chunks connected by links which offer the reader different pathways" (*Literary Machines*, 0/2). *Hypertext*, as the term is used in this work, denotes text composed of blocks of text— what Barthes terms a *lexia*—and the electronic links that join them.[4] *Hypermedia* simply extends the notion of the text in hypertext by including visual information, sound, animation, and other forms of data.[5] Since hypertext, which links one passage of verbal discourse to images, maps, diagrams, and sound as easily as to another verbal passage, expands the notion of text beyond the solely verbal, I do not distinguish between hypertext and hypermedia. *Hypertext* denotes an information medium that links verbal and nonverbal information. In this network, I shall use the terms *hypermedia* and *hypertext* interchangeably. Electronic links connect lexias "external" to a work—say, commentary on it by another author or parallel or contrasting texts—as well as within it and thereby create text that is experienced as nonlinear, or, more properly, as multilinear or multisequential. Although conventional reading habits apply within each lexia, once one leaves the shadowy bounds of any text unit, new rules and new experience apply.

The standard scholarly article in the humanities or physical sciences perfectly embodies the underlying notions of hypertext as multisequentially read text. For example, in reading an article on, say, James Joyce's *Ulysses*, one reads through what is conventionally known as the main text, encounters a number or symbol that indicates the presence of a footnote or endnote, and leaves the main text to read that note, which can contain a citation of passages in *Ulysses* that supposedly support the argument in question or information about the scholarly author's indebtedness to other authors, disagreement with them, and so on. The note can also summon up information about sources, influences, and parallels in other literary texts. In each case, the reader can follow the link to another text indicated by the note and thus move entirely outside the scholarly article itself. Having completed reading the note or having decided that it does not warrant a careful reading at the moment, one returns to the main text and continues reading until one encounters another note, at which point one again leaves the main text.

This kind of reading constitutes the basic experience and starting point of hypertext. Suppose now that one could simply touch the page where the symbol of a note, reference, or annotation appeared, and thus instantly bring into view the material contained in a note or even the entire other text—here all of *Ulysses*—to which that note refers. Scholarly articles situate themselves within a field of relations, most of which the print medium keeps out of sight and relatively difficult to follow, because in print technology the referenced (or linked) materials lie spatially distant from the references to them. Electronic hypertext, in contrast, makes individual references easy to follow and the entire field of interconnections obvious and easy to navigate. Changing the ease with which one can orient oneself within such a context and pursue individual references radically changes both the experience of reading and ultimately the nature of that which is read. For example, if one possessed a hypertext system in which our putative Joyce article was linked to all the other materials it cited, it would exist as part of a much larger system in which the totality might count more than the individual document; the article would now be woven more tightly into its context than would a printed counterpart.

As this scenario suggests, hypertext blurs the boundaries between reader and writer and therefore instantiates another quality of Barthes's ideal text. From the vantage point of the current changes in information technology, Barthes's distinction between readerly and writerly texts appears to be essentially a distinction between text based on print technology and electronic hypertext, for hypertext fulfills

the goal of literary work (of literature as work) [which] is to make the reader no longer a consumer, but a producer of the text. Our literature is characterized by the pitiless divorce which the literary institution maintains between the producer of the text and its user, between its owner and its consumer, between its author and its reader. This reader is thereby plunged into a kind of idleness—he is intransitive; he is, in short, serious: instead of functioning himself, instead of gaining access to the magic of the signifier, to the pleasure of writing, he is left with no more than the poor freedom either to accept or reject the text: reading is nothing more than a referendum. Opposite the writerly text, then, is its countervalue, its negative, reactive value: what can be read, but not written: the readerly. We call any readerly text a classic text. (*S/Z*, 4)

Compare the way the designers of Intermedia, one of the most advanced hypertext systems thus far developed, describe the active reader that hypertext requires and creates:

Both an author's tool and a reader's medium, a hypertext document system allows authors or groups of authors to link information together, create paths through a corpus of related material, annotate existing texts, and create notes that point readers to either bibliographic data or the body of the referenced text . . . Readers can browse through linked, cross-referenced, annotated texts in an orderly but nonsequential manner. (17)[6]

To get an idea of how hypertext produces Barthes's writerly text, let us examine how the print version and the hypertext version of this book would differ. In the first place, instead of encountering it in a paper copy, you would read it on a computer screen (or already have if you've read the Johns Hopkins translation of the first version into hypertext). In 1997, computer screens, which had neither the portability nor the tactility of printed books, made the act of reading somewhat more difficult than did the print version. For those people like myself who do a large portion of their reading reclining on a bed or couch, screens on desktop machines are markedly less convenient. For the past four years, however, I have worked with a series of laptops whose displays do not flicker and whose portability permits enjoyable reading in multiple locations. Of course, my Apple G4 laptop still doesn't endow the documents read on it with the pleasurable tactility of the printed book, but since my wife and I use wireless access to the Internet, we can both read Internet materials anywhere in the house or sitting outside in a recliner on the porch. Although I used to agree with people who told me that one could never read large amounts of text online, I now find that with these new displays I prefer to read the scholarly literature on my laptop; taking notes and copying passages is certainly more convenient. Nonetheless, back in the late 1980s, reading on Intermedia, the hypertext system with which I first worked, offered certain important compensations for its inconveniences.[7]

Reading an Intermedia, Storyspace, or World Wide Web version of this book, for example, you could change the size and even style of font to make reading easier. Although you could not make such changes permanently in the text as seen by others, you could make them whenever you wished. More important, since on Intermedia you would read this hypertext book on a large two-page graphics monitor, you would have the opportunity to place several texts next to one another. Thus, upon reaching the first note in the main text, which follows the passage quoted from *S/Z*, you would activate the hypertext equivalent of a reference mark (glyph, button, link marker), and this action would bring the endnote into view. A hypertext version of a note differs from that in a printed book in several ways. First, it links directly to the reference symbol and does not reside in some sequentially numbered list at the rear of the main text. Second, once opened and either superimposed on the main text or placed alongside it, it appears as an independent, if connected, document in its own right and not as some sort of subsidiary, supporting, possibly parasitic text.

Although I have since converted endnotes containing bibliographic information to in-text citations, the first version of *Hypertext* had a note containing the following information: "Roland Barthes, *S/Z*, trans. Richard Miller (New York: Hill and Wang, 1974), 5–6." A hypertext lexia equivalent to this note could include this same information, or, more likely, take the form of the quoted passage, a longer section or chapter, or the entire text of Barthes's work. Furthermore, in the various hypertext versions of this book, that passage in turn links to other statements by Barthes of similar import, comments by students of Barthes, and passages by Derrida and Foucault that also concern this notion of the networked text. As a reader, you must decide whether to return to my argument, pursue some of the connections I suggest by links, or, using other capacities of the system, search for connections I have not suggested. Reading on the World Wide Web produces this kind of reading experience. The multiplicity of hypertext, which appears in multiple links to individual blocks of text, calls for an active reader.

A full hypertext system, unlike a book and unlike some of the first approximations of hypertext available—HyperCard™, Guide™, and the current World Wide Web (except for blogs)—offers the reader and writer the same environment. Therefore, by opening the text-processing program, or editor, as it is known, you can take notes, or you can write against my interpretations, against my text. Although you cannot change my text, you can write a response and then link it to my document. You thus have read the readerly text in several ways not possible with a book: you have chosen your reading path, and since you, like all readers, will choose individualized paths, the hypertext version of this book would probably take a very different form, perhaps suggesting the values of alternate routes and probably devoting less room in the main text to quoted passages. You might have also have begun to take notes or produce responses to the text as you read, some of which might take the form of texts that either support or contradict interpretations proposed in my texts.

Very Active Readers

When one considers the history of both ancient literature and recent popular culture, the figure of the reader-as-writer hardly appears at all strange, particularly since classical and neoclassical cultural theory urged neophyte authors to learn their craft by reading the masters and then consciously trying to write like them. Anyone who's taken an undergraduate survey course will know that Vergil self-consciously read and rewrote Homer, and that Dante read and rewrote both Homer and Vergil, and Milton continued the practice. Such very active readers appear throughout the past two centuries. To an important extent, *Jane Eyre* represents a very active reading of *Pride and Prejudice,* just as *North and South* and *Aurora Leigh* represent similar readings and rewritings of the two earlier texts. In fact, all four works could have been entitled "Pride and Prejudice," and all four present women of a supposedly lower social and economic class disciplining their men; in Victorian versions of this plot the man not only has to apologize for his shortcomings but he also has to experience major punishment—bankruptcy, severe injury, blindness, or a combination of them.

Literary scholars are quite accustomed to chains of active readings that produce such rewritings. We call it a *tradition*. We also, following Harold Bloom, call it the *anxiety of influence*, the later author challenging the earlier one. Readers of, say, *Aurora Leigh* recognize that Elizabeth Barrett Browning's novel-poem simultaneously asserts the existence of a female literary tradition while also challenging its creators, the poet's predecessors, for pride of place within that tradition.

Such aggressively active reading has proved particularly popular with postcolonial and postimperial authors. Thus Jean Rhys's *Wide Sargasso Sea* offers a very different, Caribbean reading of *Jane Eyre*, telling the story almost entirely from Bertha's point of view. We encounter the empire again writing back in Peter Carey's *Jack Maggs*, a novel told from the vantage point of the Magwitch character; in this version, which includes a Dickens-like novelist, the illegally returned convict does not die with Pip at his side: realizing what a dreadful person the Pip-character has turned out to be, Maggs returns to wealth, fatherhood, and fame in Australia. Take that, Dickens!

Given the history of high culture, one is not surprised to encounter these active readings and rewritings, but such approaches also appear in so-called genre fiction, such as detective stories and science fiction. In Japser Fforde's *The Eyre Affair*, for example, we learn how Brontë's novel received its happy ending. All the examples of such very active reading thus far belong to the upper reaches of the culture industry: major commercial firms publish them, they win prestigious prizes, and they quickly earn canonicity by being taught in universities. There are, however, large numbers of very active readers who receive little notice from the publishing and academic establishments. The wide availability of low-cost information technologies—first mimeographs and photo-offset printing and later desktop publishing and finally the World Wide Web—permitted the creation of self-published rewritings of popular entertainment, such as *Star Trek*, that first appeared in books, television, and cinema. Active readings of the popular science fiction series, Constance Penley explains, have existed since the mid-1970s.

Most of the writers and readers started off in "regular" *Star Trek* fandom, and many are still involved in it, even while they pursue their myriad activities in what is called "K/S" or "slash" fandom. The slash between K(irk) and S(pock) serves as a code to those purchasing by mail amateur fanzines (or "zines") that the stories, poems, and artwork published there concern a same-sex relationship between the two men. Such a designation stands in contrast to "ST," for example, with no slash, which stands for action adventure stories based on the *Star Trek* fictional universe, or "adult ST," which refers to stories containing sexual scenes, but heterosexual ones only, say between Captain Kirk and Lieutenant Uhura, or Spock and Nurse Chapel. Other media male couples have been "slashed" in the zines, like Starsky and Hutch (S/H) . . . or *Miami Vice*'s Crockett and Tubbs. (137)

According to Penley, women produce most of these samizdat texts, and these readers-as-writers take "pride in having created both a unique, hybridized genre that ingeniously blends romance, pornography, and utopian science fiction and a comfortable social space in which women can manipulate the products of mass-produced culture to stage a popular debate around the issues of technology, fantasy, and everyday life" (137). As one might expect, the development of the World Wide Web has stimulated this active reading even more, and one can find all kinds of works by readers who want to write *their versions* of materials commercially published. The presence and productions of very active readers answer the critics of digital information technology who claim it cannot demonstrate any examples of cultural democratization. Whether one actually likes this or other kinds of cultural democratization is another matter.

Very active readers (or readers-as-writers) have tended to go unnoticed for several reasons. First, although some of these fanzines may have circulations as large as first novels published by prestigious publishers, they represent an underground culture of which mass media and educational institutions remain unaware. Another reason why the continuations and rewritings they produce receive little attention derives from some of the obvious qualities of print culture: like Carey's very active reading of Dickens's novel, these underground texts, even those that appear on the Internet, take the form of discrete works separated in time and space from the texts they rewrite. The Internet works, however, appear in a very different context than do the print ones. Anyone who stumbles upon any of these writings is likely to find them linked to a personal or group site containing biographies of the site owner, explanations of the imaginative world, and lists of links to similar stories. The link, in other words, makes immediately visible the virtual community created by these active readers.

How does such active reading-as-writing relate to the hypertext reader? First of all, this kind of print-based active reader encounters a supposedly discrete, finished text; the reader's response—writing a new text—demonstrates that this kind of reader both accepts that fact and also does not want to accept its limitations. This active reading characterizes readers of blogs: they take an existing text and add to it, but because they write in a networked computer environment the commented-on blog, employing TrackBack, can link to the active reader's text, incorporating it into the ongoing discussion.[8]

Like blogs, by-now atypical hypertext systems that permit readers to add their own links and materials (Intermedia, Storyspace in the authoring environment) or even websites that solicit reader contributions represent ways that readers can assume the role of authors. All of these forms of active reading differ from the experience of the hypertext reader in read-only systems, whose writing takes the form not of adding new texts but of establishing an order of reading in an already-written set of texts. Readers of large bodies of informational hypermedia create the document they read from the informed choices they make. It might appear that such is rarely true of readers of fictional hypertexts who may not know where particular links lead. Nonetheless, the best hyperfictions, I submit, permit the reader to deduce enough basic information, sometimes, as in Michael Joyce's *afternoon*, by retracing their steps, to make informed (thus creative) decisions when they arrive at links. Still, no matter how much power readers have to choose their ways through a hypertext, they never obtain the same degree of power—or have to expend as much effort—as those who write their texts in response to another's.

Vannevar Bush and the Memex

Writers on hypertext trace the concept to a pioneering article by Vannevar Bush in a 1945 issue of *Atlantic Monthly* that called for mechanically linked information-retrieval machines to help scholars and decision makers faced with what was already becoming an explosion of information. Struck by the "growing mountain of research" that confronted workers in every field, Bush realized that the number of publications had already "extended far beyond our present ability to make real use of the record. The summation of human experience is being expanded at a prodigious rate, and the means we use for threading through the consequent maze to the momentarily important item is the same as was used in the days of square-rigged ships" (17–18). As he emphasized, "there may be millions of fine thoughts, and the account of the experience on which they are based, all encased within stone walls of acceptable architectural form; but if the scholar can get at only one a week by diligent search, his syntheses are not likely to keep up with the current scene" (29).

According to Bush, the main problem lies with what he termed "the matter of selection"—information retrieval—and the primary reason that those who need information cannot find it lies in turn with inadequate means of storing, arranging, and tagging information:

> Our ineptitude in getting at the record is largely caused by the artificiality of systems of indexing. When data of any sort are placed in storage, they are filed alphabetically or numerically, and information is found (when it is) by tracing it down from subclass to subclass. It can be in only one place, unless duplicates are used; one has to have rules as to which path will locate it, and the rules are cumbersome. Having found one item, moreover, one has to emerge from the system and re-enter on a new path. (31)

As Ted Nelson, one of Bush's most prominent disciples, points out, "there is nothing wrong with categorization. It is, however, by its nature transient: category systems have a half-life, and categorizations begin to look fairly stupid after a few years . . . The army designation of 'Pong Balls, Ping' has a certain universal character to it" (*Literary Machines,* 2/49). According to Bush and Nelson, then, one of the greatest strengths of hypertext lies in its capacity of permitting users to find, create, and follow multiple conceptual structures in the same body of information. Essentially, they describe the technological means of achieving Derrida's concept of decentering.

In contrast to the rigidity and difficulty of access produced by present means of managing information based on print and other physical records, one needs an information medium that better accommodates the way the mind works. After describing present methods of storing and classifying knowledge, Bush complains, "The human mind does not work that way" ("As We May Think," 31) but by association. With one fact or idea "in its grasp," the mind "snaps instantly to the next that is suggested by the association of thoughts, in accordance with some intricate web of trails carried by the cells of the brain" (32).

To liberate us from the confinements of inadequate systems of classification and to permit us to follow natural proclivities for "selection by association, rather than by indexing," Bush therefore proposes a device, the "memex," that would mechanize a more efficient, more human, mode of manipulating fact and imagination. "A memex," he explains, "is a device in which an individual stores his books, records, and communications, and which is mechanized so that it may be consulted with exceeding speed and flexibility. It is an enlarged intimate supplement to his memory" (32). Writing in the days before digital computing (the first idea for a memex came to him in the mid-1930s), Bush conceived of his device as a desk with translucent screens, levers, and motors for rapid searching of microform records.

In addition to thus searching and retrieving information, the memex also permits the reader to "add marginal notes and comments, taking advantage of one possible type of dry photography, and it could even be arranged so that [an individual] can do this by a stylus scheme, such as is now employed in the telautograph seen in railroad waiting rooms, just as though he had the physical page before him" (33). Two things demand attention about this crucial aspect of Bush's conception of the memex. First, he believes that while reading, one needs to append one's own individual, transitory thoughts and reactions to texts. With this emphasis Bush in other words reconceives reading as an active process that involves writing. Second, his remark that this active, intrusive reader can annotate a text "just as though he had the physical page before him" recognizes the need for a conception of a virtual, rather than a physical, text. One of the things that is so intriguing about Bush's proposal is the way he thus allows the shortcomings of one form of text to suggest a new technology, and that leads, in turn, to an entirely new conception of text.

The "essential feature of the memex," however, lies not only in its capacities for retrieval and annotation but also in those involving "associative indexing"—what present hypertext systems term a *link*—"the basic idea of which is a provision whereby any item may be caused at will to select immediately and automatically another" (34). Bush then provides a scenario of how readers would create "endless trails" of such links:

When the user is building a trail, he names it, inserts the name in his code book, and taps it out on his keyboard. Before him are the two items to be joined, projected onto adjacent viewing positions. At the bottom of each there are a number of blank code spaces, and a pointer is set to indicate one of these on each item. The user taps a single key, and the items are permanently joined. In each code space appears the code word. Out of view, but also in the code space, is inserted a set of dots for photocell viewing; and on each item these dots by their positions designate the index number of the other item. Thereafter, at any time, when one of these items is in view, the other can be instantly recalled merely by tapping a button below the corresponding code space. (34)

Bush's remarkably prescient description of how the memex user creates and then follows links joins his major recognition that trails of such links themselves constitute a new form of textuality and a new form of writing. As he explains, "when numerous items have been thus joined together to form a trail . . . it is exactly as though the physical items had been gathered together from widely separated sources and bound together to form a new book." In fact, "it is more than this," Bush adds, "for any item can be joined into numerous trails" (34), and thereby any block of text, image, or other information can participate in numerous books.

These new memex books themselves, it becomes clear, are the new book, or one additional version of the new book, and, like books, these trail sets or webs can be shared. Bush proposes, again quite accurately, that "wholly new forms of encyclopedias will appear, ready-made with a mesh of associative trails running through them, ready to be dropped into the memex and there amplified" (35). Equally important, individual reader-writers can share document sets and apply them to new problems.

Bush, an engineer interested in technical innovation, provides the example of a memex user

studying why the short Turkish bow was apparently superior to the English long bow in the skirmishes of the Crusades. He has dozens of possibly pertinent books and articles in his memex. First he runs through an encyclopedia, finds an interesting but sketchy article, leaves it projected. Next, in a history, he finds another pertinent item, and ties the two together. Thus he goes, building a trail of many items. Occasionally he inserts a comment of his own, either linking it into the main trail or joining it by a side trail to a particular item. When it becomes evident that the elastic properties of available materials had a great deal to do with the bow, he branches off on a side trail which takes him through textbooks on elasticity and tables of physical constants. He inserts a page of longhand analysis of his own. Thus he builds a trail of his interest through the maze of materials available to him. (34–35)

And, Bush adds, his researcher's memex trails, unlike those in his mind, "do not fade," so when he and a friend several years later discuss "the queer ways in which a people resist innovations, even of vital interest" (35), he can reproduce his trails created to investigate one subject or problem and apply them to another.

Bush's idea of the memex, to which he occasionally turned his thoughts for three decades, directly influenced Nelson, Douglas Englebart, Andries van Dam, and other pioneers in computer hypertext, including the group at the Brown University's Institute for Research in Information and Scholarship (IRIS) who created Intermedia. In "As We May Think" and "Memex Revisited" Bush proposed the notion of blocks of text joined by links, and he also introduced the terms *links, linkages, trails,* and *web* to describe his new conception of textuality. Bush's description of the memex contains several other seminal, even radical, conceptions of textuality. It demands, first of all, a radical reconfiguration of the practice of reading and writing, in which both activities draw closer together than is possible with book technology. Second, despite the fact that he conceived of the memex before the advent of digital computing, Bush perceives that something like virtual textuality is essential for the changes he advocates. Third, his reconfiguration of text introduces three entirely new elements—associative indexing (or links), trails of such links, and sets or webs composed of such trails. These new elements in turn produce the conception of a flexible, customizable text, one that is open— and perhaps vulnerable—to the demands of each reader. They also produce a concept of multiple textuality, since within the memex world texts refers to individual reading units that constitute a traditional "work," those entire works, sets of documents created by trails, and perhaps those trails themselves without accompanying documents.

Perhaps most interesting to one considering the relation of Bush's ideas to contemporary critical and cultural theory is that this engineer began by rejecting some of the fundamental assumptions of the information technology that had increasingly dominated—and some would say largely created— Western thought since Gutenberg. Moreover, Bush wished to replace the essentially linear fixed methods that had produced the triumphs of capitalism and industrialism with what are essentially poetic machines—machines that work according to analogy and association, machines that capture and create the anarchic brilliance of human imagination. Bush, we perceive, assumed that science and poetry work in essentially the same way.

Forms of Linking, Their Uses and Limitations

Before showing some of the ways this new information technology shares crucial ideas and emphases with contemporary critical theory, I shall examine in more detail the link, the element that hypertext adds to writing and reading.[9] The very simplest, most basic form of linking is unidirectional lexia to lexia (Figure 1). Although this type of link has the advantage of requiring little planning, it disorients when used with long documents, since readers do not know where a link leads in the entered document. It is best used, therefore, for brief lexias or in systems that use card metaphors.

Next in complexity comes bidirectional linking of two entire lexias to one another—identical to the first form except that it includes the ability to retrace one's steps (or jump). Its advantage lies in the fact that by permitting readers to retrace their steps, it creates a simple but effective means of orientation. This mode seems particularly helpful when a reader arrives at a lexia that has only one or two links out, or when readers encounter something, say, a glossary definition or image, that they do not want to consult at that point in their reading.

Lexia to Lexia Unidirectional

Advantage: simple, requires little planning.

Disadvantage: disorients when used with long documents, since readers do not know where link leads; best used for brief lexias or in systems that use card metaphor.

Lexia to Lexia Bidirectional

Advantage: by permitting readers to retrace their steps creates simple but effective means of orientation. Particularly helpful when arriving at lexias that have only one or two departure links.

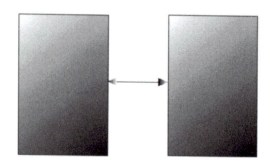

String (word or phrase) to Lexia

Advantages: (1) allows simple means of orienting readers; (2) permits longer lexias; (3) encourages different kinds of annotation and linking.

Disadvantage: disorients when used with long documents, since readers do not know where link leads; best used for brief lexias or in systems that use card metaphor.

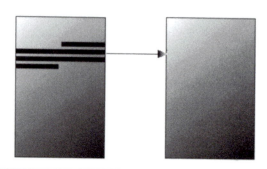

Figure 1. Three Forms of Linking

Linking a string—that is, word or phrase—to an entire lexia, the third form of linking, has three advantages. First, it permits simple means of orienting readers by allowing a basic rhetoric of departure (Figure 1). When readers see a link attached to a phrase, such as "Arminianism" or "Derrida," they have a pretty good idea that such a link will take them to information related in some obvious way to those names. Second, because string-to-lexia linking thus provides a simple means of helping readers navigate through information space, it permits longer lexias. Furthermore, since one can choose to leave the lexia at different points, one can comfortably read through longer texts. Third, this linking mode also encourages different kinds of annotation and linking, since the ability to attach links to different phrases, portions of images, and the like allows the author to indicate different kinds of link destinations. One can, for example, use icons or phrases to indicate that the reader can go to, say, another text lexia, one containing an illustration, bibliographical information, definitions, opposing arguments, and so forth.

The difficulties with string-to-lexia links, the form most characteristic of links in World Wide Web documents, arise in problems encountered at the destination lexia. Readers can find themselves disoriented when entering long documents, and therefore string-to-lexia linking works best with brief arrival lexia. The fourth form of linking occurs when one makes the link joining a string to an entire lexia bidirectional. (Most linking in HTML [HyperText Markup Language] documents takes this form in effect—"in effect," because the return function provided by most browsers creates the effect of a bidirectional link.)

The fifth form, unidirectional string-to-string linking, has the obvious advantage of permitting the clearest and easiest way to end links and thereby create a rhetoric of arrival. By bringing readers to a clearly defined point in a text, one enables them to perceive immediately the reason for a link and hence to grasp the relation between two lexias or portions of them. Readers know, in other words, why they have arrived at a particular point. The anchor feature in HTML, which is created by the <a name> tag, thus permits authors to link to a specific section of long document. The possible disadvantage of such a mode to authors—which is also a major advantage from the reader's point of view—lies in the fact that it requires more planning, or at least, more definite reasons for each link. Making such links bidirectional, our sixth category, makes navigating hyperspace even easier.

Full hypertextuality in a reading environment depends, I argue, on the multisequentiality and the reader choices created not only by attaching multiple links to a single lexia but by attaching them to a single anchor or site within a single lexia. A fully hypertextual system (or document) therefore employs a seventh form, one-to-many linking—linking that permits readers to obtain different information from the same textual site (Figure 2). One-to-many linking supports hypertextuality in several ways. First, it encourages branching and consequent reader choice. Second, attaching multiple links to a single text allows hypertext authors to create efficient overviews and directories that serve as efficient crossroad documents, or orientation points, that help the reader navigate hyperspace. Multiple overviews or sets of overviews have the additional advantage of easily permitting different authors to pro-

String to String

Advantage: permits clearest way to end links.

Disadvantage: requires more planning than do links to full lexias.

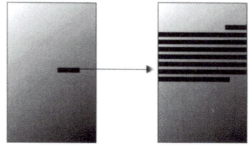

One-to-Many

Advantages: (1) encourages branching and consequent reader choice; (2) permits efficient author-generated overview and directory documents; (3) when combined with systems that provide link menus and other preview functions, helps greatly in orienting readers.

Disadvantage: can produce sense of an atomized text.

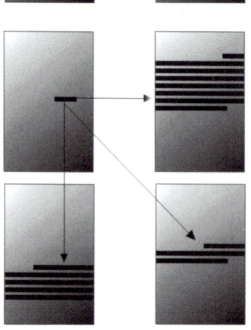

Figure 2. Two Forms of Linking

vide multiple ways through the same information space. Third, when combined with software, such as Microcosm, Storyspace, or Intermedia, that provides link menus and other so-called preview functions, one-to-many linking greatly helps in orienting readers. The major disadvantage of this kind of link, which plays a major role in most hypertext fiction, lies in its tendency to produce a sense of atomized text.

The eighth kind of link—many-to-one linking—proves particularly handy for creating glossary functions or for creating documents that make multiple references to a single text, table, image, or other data (Figure 3). DynaText,

Many-to-One Linking

Advantages: (1) handy for glossary functions or for texts that make multiple references to a single text, table, image, or other data; (2) encourages efficient reuse of important information; (3) allows simple means of producing documents for readers with differing levels of expertise.

Disadvantage: systems that create many-to-one linking automatically can produce a distracting number of identical links.

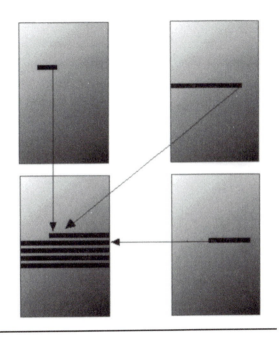

Figure 3. Many-to-One Linking

Microcosm, and the World Wide Web exemplify hypertext environments in which one can have many links lead to a single document, an arrangement that has major advantages in educational and informational applications. In particular, many-to-one linking encourages efficient reuse of important information. For example, having once created an introductory essay on, say, Charles II, Lamarckianism, or Corn Law agitation, the original (and later) authors simply use linking to provide access to it as the occasion arises. Furthermore, by providing an easy, efficient means of offering readers glossaries and other basic information, many-to-one linking also permits webs to be used easily by readers with differing levels of expertise.

The major disadvantage of such linking involves not the links themselves but the means various systems use to indicate their presence. Systems that create many-to-one linking, particularly those that create it automatically, can produce a distracting number of link markers. The World Wide Web uses colored underlining to indicate hot text (link anchors), and in the DynaText version of the first version of this book, Paul Kahn chose red text to signify the presence of links. In both cases the reader encounters distracting markup intruding into the text. Experience with these systems quickly convinces one of the need for a means of easily turning on and off such link indicators, such as one can do in Eastgate System's Storyspace. The disadvantages with many-to-one links derive not from this form of linking itself but from other aspects of individual hypertext environments, and any such disadvantages become amplified by the inexperience of readers: in the first years of the Web, for example, authors and designers generally agreed that users, many of whom had little experience with computing, required colored underlining to find links; otherwise, it was correctly reasoned, readers would not know what to do. In the very earliest days of the Web, in fact, one often encountered linked underlined text immediately beneath a linked icon because web designers knew that many neophyte users would not realize they could follow links by clicking on the icon. As people began to use the Web every day, however, they recognized that when they moved a cursor across the surface of a web browser, it changed from an arrow to a hand when placed over a link. Experienced users thus no longer required the once ubiquitous blue underlining, and many sites now do not use it.

As we shall observe shortly, some systems, such as Microcosm, include a particularly interesting and valuable extension of many-to-one linking that permits readers to obtain a menu containing two or more glossary or similar documents. While creating a hypertext version of my book on Holman Hunt and Pre-Raphaelite painting for the World Wide Web, an environment that does not permit either link menus or one-to-many links, I had to choose whether to link (connect) multiple mentions of a particular painting, say, the artist's *Finding of the Saviour in the Temple,* to an introductory discussion of the picture or to an illustration of it. In contrast, while creating a hypertext version of the same book in Microcosm, I easily arranged links so that when readers follow them from any mention of the painting, they receive a menu containing titles of the introductory text and two or more illustrations, thereby providing readers with convenient access to the kind of information they need when they need it (see Figure 6).

Typed links, our ninth category, take the form of limiting an electronic link to a specific kind of relationship, such as "exemplifies," "influences," "contrary argument," "derives from" (or "child of"), and so on (Figure 4). Software that includes such link categorization range from proposed research systems that, in attempting to help organize argument, permit only certain kinds of connections, to those like Marc and Jocelyne Nanard's MacWeb, which allows authors to create their own categories. In fact, any system, such as Intermedia, Storyspace, or Microcosm, that permits one to attach labels to individual links allows one to create typed links, since labels permit authors

Typed Links

Advantages: (1) if clearly labeled, acts as a form of link preview and aids reader comfort; (2) can produce different kinds of link behavior, including pop-up windows.

Disadvantage: can clutter reading area or confuse by producing too many different actions when one follows links.

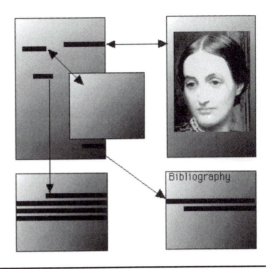

Figure 4. Typed Links

to indicate everything from document type (essay, illustration, statistics, timeline) to a particular path or trail of links that overlay a number of lexias. In fact, as the experience of the World Wide Web reveals, one can use icons or text to create what are essentially typed icons even when the system makes no provision for them. Thus, one can make clear (as I have in the *Victorian Web*) that a link leads to bibliographical information, an illustration, or an opposing argument by simply linking to a word, such as *source* or *illustration,* within parentheses.

The advantage of typed links includes the fact that, when clearly labeled, they offer a generalized kind of previewing that aids reader comfort and helps navigating information space. Such labeling can take the form of icons in the current lexia (DynaText, Voyager Expanded Book, World Wide Web), similar indications in a second window (Intermedia's Web View and similar dynamic hypergraphs, such as that created experimentally for Microcosm), and dynamic link menus (Intermedia, Storyspace). In systems that include pop-up windows overlaying the current lexia (DynaText, the proprietary one created by Cognitive Systems for the Microsoft Art Gallery, and ones created by Java for the World Wide Web), typed links can also produce different kinds of link behavior. A potential disadvantage for readers of the typed link might be confusion produced when they encounter too many different actions or kinds of information; in fact, I have never encountered hypertexts with these problems, but I'm sure some might exist. A greater danger for authors would exist in systems that prescribe the kind of links possible. My initial skepticism about typing links arose in doubts about the effectiveness of creating rules of thought in advance and a particular experience with Intermedia. The very first version of Intermedia used by faculty developers and students differentiated between annotation and commentary links, but since one person's annotation turned out to be another's commentary, no one lobbied for retaining this feature, and IRIS omitted it from later versions.

An equally basic form of linking involves the degree to which readers either activate or even create links. In contemporary hypertext jargon, the opposition is usually phrased as a question of whether links are author or wreader determined, or—putting the matter differently—whether they are hard or soft. Most writing about hypertext from Bush and Nelson to the present assumes that someone, author or reader functioning as author, creates an electronic link, a so-called hard link. Recently, workers in the field, particularly the University of Southampton's Microcosm development group, have posed the question, "Can one have hypertext 'without links'?"—that is, without the by-now traditional assumption that links have to take the form of always-existing electronic connections between anchors. This approach takes the position that the reader's actions can create on-demand links. In the late 1980s when the first conferences on hypertext convened, such a conception of hypertext might have been difficult, if not impossible, to advocate, because in those days researchers argued that information retrieval did not constitute hypertext, and the two represented very different, perhaps opposed, approaches to information. Part of the reason for such views lay in the understandable attempts of people working in a new field in computer science to distinguish their work—and thereby justify its very existence—from an established one. Although some authors, such as the philosopher Michael Heim, perceived the obvious connection between the active reader who uses search tools to probe an electronic text and the active reader of hypertext, the need of the field to constitute itself as a discrete specialty prompted many to juxtapose hypertext and information retrieval in the sharpest terms. When the late James H. Coombs created both InterLex and full-text retrieval in Intermedia, many of these oppositions immediately appeared foolish, since anyone who clicked on a word and used Intermedia's electronic version of the *American Heritage Dictionary*—whether they were aware of it or not—inevitably used a second kind of linking. After all, activating a word and following a simple sequence of keys or using a menu brought one to another text (Figure 5). Of course, Web users now have near-immediate access to the

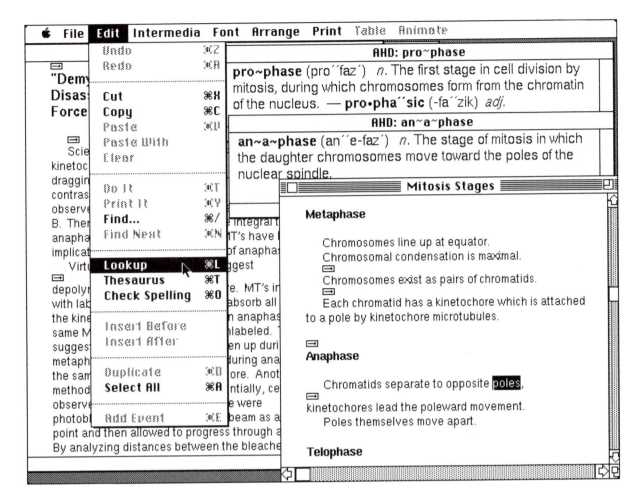

Figure 5. Hypertext Links and Information Retrieval: The InterLex Feature in Intermedia

fourth edition of *The American Heritage Dictionary* (www.bartleby.com) or to dictionaries in dozens of languages ranging from Abenaki and Armenian to Walloon and Yema on www.yourdictionary.com.

Microcosm, a system on which work began in the early days of Intermedia, has built this idea of reader-activated links into its environment in two ways. First, using the "Compute Links" function, readers activate what are essentially information-retrieval software tools to produce menus of links that take exactly the same form as menus of links created by authors. Soft links—links created on demand—appear to the reader identical to hard ones created by authors. Second, readers can activate implicit or generalized links. When readers click on hard links, they activate a connection established by a hypertext author, who, in some systems, could be a previous hypertext reader. When readers activate "Compute Links," they use what are essentially information-retrieval devices to create a dynamic relation between one text and another. In contrast to both these previous approaches, Microcosm's generalized link function produces a different form of electronic connections that we call term *soft linking*, linking activated only on demand. Essentially, Microcosm's generalized links create a link that only appears when a reader asks for it (Figure 6). No link marker, no code, indicates its existence, and nothing deforms the text in a lexia to announce its presence. In fact, only a reader's interest—a reader's energy, active interest, or aggressive relation to the text—brings such a link fully into being. Readers will recognize that this approach, this kind of linking, permits the many-to-more-than-one linking that permitted me to have readers obtain an introductory discussion and two plates of a painting by a Victorian artist by clicking on the title of one of his paintings. Microcosm's generalized linking facility, in fact, permitted me to recreate in a matter of hours links that had taken weeks to create manually in another system.

The final forms of linking—action links, warm links (or reader-activated data-exchange links), and hot linking (automatic data-exchange links)—represent, in contrast, kinds that carry the hard, author-created link in other directions. These author-created links do more, in other words, than allow readers to traverse information space or bring the document to them. They either initiate an action or they permit one to do so.

In later chapters when we examine examples of hypermedia containing animation and video, we shall observe yet other permutations of the link. Nonetheless, these preliminary remarks permit us to grasp some of the complex issues involved with adding the link to writing, with reconfiguring textuality with an element that simultaneously blurs borders, bridges gaps, and yet draws attention to them.

Linking in Open Hypermedia Systems: Vannevar Bush Walks the Web

For more than a quarter century, many computer scientists have proposed a conception of linking that differs fundamentally from the one used by HTML, Storyspace, and earlier systems, such as Guide and HyperCard. This different way of conceiving the link, not surprisingly, is also associated with a different theory of how hypermedia systems should work. HTML and Storyspace have accustomed most of us to the idea that links exist as integral parts of documents in which they appear. To anyone who has

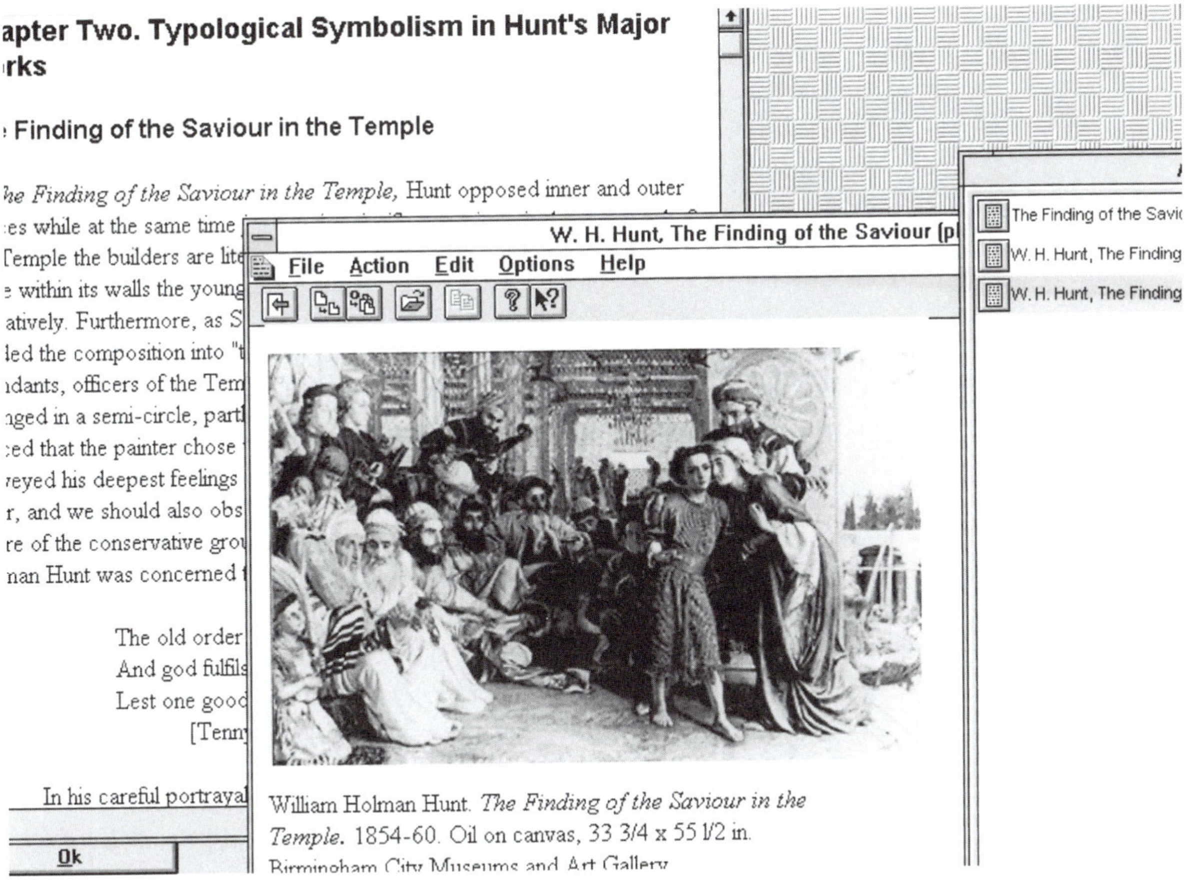

: Finding of the Saviour in the Temple

he Finding of the Saviour in the Temple, Hunt opposed inner and outer
:es while at the same time
Temple the builders are lite
e within its walls the young
atively. Furthermore, as S
led the composition into "t
idants, officers of the Tem
iged in a semi-circle, parti
:ed that the painter chose
reyed his deepest feelings
r, and we should also obs
re of the conservative grou
nan Hunt was concerned

The old order
And god fulfils
Lest one good
[Tenn

In his careful portrayal

Ok

W. H. Hunt, The Finding of the Saviour (p|

File   Action   Edit   Options   Help

William Holman Hunt. *The Finding of the Saviour in the Temple.* 1854-60. Oil on canvas, 33 3/4 x 55 1/2 in.
Birmingham City Museums and Art Gallery

The Finding of the Savio
W. H. Hunt, The Finding
W. H. Hunt, The Finding

Figure 6. Generic Linking in Microcosm. This screen shot shows the results of following a generic link either from the word "Finding" or from the phrase "The Finding of the Saviour in the Temple" in the Microcosm version of my book *William Holman Hunt and Typological Symbolism.* This action produces a menu (*at right*) with three choices: a section of my original book containing the principal discussion of this painting and two discussions of it. Since choosing "Follow Link" (or double clicking) on any word or phrase that serves as an anchor produces these three choices, this screen shot represents many-to-many linking. Furthermore, although readers experience the results of generic linking (here the menu with three destination lexias) just as if the author had manually linked each anchor to the discussion and two illustrations, in fact the links only come into existence when readers call for them. One can therefore consider this screen to exemplify soft many-to-many linking. Although Microcosm permits authors to create the usual manual form of one-to-one and one-to-many links, the generic link function takes a great deal of the work out of creating informational hypertext webs.

ever created a link in HTML that point seems obvious, and, in fact, placing links within each lexia has major benefits, including simplicity, ease of creating them, and permanence—they don't move or get lost. This conception of the link, however, represents a fundamental departure from the kind of medium proposed by Vannevar Bush. The user of the memex, we recall, created trails of associative links on top of already existing texts, saved those trails, and shared them with others. Different readers could create very different collections of links for the same texts. Links, in other words, exist outside the individual lexia in this kind of hypermedia.

Many hypertext researchers, inspired by Bush, have designed and implemented such open hypermedia systems and infrastructures, a defining characteristic of which is the link database or linkbase (see Rizk and Sutcliffe for a list of such systems). Intermedia, one such system, drew upon its separation of links and data to permit users to generate multiple webs from the same body of texts and images; depending on an individual user's access rights, he or she could view the webs created by others. In educational terms, using a linkbase had the effect of permitting students to use the main course web plus links added by students or to screen out links created by them. It also permitted instructors, as we shall see in chapter 4, to use links to incorporate materials created by those in other disciplines within their webs without affecting either the original author's text or web. In practice, readers experienced an Intermedia web, such as *Context32,* much as they do its HTML descendant, *The Victorian Web.* In fact, each hypermedia collection of documents existed only as a virtual web called into being by the linkbase and linkserver.

The linkbase and its associated server, which combine to create link services, lie at the heart of open hypermedia systems like Hyperbase, Multicard, Sun's Link Service, Microcosm, and its various later incarnations. David C. De Roure, Nigel G. Walker, and Leslie A. Carr offer the following definition of these key terms:

At its simplest, a hypermedia link server takes a source anchor in a multimedia document and returns the possible destination anchors, obtained by interrogating a link database (henceforth a *linkbase*) for links containing that anchor. The anchors might identify specific locations or objects in particular multimedia documents; alternatively they might have broader applicability, matching content rather than position (so-called *generic* linking). The linkbase query might also be refined by the user's context, perhaps based on their profile, current role, task and location. Link services may be accessed before, during or after document delivery, and they may provide an interface for link creation and maintenance as well as retrieval. (67)

The Multimedia Research Group at the University of Southhampton under the leadership of Wendy Hall and Hugh Davis stands out as the team of computer scientists that has the longest continuous experience with open hypermedia. Their articles dominate the literature in the field, and they have produced a number of commercial systems. Microcosm, at which we looked earlier, appeared in commercial form as Multicosm (1994), and as the World Wide Web became increasingly prominent, the Southampton team applied the heart of Microcosm—its link services—to the Internet, creating Distributed Link Services (1995), Multicosm (1998), and Portal Maximizer (2001). Multicosm, the company formed to provide commercial versions of the group's link-services-based applications, has recently become Active Navigation, but the open hypermedia approach remains the same.

As Hays Goodman points out about Active Navigation, "the core technology behind the company's products is the ability to insert active hyperlinks on-the-fly in almost any textual format document." The already-observed forms of linking possible with Microcosm show what an immensely powerful system it is, but that power came at considerable cost—or, rather, at two different kinds of costs. Like all open systems, Microcosm and all its descendants require a separate server for the linkbase, and the team also had to create the software to make it work. A different kind of cost appears in the way Microcosm has created anchors. At first, Microcosm recorded links solely in terms of the anchor's position—essentially counting off numbers of characters or units of spatial measurement to record where in the document a phrase (or image) begins and ends. This method proved to have enormous advantages. Initially, the Southampton team had the goal of creating the kind of hypermedia system that Vannevar Bush would love, since it could create links not just in other people's documents but also in other software: one could, for example, link a document in MS Word to another in Word Perfect to another in a PDF file. This version of Microcosm worked, and much research went into devising ways of linking among different kinds of applications; one of the most interesting of these projects involved placing links inside a very large CadCam document used by architects, and part of the difficulty included creating tiny, yet accurate, summaries of the visual data. Eventually the team discovered that some features they wished to add to the system could function only if all text had the same format, and so they turned to a more closed system. In the Microcosm version of my book on Pre-Raphaelite painting, all the text documents were created in Word and saved as in the RTF (rich text format) file format, and although the system, like current HTML, permits linking to images, in practice the need to attach captions to them resulted in placing images within text documents.

This wonderfully powerful system, which was convenient for both author and reader, permitted linking all kinds of data, but it had one Achilles heel: the computer files to which the system added links could not be modified in any way. Unlike Intermedia's linkbase, Microcosm's required freezing a document once it had links; adding or deleting words would move the link to an irrelevant phrase.

To solve this problem, one had to add a second method of identifying link anchors in the linkbase, one that required "matching content rather than position" (De Roure, Walker, and Carr, 67). This method has the great advantage of enabling powerful generic linking, but it is also much less suited to non-alphanumeric media. This seems to be the form of linkbase storage that allowed Microcosm-Multicosm to become a Web application. Goodman explains how one version marketed by Active Navigation works:

Portal Maximizer is implemented essentially as a Web proxy server. When the user requests a Web page, the browser will be directed to the Webcosm proxy. *Webcosm will fetch the page from the original location and annotate the page with extra links before passing the modified Web page back to the user's browser.* When the webmaster has activated this feature, the user will see portions of the text transformed into hyperlinks, which are derived from what is known as a linkbase. This linkbase contains at a minimum a source word or phrase, a destination URL [Uniform Resource Locator] and a description of the link. The linkbase is generated automatically by crawling the Web site at predetermined intervals, with the results fully tunable so that by moving a slider one can decide how broad or narrow particular themes can be. By making the themes broader, nearly every word in a document could theoretically be hyperlinked, but by selectively tuning that variable, more relevant results are obtained. Multiple linkbases can be used, so that different groups of users could see different results, depending on their profile or interests. (Emphasis added)

By storing links apart from text, images, and other media forms, open hypermedia systems can place links in someone else's Web document without ever affecting that document. Vannevar Bush walks the Web. Depending on the desires of those who own the server, these added links can be viewed by anyone who visits their website, or they can be screened from outsiders. The capacity of open hypermedia applications like Portal Maximizer to add links to documents coming from another site has important implications for our conceptions of authorship, intellectual property, and political rights, particularly the right of free speech.

As we have observed in the discussion of the hypermedia pioneers Bush and Nelson, they believed that one of the greatest strengths of hypertext lies in its capacity to permit users to discover or produce multiple conceptual structures in the same body of information. A respected group of computer scientists, however, reject Nelsonian style link-and-lexia hypertext represented by Intermedia, Microcosm, and the World Wide Web. In "As We Should Have Thought"—the title an obvious play on Bush's seminal essay—Peter J. Nürnberg, John J. Leggett, and Erich R. Schneider assert, for example, that "linking is more than harmful—it is downright deadly" (96). For anyone whose chief experience of hypermedia has involved reading computer help files and materials on the World Wide Web, these statements seem to come from authors dwelling in some *Through the Looking Glass* alternate universe—particularly when they explain that the "two main problems . . . with hypermedia research today" derive from "our current notion of linking. Firstly, linking implies a certain kind of structural paradigm, one in which the user (or occasionally a program) links information together for purposes of navigation . . . Secondly, linking implies the primacy of data, not structure" (96). They certainly describe what is commonly understood to be hypertext, and their reason for rejecting it becomes clear when they explain their emphasis on structure:

We should have realized that hypermedia is just a special case of a general philosophy of computing in which structure is more important than data. Structure should be the ubiquitous, atomic building block available to all systems at all times and from which all other abstractions (including data) are derived. Here we call this philosophy of the primacy of structure "structural computing." (96)

Clearly, Nürnberg, Leggett, and Schneider have an entirely different set of concerns than do Bush and Nelson.

In fact, they represent a different approach to information technology—spatial hypertext—which they point out "has always pushed the limits of our notions of hypertext . . . Structure in spatial hypertext systems is dynamic and implicit. It is defined by the placement of data objects in a space. This structure is not traversed explicitly for the purpose of navigating the information. Instead, it is traversed (by the system) for the purpose of finding higher-level compositions of atomic data objects and lower-level compositions" (97). They are concerned primarily with information systems that analyze structure computationally, and, despite their opening salvo, it turns out that they do not in fact reject link-and-lexia or navigational hypermedia at all but simply wish to place their fundamentally different approach to computing "on a par with navigation systems" (97).

Nürnberg, Leggett, and Schneider point to VIKI, a system developed by Catherine C. Marshall and Frank M. Shipman III, as an example of spatial hypertext. As Marshall and Shipman explain, VIKI, which functions as a conceptual organizer, "provides users with visual and spatial affordances for organizing and interpreting information" ("Information Triage," 125). "Spatial hypertext," they explain elsewhere, "has its origins in browser-based approaches in which the emerging hypertext network is portrayed graphically, in an overview . . . In browser-based hypertext, boxes generally symbolize nodes; lines represent the links among them. In a completely spatial view of hypertext, the lines—links—may be removed from the picture, and the nodes may move about freely against their spatial backdrop" ("Spatial Hypertext," online version). Systems like VIKI rely on our "spatial intelligence," using graphic interfaces to organize complex ideas. Boxlike icons, which may contain text, represent concepts, and users can arrange these boxes and nest one inside another to explore or express their relationship. The Storyspace view, which functions in this manner, exemplifies one feature of spatial hypertext, but to this VIKI adds "structure finding algorithms that analyze the spatial layout and the visually salient properties of the information objects," so that authors do not have to construct explicit structures themselves. Many discussions of spatial hypertext so emphasize conceptual structures that they make it seem purely an organizational tool, but Marshall and Shipman believe that both readers and writers can benefit from it: "For readers, the system provides an opportunity to read in context, with awareness of the related, nearby nodes." For writers, it supports exploring various conceptual structures. For both, the graphic display of manipulatable information hierarchies "helps keep complexity tractable" ("Spatial Hypertext," online version).

Nürnberg, Leggett, and Schneider's definition of the concept makes me suspect that what they term *spatial hypertext* has little to do with hypertext and hypermedia, though it certainly represents an important area of research in computer science. In contrast, Marshall and Shipman's description of a graphic overview in which one can hide the lines representing links (which one can do in Storyspace) suggests that such spatial display of information does play a role in specific hypertext systems, though I don't know if *by itself* that feature constitutes a form of hypermedia. Graphic sitemaps, such as *The Victorian Web*'s opening screen, the Storyspace view, and Eastgate Systems's Tinderbox all exemplify graphic presentation of conceptual structures, but they don't have VIKI's ability to analyze and represent such structures computationally. Since the computer science literature uses the term *spatial hypertext*, I shall, too: in later chapters it refers to those aspects of hypertext environments, such as Storyspace, that use the graphic arrangement of lexias to convey structural information.

The Place of Hypertext

in the History of

Information Technology

The appearance of any new information technology like hypertext provides conditions for major societal change, though any change, such as the democratizing effects of writing, which took millennia, can take a very long time to occur. Such changes of information regimes always produce both loss and gain. In fact, let's propose a fundamental law of media change: no free lunch; or, there is no gain without some loss. Thus, if writing offers us the ability to contemplate information and respond to it at our leisure, thereby permitting personal reflection and considered thought, it also lacks the immediacy of the spoken voice and the clues that we receive while observing the person to whom we are speaking. Similarly, if we gain large audiences, new forms of text preservation, and standardization of the vernacular from print, we also lose what Benjamin termed the "aura" provided by the unique object. When people find that any particular gain from a new information technology makes up for the corollary loss, they claim it represents progress; when they feel loss more than gain, they experience the new information technology as cultural decline. Printing, an information technology that has so shaped our culture that most see it as an unqualified benefit, had its bad effects, too. Late medieval and early Renaissance connoisseurs, who mourned the loss of the scribal hand and pages that integrated words and images, considered printing a crude technology that destroyed aesthetic quality and blamed it for removing an important source of beauty from the world. For this reason they paid scribes to copy printed books and create manuscripts.[10] Far more important, the printed book, as Elizabeth Eisenstein has shown, led directly to centuries of religious warfare. Historians of the printed book point out the way it has shaped our culture, influencing our notions of self, intellectual property, language, education, and scholarship, and they present it in a largely favorable light but admit it had other effects as well. So when we consider the potential of hypermedia to change the way we do things, we must ask what the gains are and how they balance the losses that any new information regime causes.

Evaluating the relative effects and values of various media in relation to one another always turns out, however, to be more than a simple matter of loss and gain, because as J. David Bolter and Richard Grushin convincingly argue, every new medium

appropriates the techniques, forms, and social significance of other media and attempts to rival or refashion them in the name of the real. A medium in our culture can never operate in isolation, because it must enter into relationships of respect and rivalry with other media. There may be or may have been cultures in which a single form of representation (perhaps painting or song) exists with little or no reference to other media. Such isolation does not seem possible for us today, when we cannot even recognize the representational power of a medium except with reference to other media. (65)[11]

For an instantly convincing example of the ways in which newer and older information technologies influence each other, we need go no farther than Bolter and Grushin's observation that CNN and other television networks have increasingly resembled webpages, and at the same time the CNN "web site borrows its sense of immediacy from the televised CNN broadcasts" (9). Some examples of remediation produce results that can strike us as very odd indeed. Take the case of the arrival of printing within a scribal culture. "Typography was no more an addition to the scribal art than the motorcar was an addition to the horse. Printing had its 'horse-less carriage' phase of being misconceived and misapplied during its first decades, when it was not uncommon for a purchaser of a printed book to take it to a scribe to have it copied and illustrated" (McLuhan, *Understanding Media,* 189).

Before examining the relation of hypertext to previous media, I propose to look briefly at the advantages and disadvantages of various forms of information technology, a term that today is often mistakenly understood to refer solely to computing. Digital information technology certainly begins with the electronic digital computer, but information technology itself has been around for millennia. It begins with spoken language, which makes possible communal or community memory that in turn permits cultural development. Unlike biological or Darwinian development, such cultural (or Lamarkian) change permits groups of people to accumulate knowledge and practice and then pass them on to later generations; writing serves as individual prosthetic memory, which in turn creates a prosthetic group or community memory.

Speech as an information technology has certain qualities, which can be experienced as advantageous or disadvantageous, depending on the specific situations in which it occurs. "Language, like currency," McLuhan reminds us in *Understanding Media,* "acts as a store of perception and as a transmitter of the perceptions and experience of one person or one generation to another. As both a translator and storehouse of experience, language is, in addition, a reducer and a distorter of experience. The very great advantage of accelerating the learning process, and of making possible the transmission of knowledge and insight across time and space, easily overrides the disadvantages of linguistic codifications of experience" (151–52). These are not the only advantages and disadvantages of spoken language, for as Derrida (following Plato) urges, it is fundamentally a technology of *presence*. Speaker and listener have to be present in the same place and time, though, as Christian Metz points out, both do not have to be in sight of each other; one can, for example, hear words spoken by someone on the other side of a door or in a darkened room. What advantages and what disadvantages, then, does such an information technology based on presence have? This question turns out to be an especially crucial one, because many info-pundits automatically assume that presence has more importance than all other qualities and effects of any particular information technology. I've often observed that many writers on media, particularly its educational applications, often react to improvements in computing, such as increased speed of Internet connections, as if the most important result of any such change lies in the possibility of affordable telepresence. They imply that speaking in sight of the listener always has more value than writing or other forms of communication. In other words, considering the possibility of sending and receiving large quantities of information over electronic networks, the first reaction of many educators and businesspersons is that we can replace written text by talking heads.

This typical reaction exemplifies Bolter and Grushin's point that the advocates of all new information technologies always claim theirs possess *immediacy,* and this claim derives from "the desire to get past the limits of representation and to achieve the real" (53). Such assumptions, which ignore the very different strengths of speech and writing, demonstrate that many people believe that being in the presence of someone trumps all the advantages, including reflection, abstraction, organization, and concision, that writing enables. Moreover, presence is itself not always a desirable, much less the most important, quality when communicating with another person. We can all think of situations in which we feel more comfortable talking on a telephone than speaking to someone face to face: when we do not look our best—say, because we've just awakened—or when we wish to fend off someone trying to solicit money for a charity or sell us something. Absence, in other words, also has great value in certain communicative situations, a crucial factor to take into account when considering the gains and losses involved in writing.

Writing, probably the most important technology human beings ever developed, exchanges presence and simultaneity for asynchronous communication—for the opportunity to respond at one's own convenience. Because it does not base the act of communication on presence, writing does not require the person communicating to be in either the same place or the same time as the person receiving the communication. The person communicating information places it in a form that permits someone else to receive it later. Writing, printing, cinema, and video are all forms of asynchronous communication, which, as McLuhan points out in *The Gutenberg Galaxy*, permits reflection, abstraction, and forms of thought impossible in an oral culture. Writing's combination of absence and asynchronisity obviously permits a new kind of education, as well as itself becoming a goal of education, since teaching reading and writing becomes a primary function of early instruction in eras in which these skills are important.

For millennia, writing, which eventually leads to silent reading, nonetheless remained a technology that oddly combined orality and literacy. The explanation for this situation lies in economic and material factors. The high cost and scarcity of writing surfaces prompted scribes to omit spaces between words and adopt a bewildering array of abbreviations, all so they could cram as many characters as possible on a scroll or page. These material conditions produced a kind of text that proved so difficult to read that it chiefly served as a mnemonic device, and readers often read aloud. Eventually, around the year 1000, cheaper writing materials led to the development of interword spacing, which in turn encouraged silent reading—a practice that tended to exchange expressive performance and a communal experience for privacy, increasing reading speed, and the sense of personal or inner place. Interword spacing, like the codex (what we generally call a book), eventually changed reading from a craft skill to an ordinary one required of every citizen.

Since the invention of writing and printing, information technology has concentrated on the problem of creating and then disseminating static, unchanging records of language. As countless authors since the inception of writing have proclaimed, such fixed records conquer time and space, however temporarily, for they permit one person to share data with other people in other times and places. As Elizabeth Eisenstein argues, printing adds the absolutely crucial element of multiple copies of the same text; this multiplicity, which preserves a text by dispersing individual copies of it, permits readers separated in time and space to refer to the same information (116). As Eisenstein, Marshall McLuhan, William M. Ivins, J. David Bolter, and other students of the history of the cultural effects of print technology have shown, Gutenberg's invention produced what we today understand as scholarship and criticism in the humanities. No longer primarily occupied by the task of preserving information in the form of fragile manuscripts that degraded with frequent use, scholars, working with books, developed new conceptions of scholarship, originality, and authorial property.

Hand-set printing with movable type permits large numbers of readers widely separated in time and space to encounter essentially the same text—and hence creates a new kind of virtual community of readers and many other things basic to modern culture. The existence of multiple copies of the same text permits readers hundreds of miles and hundreds of years apart to refer to specific passages by page number. Printing, which thus exemplifies asynchronous, silent communication, provides the conditions for the development of a humanistic and scientific culture dependent on the ability to cite and discuss specific details of individual texts. And of course it drastically changes the nature of education, which moves from dictating primary texts to the student to teaching the student modes of critical analysis. "Even in the early eighteenth century," McLuhan reminds us, "a 'textbook' was still defined as a 'Classick Author written very wide by Students, to give room for an Interpretation dictated by the Master, &c., to be inserted in the Interlines' (O.E.D.). Before printing, much of the time in school and college classrooms was spent in making such texts" (*Understanding Media*, 189).

High-speed printing, which appeared in the nineteenth century, truly acted as a democratizing force, producing many of our conceptions of self, intellectual property, and education. In addition to creating a virtual community of readers, the relatively inexpensive texts created by high-speed printing radically changed the notions of an earlier manuscript culture about how to preserve texts: with printing, one preserves texts by creating and distributing multiple copies of them rather than, as with manuscripts, which eventually degrade after many readings, protecting the text by permitting fewer people to have access to it. As we all know, the book also functions as a kind of self-teaching machine that turns out to be far more accessible and hence more quickly democratizing than manuscript texts can ever be.

Although the fixed multiple text produced by print technology has had enormous effects on modern conceptions of literature, education, and research, it still, as Bush and Nelson emphasize, confronts the knowledge worker with the fundamental problem of an information retrieval system based on physical instantiations of text—namely, that preserving information in a fixed, unchangeable linear format makes information retrieval difficult.

We may state this problem in two ways. First, no one arrangement of information proves convenient for all who need that information. Second, although both linear and hierarchical arrangements provide information in some sort of order, that order does not always match the needs of individual users of that information. Over the centuries scribes, scholars, publishers, and other makers of books have invented a range of devices to increase the speed of what today are called information processing and retrieval. Manuscript culture gradually saw the invention of individual pages, chapters, paragraphing, and spaces between words. The technology of the book found enhancement by pagination, indices, and bibliographies. Such devices have made scholarship possible, if not always easy or convenient to carry out.

The next great change in information technology—and that which most concerns us—came with the development of digital information technology. For the first time, writing, which had always been a matter of physical marks on a physical surface, instead takes the form of electronic codes, and this shift from ink to electronic code—what Jean Baudrillard calls the shift from the "tactile" to the "digital" (*Simulations,* 115)—produces an information technology that combines fixity and flexibility, order and accessibility—but at a cost.[12] Using Diane Balestri's terminology, we can say that all previous media took the form of hard text (cited in Miles, "Softvideography"); computing produces soft text, and this fundamental change, like all developments in infotech, comes with gains and losses. For example, although electronic writing has the multiplicity of print, it does not have the fixity—and hence the reliability and stability—of either written or printed texts.

As Bolter and Grushin point out, over the past half century digital computing has undergone what they call a "process of 'remediatization'" during which society understood it as having fundamentally different purposes:

■ The "programmable digital computer was invented in the 1940s as a *calculating engine* (ENIAC, EDSAC, and so on)" (66) for military and scientific application.

■ During the next decade "large corporations and bureaucracies" (66) used it, instead, for *accounting.*

■ About the same time, a few pioneers saw the computer as "a new *writing* technology" (66).

■ Turing and those involved with AI (artificial intelligence) saw the computer primarily as a *"symbol manipulator"* that could "remediate earlier technologies of arbitrary symbol manipulation, such as handwriting and printing" (66).

■ "In the 1970s, the first *word processors* appeared, and in the 1980s the desktop computer. The computer could then become a medium because it could enter into the social and economic fabric of business culture and remediate the typewriter almost out of existence" (66).

■ More recently, the computer has been seen as an *image capturer, presenter, and manipulator:* "If even ten years ago we thought of computers exclusively as numerical engines and word processors, we now think of them also as devices for generating images, reworking photographs, holding videoconferences, and providing animation and special effects for film and television" (23).

This fundamental shift from tactile to digital, physical to code, and hard to soft media produces text with distinctive qualities. First of all, since electronic text processing is a matter of manipulating computer codes, all texts that the reader-writer encounters on the screen are virtual texts. Using an analogy to optics, computer scientists speak of "virtual machines" created by an operating system that provides individual users with the experience of working on their own individual machines when they in fact share a system with as many as several hundred others.[13] According to the *Oxford English Dictionary,* "virtual" is that which "is so in essence or *effect,* although not formally or actually; admitting of being called by the name so far as the effect or result is concerned," and this definition apparently derives from the use of the term in optics, where it refers to "the apparent focus or image resulting from the effect of reflection or refraction upon rays of light." In computing, the virtual refers to something that is *"not physically existing as such but made by software to appear to do so* from the point of view of the program or the user" (emphasis added). As Marie-Laure Ryan points out, the powerful concept of virtualization "leads from the here and now, the singular, the usable once-for-all, and the solidly embodied to the timeless, abstract, general, multiple, versatile, repeatable, ubiquitous, and morphologically fluid" (37).[14]

Similarly, all texts the reader and the writer encounter on a computer screen exist as a version created specifically for them while an electronic primary version resides in the computer's memory. One therefore works on an electronic copy until such time as both versions converge when the writer commands the computer to "save" one's version of the text by placing it in memory. At this point the text on screen and in the computer's memory briefly coincide, but the reader always encounters a virtual image of the stored text and not the original version itself; in fact, in descriptions of electronic word processing, such terms and such distinctions do not make much sense.

As Bolter explains, the most "unusual feature" of electronic writing is that it is "not directly accessible to either the writer or to the reader. The bits of the text are simply not on a human scale. Electronic technology removes or abstracts the writer and reader from the text. If you hold a magnetic tape or optical disk up to the light, you will not see text at all . . . In the electronic medium several layers of sophisticated technology must intervene between the writer or reader and the coded text. There are so many levels of deferral that the reader or writer is hard put to identify the text at all: is it on the screen, in the transistor memory, or on the disk?" (*Writing Space*, 42–43). Furthermore, whereas a printed book has weight and mass, its digital form appears immaterial. "If you want to get picky about the physics," Mitchell elegantly explains, "we can say that the corpus of classical literature is now embodied electromagnetically, and, yes, electrons do have mass. But that is irrelevant at the level of everyday experience. My briefcase quickly gets weighed down if I load volumes of the Loeb Classical Library into it, but my laptop does not get any heavier if I download the *TLG* onto its hard drive" (*Me++*, 231n. 7).

The "'virtual' and the 'material,'" Ned Rossiter reminds us, "are always intimately and complexly intertwined" (177), and so emphasizing the virtuality of electronic *text and image* in no way implies that the actual reading experience involves either a disembodied reader or a nonmaterial presentation of text itself. As N. Katherine Hayles emphasizes, we have to find new ways "to think about embodiment in an age of virtuality" (193).[15] We must, for example, come to the absolutely necessary recognition that the physical, material conditions of computer devices we use affect our experience of virtual text. As I have pointed out elsewhere, the size of monitors, the change from bitmap to grayscale to color displays, the portability of computers, and our physical distance from them make dramatic differences in kinds of texts we can read and write ("What's a Critic to Do?" and "Connected Images," 82).[16] Computer text may be virtual, but we who read it are still physical, to read it we rely on physical devices, and it has effects on the physical world. "Bits just don't sit out there in cyberspace," Mitchell reminds us, and therefore "it makes more sense to recognize that invisible, intangible, electromagnetically encoded information establishes new types of relationships among *physical* events occurring in *physical* places" (*Me++*, 4).

The code-based existence of electronic text that makes it virtual also makes it infinitely variable. If one changes the code, one changes the text. As Hayles has pointed out, "When a text presents itself as a constantly refreshed image rather than as a durable inscription, transformations can occur that would be unthinkable if matter and energy, rather than informational patterns, formed the primary basis for the systemic exchanges" (30). Furthermore, since digital information technology stores both alphanumeric text (words) and images as codes, it sees no essential difference between them. With images, as with words, if one manipulates the code, one manipulates the text that this code preserves and produces. Furthermore, as anyone who has ever resized a web browser window or enlarged a font in a Microsoft Word or PDF file knows, this text-as-code is always adaptable. Because users only experience a virtual image of the text, they can manipulate the version they see without affecting the source. Many forms of computer text, in other words, grant the reader more power than does any example of writing or print, though occasionally at the cost of a loss of powerful graphic design. E-text documents also have permeable limits: borders and edges, like spaces, are matters of physicality, materiality, embodiment, but digital text—text woven of codes— does not have and cannot have such unity, such closure. The digital text, which exists independent of the place in which we experience it, e-merges as dispersed text. When we discuss hypertext later, we shall see that hypertextual linking relates in important ways to this property of electronic text.

The coded basis of digital text permits it to be processed in various ways, producing documents, for example, that are both searchable and analyzable. Thus users can search electronic texts for letters and other characters, words, or various groups of them. Users can also take advantage of such code-based textuality to check the spelling, grammar, and style of digital text. Processable text also permits text as simulation since changing the code makes the text move to show things impossible to present with a static image or text. As we shall see when we examine examples of animation in chapter 3, such capacities permit one to argue by demonstrating things often too difficult to show easily with linguistic argument.

Digital text can be infinitely duplicated at almost no cost or expenditure of energy. Duplicate the code, duplicate the text—a fact true for images (including images of text, as above) or alphabetic text. As Mitchell explains with characteristic clarity, "Digital texts, images, and other artifacts begin to behave differently from their heavier, materially embedded predecessors. They become nontrivial assets—they are neither depleted not divided when shared, they can be reproduced indefinitely without cost or loss of quality, and they can be given away without loss to the giver" (*Me++*, 83). One can just duplicate the code and thereby repeat—reproduce—the text, thereby affecting the cost (and value) of the text and the potential size of one's audience.

Because the codes that constitute electronic text can move at enormous speed over networks, either locally within organizations or on the Internet, they create the conditions for new forms of scholarly and other communication. Before networked computing, scholarly communication relied chiefly on moving physical marks on a surface from one place to another with whatever cost in time and money such movement required. Networked electronic communication so drastically reduces the time scale of moving textual information that it produces new forms of textuality. Just as transforming print text to electronic coding radically changed the temporal scale involved in manipulating texts, so too has it changed the temporal scale of sharing them. Networked electronic communication has both dramatically speeded up scholarly communication and created quickly accessible versions of older forms of it, such as online, peer-reviewed scholarly journals, and new forms of it, such as discussion lists, chat groups, blogs, and IRC (Internet Relay Chat) (Landow, "Electronic Conferences," 350). In networked environments users also experience electronic text as location independent, since wherever the computer storing the text may reside in physical reality, users experience it as *being here,* on their machines. When one moves the text-as-code, it moves fast enough that it doesn't matter where it "is" because it can be everywhere . . . and nowhere.[17] Finally, electronic text is *net-work-able,* always capable of being joined in electronic networks. Thus, hypertext and the World Wide Web.

Like many features of digital textuality, the sheer speed of obtaining information has its good and bad sides. Its advantages include increasingly sophisticated World Wide Web search tools, such as Google, that can provide needed information nearly instantaneously. For example, as part of the process of writing *Hypertext 3.0,* I wanted to look up some technical terms (RSS, Atom feed) related to blogs. Typing one of these terms into Google, I pressed the "return" key and received a list of relevant web documents in less than a second—0.22 second, to be exact; the information I found most useful occurred in the first and third listed items. The convenience of such information retrieval has increasingly led students and faculty to use such search tools instead of physical libraries. Indeed, "'one of the rarest things to find is a member of the faculty in the library stacks,'"—so Katie Hafner's article in the *New York Times* quotes an instructor at a major research university.

True, Hafner slightly sensationalizes the use of Google in research by not clarifying the difference between Internet searches and online resources, such as large collections of scholarly journals that originally appeared in print. Faculty and students devote a good deal of their research time to locating and reading these scholarly journals, so online versions of them are enormously convenient: one can locate individual articles in a few minutes at most, multiple users can read them at the same time, and one can obtain them when the library is closed; some journals are actually more to pleasant to read online, since one can increase the size of print in the online copy. Nonetheless, not all research involves back issues of specialist periodicals, and depending on Internet search tools at the present time might cause one to miss a good deal:

> The biggest problem is that search engines like Google skim only the thinnest layers of information that has been digitized. Most have no access to the so-called deep Web, where information is contained in isolated databases like online library catalogs. Search engines seek so-called static Web pages, which generally do not have search functions of their own. Information on the deep Web, on the other hand, comes to the surface only as the result of a database query from within a particular site. Use Google, for instance, to research Upton Sinclair's 1934 campaign for governor of California, and you will miss an entire collection of pamphlets accessible only from the University of California at Los Angeles's archive of digitized campaign literature.

With an estimated 500 billion webpages hidden from search engines, companies like Google and Yahoo have entered into agreements with major libraries to index their collections. Still, as many observers have pointed out, researchers who only Google for their information—yes, it's actually become a much-used verb—miss not only a good deal of valuable material but the pleasures of working with printed books and materials, including the delightful serendipity of stumbling onto something particularly interesting while looking for something else. If history provides any lessons, then the marked convenience of Internet resources will increasingly dominate both scholarly research and far more common everyday searches for information: the appearance of the printed book did not make individual manuscripts any more difficult to use than they had been before Gutenberg, but eventually the rapidly growing number of texts in print, their standardized vernacular, and their increased legibility made them so convenient that only scholars with very specific interests consulted manuscripts, or still do.

Frankly, I think the consequences for literary education, criticism, and scholarship are vastly exaggerated for two simple reasons. First, comparatively little—indeed, almost no—*literary* research requiring this kind of inaccessible information takes place in colleges and universities. Much of what is now termed *literary research* simply takes the form of reading secondary materials, and the rest involves working with materials contemporary with the texts one is studying—materials almost always catalogued. As far as undergraduate education is concerned, I believe electronic resources like JSTOR provide a far greater range of information than do *Twentieth-Century Views* and other prepackaged collections of secondary materials.

Far more important, libraries frequently do not have that kind of information missed by Internet search tool in handwritten, typed, or printed form because many of these materials are out of fashion and hence fall beneath the radar. As an old-fashioned hide-bound scholar whose first books depended on manuscripts and extremely rare printed material, I quickly discovered that my own university library and others had neither the information I needed nor the information about the information. For example, looking for the published transcripts of sermons John Ruskin commented on in his diary while in the midst of an agonizing religious crisis, I discovered that major New York libraries had no record of what was once an extremely popular and profitable genre (at least three weekly British periodicals dispatched stenographers to take down the sermons of popular preachers; I stumbled onto the fact of their existence when Ruskin quoted from one in a famous letter to *The Times;* the great Victorian scholar Geoffrey Tillotson, then my Fulbright advisor, told me: "I think you're on to something important. Follow it up."). I finally found un-catalogued copies stored in a carton in the basement of a theological seminary. Another example: the catalogue of the Beineke rare book library at Yale—"accessed" by snailmail and the good offices of a librarian—listed the manuscript of one of Ruskin's own childhood notes on sermons (his mother made him do it), but I unexpectedly discovered the valuable first draft of these notes in a display case in the tiny museum in Coniston, where Ruskin lived for many years. Even if one knows where materials are located through the scholarly grapevine, they may not be maintained in easily searchable form. After traveling to the Isle of Wight to work with the vast collection of Ruskin letters and diaries at the Bembridge School, I discovered they were uncatalogued. Even locating the catalogue entry for an item (the information about the information) doesn't mean you will find it. Thus, when I thought I had located in the then-British Museum Library a crucial anonymous exhibition pamphlet in fact written by the artist W. Holman Hunt himself, I submitted my call-slip, waited forty-five minutes, and discovered that it had been "destroyed by enemy action" during the Blitz; I unexpectedly bumbled onto a copy at the bottom of a trunk when, as I was leaving his adopted granddaughter's home, she asked, "Would you like to look through some things in the garage?" That's enough of what van Dam called van "barefoot-in-the-snow stories" (889) in his keynote address at the world's first hypertext conference. Two points: first, it's obviously better to be lucky than good, and, second, digitizing all the library catalogues and deep Web material in the world does not help if the information you need is not there in the first place—and for much of the most interesting kind of research that cataloguing information does not exist.

Far more important a problem with digital searches, as Eugene Provenzo warned two decades ago, is not what necessary information we can't find but what personal information governments and corporations can near-instantly discover about us. One example will suffice. Google, which has so shaped the world of education and scholarship, is currently offering free e-mail accounts with enormous storage (1 gigabyte), and the company urges users never to discard anything: "You never know when you might need a message again, but with traditional webmail services, you delete it and it's gone forever. With Gmail, you can easily archive your messages instead, so they'll still be accessible when you need them." According to a website whose URL is gmail-is-too-creepy.com, "Google admits that even deleted messages will remain on their system, and may also be accessible internally at Google, for an indefinite period of time." The danger, according to Public Information Research, which created the site, is that the company pools its information, keeps it indefinitely, and can share it with anyone they wish. "All that's required is for Google to 'have a good faith belief that access, preservation or disclosure of such information is reasonably necessary to protect the rights, property or safety of Google, its users or the public." These privacy advocates claim that the company's statements about terms of use and privacy "mean that all Gmail account holders have consented to allow Google to show any and all email in their Gmail accounts to any official from any government whatsoever, even when the request is informal or extralegal, at Google's sole discretion." Moreover, nothing in Gmail's stated policy clarifies if it will save and index incoming mail from those who have not agreed to use their system. When one uses Google as a search tool, its software, like that of many other sites, places a so-called cookie with a unique ID number on your computer that does not expire until 2038. By that means it keeps track of any search you have ever made. According to various privacy advocates and consumer groups, connecting e-mail to this powerful tool creates the inevitability of enormous abuses by corporate and government interests, many of whom are not subject to U.S. law—this last a particularly relevant point since two-thirds of Google users live outside the United States, many in countries without privacy laws. No free lunch.

Interactive or Ergodic?

Readers may have noticed that in the preceding discussions of electronic media I have not employed the words *interactive* and *interactivity*. As many commentators during the past decade and a half have observed, these words have been used so often and so badly that they have little exact meaning anymore. Just as chlorophyll was used to sell toothpaste in the 1950s and aloe was used to sell hand lotion and other cosmetic products in the 1970s and 1980s, *interactive* has been used to sell anything to do with computing, and the word certainly played a supporting role in all the hype that led to the dotcom bust. The first time I heard the two terms criticized, I believe, was in 1988, when a speaker at a conference, who was satirizing false claims that computers always give users choices, projected a slide of a supposed dialogue box. To the question, "Do you want me to erase all your data?" the computer offered two choices: "Yes" and "OK."[18]

Espen Aarseth, who has particular scorn for *interactive* and *interactivity*, quite rightly points out that "to declare a system is interactive is to endorse it with a magic power" (48). He proposes to replace it by *ergotic*, "using a term appropriated from physics that derives from the Greek words *ergon* and *hodos*, meaning 'work' and 'path.' In ergodic literature, nontrivial effort is required to allow the reader to traverse the text. If ergodic literature is to make sense as a concept, there must also be nonergodic literature, where the effort to traverse the text is trivial, with no extranoematic responsibilities placed on the reader except (for example) eye movement and the periodic or arbitrary turning of pages" (1–2). *Ergodic*, which has the particular value of being new and thus far not used in false advertising, has received wide acceptance, particularly by those who study computer games as cultural forms. Still, Marie-Laure Ryan's *Narrative as Virtual Reality* (2001), one of the most important recent books on digital culture, retains "interactive," and a glance through the proceedings of 2003 Melbourne Digital Arts Conference reveals that people working with film and video also prefer the term.[19]

Ergodic, when used as a technical term, has its problems, too, since it's not clear that the reader's "eye movement" and turning pages, which result from intellectual effort, are in fact trivial—a point Aarseth himself seems to accept when he emphasizes Barthes's point that readers can skip about a page (78). *Ergodic* nonetheless appears a useful coinage, and so is the word *interactive* when used, as in Ted Nelson's writings, to indicate that the computer user has power to intervene in processes while they take place, as opposed to the power to act in a way that simply produces an effect, such as flipping a switch to turn on a light. The wide misuse of an important term is hardly uncommon. After all, *deconstruction* has been used in academic writing and newspapers to mean everything from "ordinary interpretation" to "demolition" while the term *classical* has meant everything from a "historical period," to an "aesthetic style," to an "eternal principle found throughout human culture." Before writing these paragraphs I checked the earlier version of *Hypertext* and found only four uses of *interactive* other than in quoted material; this one uses six. Even though I do not employ it very much, I think *interactive,* like *ergodic,* has its uses.

Baudrillard, Binarity, and the Digital

Jean Baudrillard, who presents himself as a follower of Walter Benjamin and Marshall McLuhan, is someone who seems both fascinated and appalled by what he sees as the all-pervading effects of digital encoding, though his examples suggest that he is often confused about which media actually employ it.[20] The strengths and weaknesses of Baudrillard's approach appear in his remarks on the digitization of knowledge and information. Baudrillard correctly perceives that movement from the tactile to the digital is the primary fact about the new information technology, but then he misconceives—or rather only partially perceives—the implications of his point. According to him, digitality involves binary opposition: "Digitality is with us. It is that which haunts all the messages, all the signs of our societies. The most concrete form you see it in is that of the test, of the question/answer, of the stimulus/response" (*Simulations,* 115). Baudrillard most clearly posits this equivalence, which he mistakenly takes to be axiomatic, in his statement that "the true generating formula, that which englobes all the others, and which is somehow the stabilized form of the code, is that of binarity, of digitality" (145). From this he concludes that the primary fact about digitality is its connection to "cybernetic control . . . the new operational configuration," since "digitalization is its metaphysical principle (the God of Leibnitz), and DNA its prophet" (103).[21]

True, at the most basic level of machine code and at the far higher one of program languages, digitalization, which constitutes a fundamental of electronic computing, does involve binarity. But from this fact one cannot so naively extrapolate, as Baudrillard does, a complete thought-world or *episteme.* Baudrillard, of course, may well have it partially right: he might have perceived one key connection between the stimulus/response model and digitality. The fact of hypertext, however, demonstrates quite clearly that digitality does not necessarily lock one into either a linear world or one of binary oppositions.

Unlike Derrida, who emphasizes the role of the book, writing, and writing technology, Baudrillard never considers verbal text, whose absence glaringly runs through his argument and reconstitutes it in ways that he obviously did not expect. Part of Baudrillard's theoretical difficulty, I suggest, derives from the fact that he bypasses digitized verbal text and moves with too easy grace directly from the fact of digital encoding of information in two directions: (1) to his stimulus/response, either/or model, and (2) to other nonalphanumeric (or nonwriting) media, such as photography, radio, and television. Interestingly enough, when Baudrillard correctly emphasizes the role of digitality in the postmodern world, he generally derives his examples of digitalization from media that, particularly at the time he wrote, for the most part depended on analogue rather than digital technology—and the different qualities and implications of each are great. Whereas analogue recording of sound and visual information requires serial, linear processing, digital technology removes the need for sequence by permitting one to go directly to a particular bit of information. Thus, if one wishes to find a particular passage in a Bach sonata on a tape cassette, one must scan through the cassette sequentially, though modern tape decks permit one to speed the process by skipping from space to space between sections of music. In contrast, if one wishes to locate a passage in digitally recorded music, one can instantly travel to that passage, note it for future reference, and manipulate it in ways impossible with analogue technologies—for example, one can instantly replay passages without having to scroll back through them.

In concentrating on nonalphanumeric media, and in apparently confusing analogue and digital technology, Baudrillard misses the opportunity to encounter the fact that digitalization also has the potential to prevent, block, and bypass linearity and binarity, which it replaces with multiplicity, true reader activity and activation, and branching through networks. Baudrillard has described one major thread or constituent of contemporary reality that is potentially at war with the multilinear, hypertextual one.

In addition to hypertext, several aspects of humanities computing derive from virtuality of text. First of all, the ease of manipulating individual alphanumeric symbols produces simpler word processing. Simple word processing in turn makes vastly easier old-fashioned, traditional scholarly editing—the creation of reliable, supposedly authoritative texts from manuscripts or published books—at a time when the very notion of such single, unitary, univocal texts may be changing or disappearing.

Second, this same ease of cutting, copying, and otherwise manipulating texts permits different forms of scholarly composition, ones in which the researcher's notes and original data exist in experientially closer proximity to the scholarly text than ever before. According to Michael Heim, as electronic textuality frees writing from the constraints of paper-print technology, "vast amounts of information, including further texts, will be accessible immediately below the electronic surface of a piece of writing . . . By connecting a small computer to a phone, a professional will be able to read 'books' whose footnotes can be expanded into further 'books' which in turn open out onto a vast sea of data bases systemizing all of human cognition" (*Electric Language*, 10–11). The manipulability of the scholarly text, which derives from the ability of computers to search databases with enormous speed, also permits full-text searches, printed and dynamic concordances, and other kinds of processing that allow scholars in the humanities to ask new kinds of questions. Moreover, as one writes, "the text in progress becomes interconnected and linked with the entire world of information" (*Electric Language,* 161).

Third, the electronic virtual text, whose appearance and form readers can customize as they see fit, also has the potential to add an entirely new element—the electronic or virtual link that reconfigures text as we who have grown up with books have experienced it. Electronic linking creates hypertext, a form of textuality composed of blocks and links that permits multilinear reading paths. As Heim has argued, electronic word processing inevitably produces linkages, and these linkages move text, readers, and writers into a new writing space:

The distinctive features of formulating thought in the psychic framework of word processing combine with the automation of information handling and produce an unprecedented linkage of text. By *linkage* I mean not some loose physical connection like discrete books sharing a common physical space in the library. Text derives originally from the Latin word for weaving and for interwoven material, and it has come to have extraordinary accuracy of meaning in the case of word processing. Linkage in the electronic element is interactive, that is, texts can be brought instantly into the same psychic framework. (*Electric Language,* 160–61)

The presence of multiple reading paths, which shift the balance between reader and writer, thereby creating Barthes's writerly text, also creates a text that exists far less independently of commentary, analogues, and traditions than does printed text. This kind of democratization not only reduces the hierarchical separation between the so-called main text and the annotation, which now exist as independent texts, reading units, or lexias, but it also blurs the boundaries of individual texts. In so doing, electronic linking reconfigures our experience of both author and authorial property, and this reconception of these ideas promises to affect our conceptions of both the authors (and authority) of texts we study and of ourselves as authors.

Equally important, all these changes take place in an electronic environment, the Nelsonian docuverse, in which publication changes meaning. Hypertext, far more than any other aspect of computing, promises to make publication a matter of gaining access to electronic networks. For the time being scholars will continue to rely on books, and one can guess that continuing improvements in desktop publishing and laser printing will produce a late efflorescence of the text as a physical object. Nonetheless, these physical texts will be produced (or rather reproduced) from electronic texts, and as readers increasingly become accustomed to the convenience of electronically linked texts, books, which now define the scholar's tools and end-products, will gradually lose their primary role in humanistic scholarship.

Books Are Technology, Too

We find ourselves, for the first time in centuries, able to see the book as unnatural, as a near-miraculous technological innovation and not as something intrinsically and inevitably human. We have, to use Derridean terms, decentered the book. We find ourselves in the position, in other words, of perceiving the book *as technology*. I think it no mere coincidence that it is at precisely this period in human history we have acquired crucial intellectual distance from the book as object and as cultural product. First came distant writing (the telegraph), next came distant hearing (the telephone), which was followed by the cinema and then the distant seeing of television. It is only with the added possibilities created by these new information media and computing that Harold Innis, Marshall McLuhan, Jack Goody, Elizabeth Eisenstein, Alvin Kernan, Roger Chartier, and the European scholars of *Lesengeshichte* could arise.

Influential as these scholars have been, not all scholars willingly recognize the power of information technologies on culture. As Geert Lovink, the Dutch advocate of the sociopolitical possibilities of the Internet, has wryly observed, "By and large, [the] humanities have been preoccupied with the impact of technology from a quasi-outsider's perspective, as if society and technology can still be separated" (*Dark Fiber*, 13). This resistance appears in two characteristic reactions to the proposition that information technology constitutes a crucial cultural force. First, one encounters a tendency among many humanists contemplating the possibility that information technology influences culture to assume that before now, before computing, our intellectual culture existed in some pastoral nontechnological realm. *Technology*, in the lexicon of many humanists, generally means "only that technology of which I am frightened." In fact, I have frequently heard humanists use the word *technology* to mean "some intrusive, alien force like computing," as if pencils, paper, typewriters, and printing presses were in some way *natural*. Digital technology may be new, but technology, particularly information technology, has permeated all known culture since the beginnings of human history. If we hope to discern the fate of reading and writing in digital environments, we must not treat all previous information technologies of language, rhetoric, writing, and printing as nontechnological.

As John Henry Cardinal Newman's *Idea of a University* reminds us, writers on education and culture have long tended to perceive only the negative effects of technology. To us who live in an age in which educators and pundits continually elevate reading books as an educational ideal and continually attack television as a medium that victimizes a passive audience, it comes as a shock to encounter Newman claiming that cheap, easily available reading materials similarly victimized the public. According to him,

What the steam engine does with matter, the printing press is to do with mind; it is to act mechanically, and the population is to be passively, almost unconsciously enlightened, by the mere multiplication and dissemination of volumes. Whether it be the school boy, or the school girl, or the youth at college, or the mechanic in the town, or the politician in the senate, all have been the victims in one way or other of this most preposterous and pernicious of delusions. (108)

Part of Newman's rationale for thus denouncing cheap, abundant reading materials lies in the belief that they supposedly advance the dangerous fallacy that "learning is to be without exertion, without attention, without toil; without grounding, without advance, without finishing"; but, like any conservative elitist in our own day, he fears the people unsupervised, and he cannot believe that reading without proper guidance—guidance, that is, from those who know, from those in institutions like Oxford—can produce any sort of valid education, and, one expects, had Newman encountered the self-taught mill-workers and artisans of Victorian England who made discoveries in chemistry, astronomy, and geology after reading newly available books, he would not have been led to change his mind.

Like Socrates, who feared the effects of writing, which he took to be an anonymous, impersonal denaturing of living speech, Newman also fears an "impersonal" information technology that people can use without supervision. And also like Socrates, he desires institutions of higher learning—which for the ancient took the form of face-to-face conversation in the form of dialectic—to be sensitive to the needs of specific individuals. Newman therefore argues that "a University is, according to the usual designation, an Alma Mater, knowing her children one by one, not a foundry, or a mint, or a treadmill."

Newman's criticism of the flood of printed matter produced by the new technology superficially echoes Thomas Carlyle, whose "Signs of the Times"

(1829) had lambasted his age for being a mechanical one whose "true Deity is Mechanism." In fact, claims this first of Victorian sages,

not the external and physical alone is managed by machinery, but the internal and spiritual also. Here too nothing follows its spontaneous course, nothing is left to be accomplished by old, natural methods . . . Instruction, that mysterious communing of Wisdom with Ignorance, is no longer an indefinable tentative process, requiring a study of individual aptitudes, and a perpetual variation of means and methods, to attain the same end; but a secure, universal, straightforward business, to be conducted in the gross, by proper mechanism, with such intellect as comes to hand. (101)

Several things demand remark in this passage, the first and most obvious of which is that it parallels and might have provided one of the major inspirations for Newman's conceptions of education. The second recognition, which certainly shocks us more than does the first, is that Carlyle attacks those like Newman who propose educational systems and design institutions.

In sentences that I have omitted from the quoted passage, Carlyle explained that everything, with his contemporaries, "has its cunningly devised implements, its preestablished apparatus; it is not done by hand but by machinery. Thus we have machines for education: Lancasterian machines; Hamiltonian machines; monitors, maps, and emblems." Or, as Carlyle might say today, we have peer tutoring, core curricula, distribution requirements, work-study programs, and junior years abroad.

What is not at issue here is the practicality of Carlyle's criticisms of the mechanization of education and other human activities—after all, it would seem that he would attack any organizational change on the same grounds. No, what is crucial here is that Carlyle, who apparently denies all possibilities for reforming existing institutions, recognizes something crucial about them that Newman, the often admirable theorist of education, does not. Carlyle, in other words, recognizes that all institutions and forms of social organization are properly to be considered technologies. Carlyle, who pointed out elsewhere that gunpowder and the printing press destroyed feudalism, recognized that writing, printing, pedagogical systems, and universities are all technologies of cultural memory. Newman, like most academics of the past few hundred years, considers them, more naively, as natural and inevitable, and consequently notices the effects of only those institutions new to him or that he does not like.

The great value of such a recognition to our project here lies in the fact that it reminds us that electronifying universities does not take the form of technologizing them or adding technology to them in some way alien to their essential spirit. Digital information technology, in other words, is only the latest to shape an institution that, as Carlyle reminds us, is both itself a form of technology, a mechanism, and has also long been influenced by those technologies on which it relies.

A second form of resistance to recognizing the role of information technology in culture appears in implicit claims that technology, particularly information technology, can *never* have cultural effects. Almost always presented by speakers and writers as evidence of their own sophistication and sensitivity, this strategy of denial has an unintended effect: denying that Gutenberg's invention or television can exist in a causal connection to any other aspect of culture immediately transforms technology—whatever the author means by that term—into a kind of intellectual monster, something so taboo that civilized people cannot discuss it in public. In other words, it takes technology, which is both an agent and effect of our continuing changing culture(s), and denies its existence as an element of human culture. One result appears in the strategies of historical or predictive studies that relate cultural phenomena to all sorts of economic, cultural, and ideological factors but avert their eyes from any technological causation, as if it, and only it, were in some way reductive. The effect, of course, finally is to deny that this particular form of cultural product can have any effect.

We have to remind ourselves that if, how, and whenever we move beyond the book, that movement will not embody a movement from something natural or human to something artificial—from nature to technology—since writing, and printing, and books are about as technological as one can get. Books, after all, are teaching and communicating *machines*. Therefore, if we find ourselves in a period of fundamental technological and cultural change analogous to the Gutenberg revolution, one of the first things we should do is remind ourselves that printed books are technology, too.

Analogues to the Gutenberg Revolution

What can we predict about the future by understanding the "logic" of a particular technology or set of technologies? According to Kernan, "the 'logic' of a technology, an idea, or an institution is its tendency consistently to shape whatever it affects in a limited number of definite forms or directions" (49). The work of Kernan and others like Chartier and Eisenstein who have studied the complex transitions from manuscript to print culture suggest three clear lessons or rules for anyone anticipating similar transitions.

First of all, such transitions take a long time, certainly much longer than early studies of the shift from manuscript to print culture led one to expect.

Students of technology and reading practice point to several hundred years of gradual change and accommodation, during which different reading practices, modes of publication, and conceptions of literature obtained. According to Kernan, not until about 1700 did print technology "transform the more advanced countries of Europe from oral into print societies, reordering the entire social world, and restructuring rather than merely modifying letters" (9). How long, then, will it take computing, specifically computer hypertext, to effect similar changes? How long, one wonders, will the change to electronic language take until it becomes culturally pervasive? And what byways, transient cultural accommodations, and the like will intervene and thereby create a more confusing, if culturally more interesting, picture?

The second chief rule is that studying the relations of technology to literature and other aspects of humanistic culture does not produce any mechanical reading of culture, such as that feared by Jameson and others. As Kernan makes clear, understanding the logic of a particular technology cannot permit simple prediction because under varying conditions the same technology can produce varying, even contradictory, effects. J. David Bolter and other historians of writing have pointed out, for example, that initially writing, which served priestly and monarchical interests in recording laws and records, appeared purely elitist, even hieratic; later, as the practice diffused down the social and economic scale, it appeared democratizing, even anarchic. To a large extent, printed books had similarly diverse effects, though it took far less time for the democratizing factors to triumph over the hieratic—a matter of centuries, perhaps decades, instead of millennia!

Similarly, as Marie-Elizabeth Ducreux and Roger Chartier have shown, both printed matter and manuscript books functioned as instruments of "religious acculturation controlled by authority, but under certain circumstances [they] also supported resistance to a faith rejected, and proved an ultimate and secret recourse against forced conversion." Books of hours, marriage charters, and so-called evangelical books all embodied a "basic tension between public, ceremonial, and ecclesiastical use of the book or other print object, and personal, private, and internalized reading."[22]

Kernan himself points out that "knowledge of the leading principles of print logic, such as fixity, multiplicity, and systematization, makes it possible to predict the tendencies but not the *exact* ways in which they were to manifest themselves in the history of writing and in the world of letters. The idealization of the literary text and the attribution to it of a stylistic essence are both developments of latent print possibilities, but there was, I believe, no precise necessity beforehand that letters would be valorized in these particular ways" (181). Kernan also points to the "tension, if not downright contradiction, between two of the primary energies of print logic, multiplicity and fixity—what we might call 'the remainder house' and the 'library' effects" (55), each of which comes into play, or becomes dominant, only under certain economic, political, and technological conditions.

The third lesson or rule one can derive from the work of Kernan and other historians of the relations among reading practice, information technology, and culture is that transformations have political contexts and political implications. Considerations of hypertext, critical theory, and literature have to take into account what Jameson calls the basic "recognition that there is nothing that is not social and historical—indeed, that everything is 'in the last analysis' political" (*Political Unconscious,* 20).

If the technology of printing radically changed the world in the manner that Kernan convincingly explains, what, then, will be the effects of the parallel shift from print to computer hypertext? Although the changes associated with the transition from print to electronic technology may not parallel those associated with that from manuscript to print, paying attention to descriptions of the most recent shift in the technology of alphanumeric text provides areas for investigation.

One of the most important changes involved fulfilling the democratizing potential of the new information technology. During the shift from manuscript to print culture "an older system of polite or courtly letters—primarily oral, aristocratic, authoritarian, court-centered—was swept away . . . and gradually replaced by a new print-based, market-centered, democratic literary system" whose fundamental values "were, while not strictly determined by print ways, still indirectly in accordance with the actualities of print" (Kernan, *Printing Technologies,* 4). If hypertextuality and associated electronic information technologies have similarly pervasive effects, what will they be? Nelson, Miller, and almost all authors on hypertext who touch upon the political implications of hypertext assume that the technology is essentially democratizing and that it therefore supports some sort of decentralized, liberated existence. As our earlier brief glance at Internet search technology shows, networked electronic media have at least two contradictory logics— empowerment of individual readers and their vastly increased vulnerability to surveillance and consequent loss of privacy and security.

Kernan offers numerous specific instances of ways that technology "actually affects individual and social life." For example, "by changing their work and their writing, [print] forced the writer, the scholar, and the teacher—the standard literary roles—to redefine themselves, and if it did not entirely create, it noticeably increased the importance and number of critics, editors, bibliographers, and literary historians." Print technology similarly redefined the audience for literature by transforming it from

> a small group of manuscript readers or listeners . . . to a group of readers . . . who bought books to read in the privacy of their homes. Print also made literature objectively real for the first time, and therefore subjectively conceivable as a universal fact, in great libraries of printed books containing large collections of the world's writing . . . Print also rearranged the relationship of letters to other parts of the social world by, for example, freeing the writer from the need for patronage and the consequent subservience to wealth, by challenging and reducing established authority's control of writing by means of state censorship, and by pushing through a copyright law that made the author the owner of his own writing. (4–5)

Electronic linking shifts the boundaries between one text and another as well as between the author and the reader and between the teacher and the student. It also has radical effects on our experience of author, text, and work, redefining each. Its effects are so basic, so radical, that it reveals that many of our most cherished, most commonplace, ideas and attitudes toward literature and literary production turn out to be the result of that particular form of information technology and technology of cultural memory that has provided the setting for them. This technology—that of the printed book and its close relations, which include the typed or printed page—engenders certain notions of authorial property, authorial uniqueness, and a physically isolated text that hypertext makes untenable. The evidence of hypertext, in other words, historicizes many of our most commonplace assumptions, thereby forcing them to descend from the ethereality of abstraction and appear as corollary to a particular technology rooted in specific times and places. In making available these points, hypertext has much in common with some major points of contemporary literary and semiological theory, particularly with Derrida's emphasis on decentering and with Barthes's conception of the readerly versus the writerly text. In fact, hypertext creates an almost embarrassingly literal embodiment of both concepts, one that in turn raises questions about them and their interesting combination of prescience and historical relations (or embeddedness).

2 Hypertext and Critical Theory

Like Barthes, Foucault, and Mikhail Bakhtin, Jacques Derrida continually uses the terms *link (liasons), web (toile), network (rèseau),* and *interwoven (s'y tissent)*, which cry out for hypertextuality; but in contrast to Barthes, who emphasizes the writerly text and its nonlinearity, Derrida emphasizes textual openness, intertextuality, and the irrelevance of distinctions between inside and outside a particular text. These emphases appear with particular clarity when he claims that "like any text, the text of 'Plato' couldn't not be involved, or at least in a virtual, dynamic, lateral manner, with all the worlds that composed the system of the Greek language" (*Dissemination*, 129). Derrida in fact here describes extant hypertext systems in which the active reader in the process of exploring a text, probing it, can call into play dictionaries with morphological analyzers that connect individual words to cognates, derivations, and opposites. Here again something that Derrida and other critical theorists describe as part of a seemingly extravagant claim about language turns out precisely to describe the new economy of reading and writing with electronic virtual, rather than physical, forms.

Derrida properly recognizes (in advance, one might say) that a new, freer, richer form of text, one truer to our potential experience, perhaps to our actual if unrecognized experience, depends on discrete reading units. As he explains, in what Gregory Ulmer terms "the fundamental generalization of his writing" (*Applied Grammatology*, 58), there also exists "the possibility of disengagement and citational graft which belongs to the structure of every mark, spoken and written, and which constitutes every mark in writing before and outside of every horizon of semiolinguistic communication . . . Every sign, linguistic or non-linguistic, spoken or written . . . can be cited, put between quotation marks." The implication of such citability, separability, appears in the fact, crucial to hypertext, that, as Derrida adds, "in so doing it can break with every given context, engendering an infinity of new contexts in a manner which is absolutely illimitable" ("Signature," 185).

Like Barthes, Derrida conceives of text as constituted by discrete reading units. Derrida's conception of text relates to his "methodology of decomposition" that might transgress the limits of philosophy. "The organ of this new philospheme," as Ulmer points out, "is the mouth, the mouth that bites, chews, tastes . . . The first step of decomposition is the bite" (*Applied Grammatology*, 57). Derrida, who describes text in terms of something close to Barthes's lexias, explains in *Glas* that "the object of the present work, its style too, is the 'mourceau,'" which Ulmer translates as "bit, piece, morsel, fragment; musical composition; snack, mouthful." This mourceau, adds Derrida, "is always detached, as its name indicates and so you do not forget it, with the teeth," and these teeth, Ulmer explains, refer to "quotation marks, brackets, parentheses: when language is cited (put between quotation marks), the effect is that of releasing the grasp or hold of a controlling context" (58).

Derrida's groping for a way to foreground his recognition of the way text operates in a print medium—he is, after all, the fierce advocate of writing as against orality—shows the position, possibly the dilemma, of the thinker working with print who sees its shortcomings but for all his brilliance cannot think his way outside this *mentalité*. Derrida, the experience of hypertext shows, gropes toward a new kind of text: he describes it, he praises it, but he can only present it in terms of the devices—here those of punctuation—associated with a particular kind of writing. As the Marxists remind us, thought derives from the forces and modes of production, though, as we shall see, few Marxists or Marxians ever directly confront the most important mode of literary production—that dependent on the *techne* of writing and print.

From this Derridean emphasis on discontinuity comes the conception of hypertext as a vast assemblage, what I have elsewhere termed the *metatext* and what Nelson calls the *docuverse*. Derrida in fact employs the word *assemblage* for cinema, which he perceives as a rival, an alternative, to print. Ulmer points out that "the gram or trace provides the 'linguistics' for collage/montage" (*Applied Grammatology*, 267), and he quotes Derrida's use of *assemblage* in *Speech and Phenomena*: "The word 'assemblage' seems more apt for suggesting that the kind of bringing-together proposed here has the structure of an interlacing, a weaving, or a web, which would allow the different threads and different lines of sense or force to separate again, as well as being ready to bind others together" (131). To carry Derrida's instinctive theorizing of hypertext further, one may also point to his recognition that such a montage-like textuality marks or foregrounds the writing process and therefore rejects a deceptive transparency.

Hypertext and Intertextuality

Hypertext, which is a fundamentally intertextual system, has the capacity to emphasize intertextuality in a way that page-bound text in books cannot. As we have already observed, scholarly articles and books offer an obvious example of *explicit* hypertextuality in nonelectronic form. Conversely, any work of literature—which for the sake of argument and economy I shall here confine in a most arbitrary way to mean "high" literature of the sort we read and teach in universities—offers an instance of *implicit* hypertext in nonelectronic form. Again, take Joyce's *Ulysses* for an example. If one looks, say, at the Nausicaa section, in which Bloom watches Gerty McDowell on the beach, one notes that Joyce's text here "alludes" or "refers" (the terms we usually employ) to many other texts or phenomena that one can treat as texts, including the Nausicaa section of the *Odyssey*, the advertisements and articles in the women's magazines that suffuse and inform Gerty's thoughts, facts about contemporary Dublin and the Catholic Church, and material that relates to other passages within the novel. Again, a hypertext presentation of the novel links this section not only to the kinds of materials mentioned but also to other works in Joyce's career, critical commentary, and textual variants. Hypertext here permits one to make explicit, though not necessarily intrusive, the linked materials that an educated reader perceives surrounding it.

Thaïs Morgan suggests that intertextuality, "as a structural analysis of texts in relation to the larger system of signifying practices or uses of signs in culture," shifts attention from the triad constituted by author/work/tradition to another constituted by text/discourse/culture. In so doing, "intertextuality replaces the evolutionary model of literary history with a structural or synchronic model of literature as a sign system. The most salient effect of this strategic change is to free the literary text from psychological, sociological, and historical determinisms, opening it up to an apparently infinite play of relationships" (1–2). Morgan well describes a major implication of hypertext (and hypermedia) intertextuality: such opening up, such freeing one to create and perceive interconnections, obviously occurs. Nonetheless, although hypertext intertextuality would seem to devalue any historic or other reductionism, it in no way prevents those interested in reading in terms of author and tradition from doing so. Scholarship and criticism in hypertext from Intermedia and HyperCard to Weblogs demonstrates that hypertext does not necessarily turn one's attention away from such approaches. What is perhaps most interesting about hypertext, though, is not that it may fulfill certain claims of structuralist and poststructuralist criticism but that it provides a rich means of testing them.

Hypertext and Multivocality

In attempting to imagine the experience of reading and writing with (or within) this new form of text, one would do well to pay heed to what Mikhail Bakhtin has written about the dialogic, polyphonic, multivocal novel, which he claims "is constructed not as the whole of a single consciousness, absorbing other consciousnesses as objects into itself, but as a whole formed by the interaction of several consciousnesses, none of which entirely becomes an object for the other" (18). Bakhtin's description of the polyphonic literary form presents the Dostoevskian novel as a hypertextual fiction in which the individual voices take the form of lexias.

If Derrida illuminates hypertextuality from the vantage point of the "bite" or "bit," Bakhtin illuminates it from the vantage point of its own life and force—its incarnation or instantiation of a voice, a point of view, a Rortyian conversation.[1] Thus, according to Bakhtin, "in the novel itself, nonparticipating 'third persons' are not represented in any way. There is no place for them, compositionally or in the larger meaning of the work" (18). In terms of hypertextuality this points to an important quality of this information medium: complete read-write hypertext (exemplified by blogs and Intermedia) does not permit a tyrannical, univocal voice. Rather, the voice is always that distilled from the combined experience of the momentary focus, the lexia one presently reads, and the continually forming narrative of one's reading path.

Hypertext and Decentering

As readers move through a web or network of texts, they continually shift the center—and hence the focus or organizing principle—of their investigation and experience. Hypertext, in other words, provides an infinitely recenterable system whose provisional point of focus depends on the reader, who becomes a truly active reader in yet another sense. One of the fundamental characteristics of hypertext is that it is composed of bodies of linked texts that have no primary axis of organization. In other words, the metatext or document set—the entity that describes what in print technology is the book, work, or single text—has no center. Although this absence of a center can create problems for the reader and the writer, it also means that anyone who uses hypertext makes his or her own interests the de facto organizing principle (or center) for the investigation at the moment. One experiences hypertext as an infinitely decenterable and recenterable system, in part because hypertext transforms any document that has more than one link into a transient center, a partial sitemap that one can employ to orient oneself and to decide where to go next.

Western culture imagined quasi-magical entrances to a networked reality long before the development of computing technology. Biblical typology, which played such a major role in English culture during the seventeenth and nineteenth centuries, conceived sacred history in terms of types and shadows of Christ and his dispensation. Thus, Moses, who existed in his own right, also existed as Christ, who fulfilled and completed the prophet's meaning. As countless seventeenth-century and Victorian sermons, tracts, and commentaries demonstrate, any particular person, event, or phenomenon acted as a magical window into the complex semiotic of the divine scheme for human salvation. Like the biblical type, which allows significant events and phenomena to participate simultaneously in many realities or levels of reality, the individual lexia inevitably provides a way into the network of connections. Given that Evangelical Protestantism in America preserves and extends these traditions of biblical exegesis, one is not surprised to discover that some of the first applications of hypertext involved the Bible and its exegetical tradition.[2]

Not only do lexias work much in the manner of types, they also become Borgesian Alephs, points in space that contain all other points, because from the vantage point each provides one can see everything else—if not exactly simultaneously, then a short way distant, one or two jumps away, particularly in systems that have full text searching. Unlike Jorge Luis Borges's Aleph, one does not have to view it from a single site, neither does one have to sprawl in a cellar resting one's head on a canvas sack.[3] The hypertext document becomes a traveling Aleph.

As Derrida points out in "Structure, Sign, and Play in the Discourse of the Human Sciences," the process or procedure he calls decentering has played an essential role in intellectual change. He says, for example, that "ethnology could have been born as a science only at the moment when a de-centering had come about: at the moment when European culture—and, in consequence, the history of metaphysics and of its concepts—had been dislocated, driven from its locus, and forced to stop considering itself as the culture of reference" (251). Derrida makes no claim that an intellectual or ideological center is in any way bad, for, as he explains in response to a query from Serge Doubrovsky, "I didn't say that there was no center, that we could get along without a center. I believe that the center is a function, not a being—a reality, but a function. And this function is absolutely indispensable" (271).

All hypertext systems permit the individual reader to choose his or her own center of investigation and experience. What this principle means in practice is that the reader is not locked into any kind of particular organization or hierarchy. Experiences with various hypertext systems reveal that for those who choose to organize a session on the system in terms of authors—moving, say, from Keats to Tennyson—the system represents an old-fashioned, traditional, and in many ways still useful author-centered approach. On the other hand, nothing constrains the reader to work in this manner, and readers who wish to investigate the validity of period generalizations can organize their sessions in terms of such periods by using the Victorian and Romantic overviews as starting points or midpoints while yet others can begin with ideological or critical notions, such as feminism or the Victorian novel. In practice most readers employ the materials in *The Victorian Web* as a text-centered system, since they tend to focus on individual works, with the result that even if they begin sessions by looking for information about an individual author, they tend to spend most time with lexias devoted to specific texts, moving between poem and poem (Swinburne's "Laus Veneris" and Keats's "La Belle Dame Sans Merci" or works centering on Ulysses by Joyce, Tennyson, and Soyinka) and between poem and informational texts ("Laus Veneris" and files on chivalry, medieval revival, courtly love, Wagner, and so on).

Hypertext as Rhizome

Shortly after I began to teach hypertext and critical theory, Tom Meyer, a member of my first class, advised me that Gilles Deleuze and Félix Guattari's *1,000 Plateaus* demanded a place in *Hypertext.* And he is clearly right. Anyone considering the subject of this book has to look closely at their discussion of rhizomes, plateaus, and nomadic thought for several obvious reasons, only the most obvious of which is that they present *1,000 Plateaus* as a print protohypertext. Like Julio Cortzàr's *Hopscotch,* their volume comes with instructions to read it in various reader-determined orders, so that, as Stuart Moulthrop explains, their "rhizome-book may itself be considered an incunabular hypertext . . . designed as a matrix of independent but cross-referential discourses which the reader is invited to enter more or less at random (Deleuze and Guattari, xx)" and read in any order. "The reader's implicit task," Moulthrop explains, "is to build a network of virtual connections (which more than one reader of my acquaintance has suggested operationalizing as a web of hypertext links)" ("Rhizome and Resistance," 300–301).

Certainly, many of the qualities Deleuze and Guattari attribute to the rhizome require hypertext to find their first approximation if not their complete answer or fulfillment. Thus, their explanation of a plateau accurately describes the way both individual lexias and clusters of them participate in a web. "A plateau," they explain, "is always in the middle, not at the beginning or the end. A rhizome is made of plateaus. Gregory Bateson uses the word 'plateau' to designate something very special: a continuous, self-vibrating region of intensities whose development avoids any orientation toward a culmination point or external end" (21–22), such as orgasm, victory in war, or other point of culmination. Deleuze and Guattari, who criticize the "Western mind" for relating "expressions and actions to exterior or transcendent ends, instead of evaluating them on a plane of consistency on the basis of their intrinsic value," take the printed book to exemplify such characteristic climactic thought, explaining that "a book composed of chapters has culmination and termination points" (22).

Like Derrida and like the inventors of hypertext, they propose a newer form of the book that might provide a truer, more efficient information technology, asking: "What takes place in a book composed instead of plateaus that communicate with one another across microfissures, as in a brain? We call a 'plateau' any multiplicity connected to other multiplicities by superficial underground stems in such a way as to form or extend a rhizome" (22). Such a description, I should add, perfectly matches the way clusters or subwebs organize themselves in large networked hypertext environments, such as the World Wide Web. In fact, reducing Deleuze and Guattari's grand prescription to relatively puny literal embodiment, one could take the sections concerning Gaskell and Trollope in *The Victorian Web,* or the individual diary entries in Phil Gyford's Weblog version of Samuel Pepys's *Diaries,* as embodiments of plateaus. Indeed, one of the principles of reading and writing hypermedia— as in exploring a library of printed books—lies in the fact that one can begin anywhere and make connections, or, as Deleuze and Guattari put it, "each plateau can be read starting anywhere and can be related to any other plateau."

Such a characteristic organization (or lack of it) derives from the rhizome's fundamental opposition to hierarchy, a structural form whose embodiment Deleuze and Guattari find in the arborescent: "unlike trees or their roots, the rhizome connects any point to any other point, and its traits are not necessarily linked to traits of the same nature; it brings into play very different regimes of signs, and even nonsign states" (21). As Meyer explains in *Plateaus,* a Storyspace web that has since been published as part of *Writing at the Edge,* we generally rely on "arborescent structures," such as binary thought, genealogies, and hierarchies, to divide the "seemingly endless stream of information about the world into more easily assimilable bits. And, for this purpose, these structures serve admirably." Unfortunately, these valuable "organizational tools end up becoming the only methods of understanding," and limit instead of enhance or liberate our thought. "In contrast, Deleuze and Guattari propose the rhizome as a useful model for analysing structures—the potato, the strawberry plant, with their thickenings and shifting connections, with their network-like structure instead of a tree-like one" ("Tree/Rhizome").

This fundamental network structure explains why

the rhizome is reducible neither to the One nor the multiple . . . It has neither beginning nor end, but always a middle (milieu) from which it grows and which it overspills . . . When a multiplicity of this kind changes dimension, it necessarily changes in nature as well, undergoes a metamorphosis . . . The rhizome is an antigenealogy. It is a short-term memory, or antimemory. The rhizome operates by variation, expansion, conquest, capture, offshoots. Unlike the graphic arts, drawing, or photography, unlike tracings, the rhizome pertains to a map that must be produced, constructed, a map that is always detachable, connectable, reversible, modifiable, and has multiple entryways and exits and its own lines of flight . . . In contrast to centered (even polycentric) systems with hierarchical modes of communication and preestablished paths, the rhizome is an acentered, nonhierarchical, nonsignifying system without a General and without an organizing memory or central automaton, defined solely by a circulation of states. (21)

As we explore hypertext in the following pages, we shall repeatedly encounter the very qualities and characteristics Deleuze and Guattari here specify: like the rhizome, hypertext, which has "has multiple entryways and exits," embodies something closer to anarchy than to hierarchy, and it "connects any point to any other point," often joining fundamentally different kinds of information and often violating what we understand to be both discrete print texts and discrete genres and modes.

Any reader of hypertext who has experienced the way our own activities within the networked text produce multiple versions and approaches to a single lexia will see the parallel to hypertext in Deleuze and Guattari's point that "multiplicities are rhizomatic, and expose arborescent pseudomultiplicities for what they are. There is no unity to serve as a pivot in the object, or to divide in the subject" (8). Therefore, like hypertext considered in its most general sense, "a rhizome is not amenable to any structural or generative model. It is a stranger to any idea of genetic axis or deep structure" (12). As Deleuze and Guattari explain, a rhizome is "a map and not a tracing. Make a map, not a tracing. The orchid does not reproduce the tracing of the wasp; it forms a map with the wasp, in a rhizome. What distinguishes the map from the tracing is that it is entirely oriented toward an experimentation in contact with the real" (12). Maps and hypertexts both, in other words, relate directly to performance, to interaction.

Like some statements by Derrida, some of Deleuze and Guattari's more cryptic discussions of the rhizome often become clearer when considered from the vantage point of hypertext. For example, when they state that the rhizome is a "a short-term memory, or antimemory," something apparently in complete contrast with any information technology or technology of cultural memory, they nonetheless capture the provisional, temporary, changing quality in which readers make individual lexias the temporary center of their movement through an information space.

Perhaps one of the most difficult portions of *A Thousand Plateaus* involves the notion of nomadic thought, something, again, much easier to convey and experience in a fluid electronic environment than from within the world of print. According to Michael Joyce, the first important writer of hypertext fiction and one of the creators of Storyspace, Deleuze and Guattari reject "the word and world fully mapped as logos," proposing instead that "we write ourselves in the gap of nomos, the nomadic" (*Of Two Minds*, 207). They offer or propose, he explains, "being-for space against being-in space. We are in the water, inscribing and inscribed by the flow in our sailing. We write ourselves in oscillation between the smooth space of being for-time (what happens to us as we go as well as what happens to the space in which we do so) and the striated space of in-time (what happens outside the space and us)" (207).

Those who find the ruptures and seams as important to hypertext as the link that bridges such gaps find that the rhizome has yet another crucial aspect of hypertextuality. Moulthrop, for example, who "describes hypertexts as composed of nodes and links, local coherences and linearities broken across the gap or synapse of transition," takes this approach: "In describing the rhizome as a model of discourse, Deleuze and Guattari invoke the 'principle of asignifying rupture' (9), a fundamental tendency toward unpredictability and discontinuity. Perhaps then hypertext and hypermedia represent the expression of the rhizome in the social space of writing" ("Rhizome and Resistance," 304).

We must take care not to push the similarity too far and assume that their descriptions of rhizome, plateau, and nomadic thought map one to one onto hypertext, since many of their descriptions of the rhizome and rhizomatic thought appear impossible to fulfill in any information technology that uses words, images, or limits of any sort. Thus when Deleuze and Guattari write that a rhizome "has neither beginning nor end, but always a middle (milieu) from which it grows and which it overspills," they describe something that has much in common with the kind of quasi-anarchic networked hypertext one encounters on the World Wide Web, but when the following sentence adds that the rhizome "is composed not of units but of dimensions, or rather directions in motion" (21), the parallel seems harder to complete. The rhizome is essentially a counterparadigm, not something realizable in any time or culture, but it can serve as an ideal for hypertext, and hypertext, at least Nelsonian, ideal hypertext, approaches it as much as can any human creation.

The Nonlinear Model of

the Network in Current

Critical Theory

Discussions and designs of hypertext share with contemporary critical theory an emphasis on the model or paradigm of the network. At least four meanings of *network* appear in descriptions of actual hypertext systems and plans for future ones. First, individual print works when transferred to hypertext take the form of blocks, nodes, or lexias joined by a network of links and paths. Network, in this sense, refers to one kind of electronically linked electronic equivalent to a printed text. Second, any gathering of lexias, whether assembled by the original author of the verbal text, or by someone else gathering together texts created by multiple authors, also takes the form of a network; thus document sets, whose shifting borders make them in some senses the hypertextual equivalent of a work, are called in some present systems a web.

Third, the term *network* also refers to an electronic system involving additional computers as well as cables or wire connections that permit individual machines, workstations, and reading-and-writing sites to share information. These networks can take the form of contemporary Local Area Networks (LANs), such as Ethernet, that join sets of machines within an institution or a part of one, such as a department or administrative unit. Networks also take the form of Wide Area Networks (WANs) that join multiple organizations in widely separated geographical locations. Early versions of such wide-area national and international networks include JANET (in the United Kingdom), ARPANET (in the United States), the proposed National Research and Education Network (NREN), and BITNET, which linked universities, research centers, and laboratories in North America, Europe, Israel, and Japan.[4] Such networks, which until the arrival of the World Wide Web had been used chiefly for electronic mail and transfer of individual files, have also supported international electronic bulletin boards, such as Humanist. More powerful networks that transfer large quantities of information at great speed were necessary before such networks could fully support hypertext.

The fourth meaning of network in relation to hypertext comes close to matching the use of the term in critical theory. Network in this fullest sense refers to the entirety of all those terms for which there is no term and for which other terms stand until something better comes along, or until one of them gathers fuller meanings and fuller acceptance to itself: *literature, infoworld, docuverse,* in fact, *all writing* in the alphanumeric as well as Derridean senses. The future wide area networks necessary for large-scale, interinstitutional and intersite hypertext systems will instantiate and reify the current information worlds, including that of literature. To gain access to information, in other words, will require access to some portion of the network. To publish in a hypertextual world requires gaining access, however limited, to the network.

The analogy, model, or paradigm of the network so central to hypertext appears throughout structuralist and poststructuralist theoretical writings. Related to the model of the network and its components is a rejection of linearity in form and explanation, often in unexpected applications. One example of such antilinear thought will suffice. Although narratologists have almost always emphasized the essential linearity of narrative, critics have recently begun to find it to be nonlinear. Barbara Herrnstein Smith, for example, argues that "by virtue of the very nature of discourse, nonlinearity is the rule rather than the exception in narrative accounts" ("Narrative Versions, Narrative Theories," 223). Since I shall return to the question of linear and nonlinear narrative in a later chapter, I wish here only to remark that nonlinearity has become so important in contemporary critical thought, so fashionable, one might say, that Smith's observation, whether accurate or not, has become almost inevitable.

The general importance of non- or antilinear thought appears in the frequency and centrality with which Barthes and other critics employ the terms *link, network, web,* and *path*. More than almost any other contemporary theorist, Derrida uses the terms *link, web, network, matrix,* and *interweaving* associated with hypertextuality; and Bakhtin similarly employs *links* (*Problems,* 9, 25), *linkage* (9), *interconnectedness* (19), and *interwoven* (72).

Like Barthes, Bakhtin, and Derrida, Foucault conceives of text in terms of the network, and he relies precisely on this model to describe his project, "the archaeological analysis of knowledge itself." Arguing in *The Order of Things* that his project requires rejecting the "celebrated controversies" that occupied contemporaries, he claims that "one must reconstitute the general system of thought whose network, in its positivity, renders an interplay of simultaneous and apparently contradictory opinions possible. It is this network that defines the conditions that make a controversy or problem possible, and that bears the historicity of knowledge" (75). Order, for Foucault, is in part "the inner law, the hidden network" (xx); and according to him a "network" is the phenomenon "that is able to link together" (127) a wide range of often contradictory taxonomies, observations, interpretations, categories, and rules of observation.

Heinz Pagels's description of a network in *The Dreams of Reason* suggests why it has such appeal to those leery of hierarchical or linear models. According to Pagels, "A network has no 'top' or 'bottom.' Rather it has a plurality of connections that increase the possible interactions between the components of the network. There is no central executive authority that oversees the system" (20). Furthermore, as Pagels also explains, the network functions in various physical sciences as a powerful theoretical model capable of describing—and hence offering research agenda for—a range of phenomena at enormously different temporal and spatial scales. The model of the network has captured the imaginations of those working on subjects as apparently diverse as immunology, evolution, and the brain.

The immune system, like the evolutionary system, is thus a powerful pattern-recognition system, with capabilities of learning and memory. This feature of the immune system has suggested to a number of people that a dynamical computer model, simulating the immune system, could also learn and have memory . . . The evolutionary system works on the time scale of hundreds of thousands of years, the immune system in a matter of days, and the brain in milliseconds. Hence if we understand how the immune system recognizes and kills antigens, perhaps it will teach us about how neural nets recognize and can kill ideas. After all, both the immune system and the neural network consist of billions of highly specialized cells that excite and inhibit one another, and they both learn and have memory. (134–35)

Terry Eagleton and other Marxist theorists who draw on poststructuralism frequently employ the kind of network model or image to which the connectionists subscribe (see Eagleton, *Literary Theory*, 14, 33, 78, 104, 165, 169, 173, 201). In contrast, more orthodox Marxists, who have a vested interest (or sincere belief) in linear narrative and metanarrative, tend to use *network* and *web* chiefly to characterize error. Pierre Macherey might therefore at first appear slightly unusual in following Barthes, Derrida, and Foucault in situating novels within a network of relations to other texts. According to Machery, "The novel is initially situated in a network of books which replaces the complexity of real relations by which a world is effectively constituted." Machery's next sentence, however, makes clear that unlike most poststructuralists and postmodernists who employ the network as a paradigm of an open-ended, non-confining situation, he perceives a network as something that confines and limits: "Locked within the totality of a corpus, within a complex system of relationships, the novel is, in its very letter, allusion, repetition, and resumption of an object which now begins to resemble an inexhaustible world" (268).

Frederic Jameson, who attacks Louis Althusser in *The Political Unconscious* for creating impressions of "facile totalization" and of "a seamless web of phenomena" (27), himself more explicitly and more frequently makes these models the site of error. For example, when he criticizes the "antispeculative bias" of the liberal tradition in *Marxism and Form,* he notes "its emphasis on the individual fact or item at the expense of the network of relationships in which that item may be imbedded" as liberalism's means of keeping people from "drawing otherwise unavoidable conclusions at the political level" (x). The network model here represents a full, adequate contextualization, one suppressed by an other-than-Marxist form of thought, but it is still only necessary in describing pre-Marxian society. Jameson repeats this paradigm in his chapter on Herbert Marcuse when he explains that "genuine desire risks being dissolved and lost in the vast network of pseudosatisfactions which make up the market system" (100–101). Once again, network provides a paradigm apparently necessary for describing the complexities of a fallen society. It does so again when in the Sartre chapter he discusses Marx's notion of fetishism, which, according to Jameson, presents "commodities and the 'objective' network of relationships which they entertain with each other" as the illusory appearance masking the "reality of social life," which "lies in the labor process itself" (296).

Cause or Convergence, Influence or Confluence?

What relation obtains between electronic computing, hypertext in particular, and literary theory of the past three or four decades? J. Hillis Miller proposes that "the relation . . . is multiple, non-linear, non-causal, non-dialectical, and heavily overdetermined. It does not fit most traditional paradigms for defining 'relationship'" ("Literary Theory," 11). Miller himself provides a fine example of the convergence of critical theory and technology. Before he discovered computer hypertext, he wrote about text and (interpretative) text processing in ways that sound very familiar to anyone who has read or worked with hypertext. Here, for example, is the way *Fiction and Repetition* describes the way he reads a novel by Hardy in terms of what I would term a *Bakhtinian hypertextuality:* "Each passage is a node, a point of intersection or focus, on which converge lines leading from many other passages in the novel and ultimately including them all." No passage has any particular priority over the others, in the sense of being more important or as being the "origin or end of the others" (58).

Similarly, in providing "an 'example' of the deconstructive strategy of interpretation," in "The Critic as Host" (1979), he describes the dispersed, linked text block whose paths one can follow to an ever-widening, enlarging metatext or universe. He applies deconstructive strategy "to the cited fragment of a critical essay containing within itself a citation from another essay, like a parasite within its host." Continuing the microbiological analogy, Miller next explains that "the 'example' is a fragment like those miniscule bits of some substance which are put into a tiny test tube and explored by certain techniques of analytical chemistry. [One gets] so far or so much out of a little piece of language, context after context widening out from these few phrases to include as their necessary milieux all the family of Indo-European languages, all the literature and conceptual thought within these languages, and all the permutations of our social structures of household economy, gift-giving and gift receiving" (223).

Miller does point out that Derrida's "*Glas* and the personal computer appeared at more or less the same time. Both work self-consciously and deliberately to make obsolete the traditional codex linear book and to replace it with the new multilinear multimedia hypertext that is rapidly becoming the characteristic mode of expression both in culture and in the study of cultural forms. The 'triumph of theory' in literary studies and their transformation by the digital revolution are aspects of the same sweeping change" ("Literary Theory," 20–21). This sweeping change has many components, to be sure, but one theme appears both in writings on hypertext (and the memex) and in contemporary critical theory—the limitations of print culture, the culture of the book. Bush and Barthes, Nelson and Derrida, like all theorists of these perhaps unexpectedly intertwined subjects, begin with the desire to enable us to escape the confinements of print. This common project requires that one first recognize the enormous power of the book, for only after we have made ourselves conscious of the ways it has formed and informed our lives can we seek to pry ourselves free from some of its limitations.

Looked at within this context, Claude Lèvi-Strauss's explanations of preliterate thought in *The Savage Mind* and in his treatises on mythology appear in part as attempts to decenter the culture of the book—to show the confinements of our literate culture by getting outside of it, however tenuously and however briefly. In emphasizing electronic, noncomputer media, such as radio, television, and film, Baudrillard, Derrida, Jean-François

Lyotard, McLuhan, and others similarly argue against the future importance of print-based information technology, often from the vantage point of those who assume analogue media employing sound and motion as well as visual information will radically reconfigure our expectations of human nature and human culture.

Among major critics and critical theorists, Derrida stands out as the one who most realizes the importance of free-form information technology based on digital, rather than analogue, systems. As he points out, "the development of *practical methods* of information retrieval extends the possibilities of the 'message' vastly, to the point where it is no longer the 'written' translation of a language, the transporting of a signified which could remain spoken in its integrity" (10). Derrida, more than any other major theorist, understands that electronic computing and other changes in media have eroded the power of the linear model and the book as related culturally dominant paradigms. "The end of linear writing," Derrida declares, "is indeed the end of the book," even if, he continues, "it is within the form of a book that the new writings—literary or theoretical—allow themselves to be, for better or worse, encased" (*Of Grammatology*, 86). Therefore, as Ulmer points out, "grammatalogical writing exemplifies the struggle to break with the investiture of the book" (13).

According to Derrida, "the form of the 'book' is now going through a period of general upheaval, and while that form appears less natural, and its history less transparent, than ever . . . the book form alone can no longer settle . . . the case of those writing processes which, in *practically* questioning that form, must also dismantle it." The problem, too, Derrida recognizes, is that "one cannot tamper" with the form of the book "without disturbing everything else" (*Dissemination,* 3) in Western thought. Always a tamperer, Derrida does not find that much of a reason for not tampering with the book, and his questioning begins in the chain of terms that appear as the more-or-less title at the beginning pages of *Dissemination:* "Hors Livres: Outwork, Hors D'oeuvre, Extratext, Foreplay, Bookend, Facing, and Prefacing." He does so willingly because, as he announced in *Of Grammatology,* "All appearances to the contrary, this death of the book undoubtedly announces (and in a certain sense always has announced) nothing but a death of speech (of a *so-called* full speech) and a new mutation in the history of writing, in history as writing. Announces it at a distance of a few centuries. It is on that scale that we must reckon it here" (8).

In conversation with me, Ulmer mentioned that since Derrida's gram equals link, grammatology is the art and science of linking—the art and science, therefore, of hypertext.[5] One may add that Derrida also describes dissemination as a description of hypertext: "Along with an ordered extension of the concept of text, dissemination inscribes a different law governing the effects of sense or reference (the interiority of the 'thing,' reality, objectivity, essentiality, existence, sensible or intelligible presence in general, etc.), a different relation between writing, in the metaphysical sense of the word, and its 'outside' (historical, political, economical, sexual, etc.)" (*Dissemination*, 42).

3

Reconfiguring the Text

Although in some distant, or not-so-distant, future all individual texts will electronically link to one another, thus creating metatexts and metametatexts of a kind only partly imaginable at present, less far-reaching forms of hypertextuality have already appeared. Translations into hypertextual form already exist of poetry, fiction, and other materials originally conceived for book technology. The simplest, most limited form of such translation preserves the linear text with its order and fixity and then appends various kinds of texts to it, including critical commentary, textual variants, and chronologically anterior and later texts.[1]

Hypertext corpora that employ a single text, originally created for print dissemination, as an unbroken axis off which to hang annotation and commentary appear in the by-now common educational and scholarly presentations of canonical literary texts (Figure 7). At Brown University my students and I first used Intermedia and Storyspace to provide annotated versions of stories by Kipling and Lawrence, and I have since created more elaborate World Wide Web presentations of Carlyle's "Hudson's Statue" and other texts. *The Dickens Web*, a corpus of materials focused on *Great Expectations* published in Intermedia (IRIS, 1990) and Storyspace (Eastgate, 1992), differs from these projects in not including the primary text, as does Christiane Paul's *Unreal City: A Hypertext Guide to T. S. Eliot's "The Waste Land"* (1994).

A second case appears when one adapts for hypertextual presentation material originally conceived for book technology that divides into discrete lexias, particularly if it has multilinear elements that call for the kind of multisequential reading associated with hypertext. An early example of this form of hypertext appears in Brian Thomas's early HyperCard version of *Imitatio*

Axial structure characteristic of electronic books and scholarly books with foot- or endnotes

versus

Network structure of hypertext

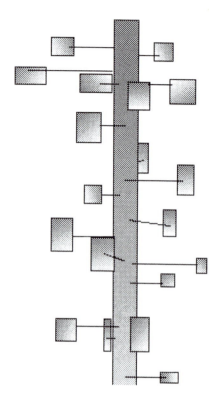

1. Where does the reader enter the text?

2. Where does the reader leave the text?

3. Where are the borders of the text?

Figure 7. Axial versus Network Structure in Hypertext

Cristi, and another is the electronic edition of the *New Oxford Annotated Bible* (1995), a hypertext presentation of the Revised Standard Version that uses AND Software's CompLex system. Like many commercially available electronic texts, the *New Oxford Annotated Bible* appears more a digitized book than a true hypertext, though it is nonetheless valuable for that. Readers can supplement the biblical text with powerful search tools and various indices, including ones for Bible and annotation topics, and substantial supplementary essays, including those on approaches to Bible study, literary forms in the Gospels, and the characteristics of Hebrew poetry. The *New Oxford Annotated Bible*'s hypertextuality consists largely of variant readings (indicated by link icons in the form of red crosses) and the fact that readers can add both book-marks and their own annotations.

A more elaborate form of hypertextuality appears in the earlier *CD Word: The Interactive Bible Library*, which a team based at Dallas Theological Semi-nary created using an enhanced version of Guide™. This hypertext Bible cor-pus, "intended for the student, theologian, pastor, or lay person" rather than for the historian of religion, includes the King James, New International, New American Standard, and Revised Standard versions of the Bible, as well as Greek texts for the New Testament and Septuagint. These materials are sup-plemented by three Greek lexica, two Bible dictionaries, and three Bible com-mentaries (DeRose, *CD Word*, 1, 117–26). Using this system, which stores the electronic texts on a compact disc, the Bible reader can juxtapose passages from different versions and compare variants, examine the original Greek, and receive rapid assistance on Greek grammar and vocabulary.

A similar kind of corpus that uses a more sophisticated hypertext system is Paul D. Kahn's pioneering *Chinese Literature* Intermedia web, which offers different versions of the poetry of Tu Fu (712–770), ranging from the Chinese text, Pin-yin transcriptions, and literal translations to much freer ones by Ken-neth Rexroth and others. *Chinese Literature* also includes abundant second-ary materials that support interpreting Tu Fu's poetry. Like *CD Word*, Kahn's Intermedia corpus permits both beginning and advanced students to approach a canonical text in a foreign language through various versions, and like the hypertext Bible on compact disc, it also situates its primary text within a net-work of links to both varying translations and reference materials.

Before considering other kinds of hypertext, we should note the implicit justifications or rationales for these two successful projects. *CD Word* pres-ents its intended readers with a particularly appropriate technological pres-entation of the Bible because they habitually handle this text in terms of brief passages—or, as writers on hypertext might put it, as if it had "fine granu-larity." Because the individual poems of Tu Fu are fairly brief, a body of them invites similar conversion to hypertext.

The *In Memoriam* Web

In contrast to the *CD Word Bible* and the *Chinese Literature Web*, which support study chiefly by electronically linking multiple parallel texts, the *In Memoriam* web (Figure 8), an-other Intermedia corpus created at Brown University and since published in Storyspace (Figure 9) after an extensive expansion by Jon Lanestedt and me (Eastgate Systems, 1992), uses electronic links to map and hence reify a text's internal and external allusions and references—its *inter-* and *intra*textuality.[2]

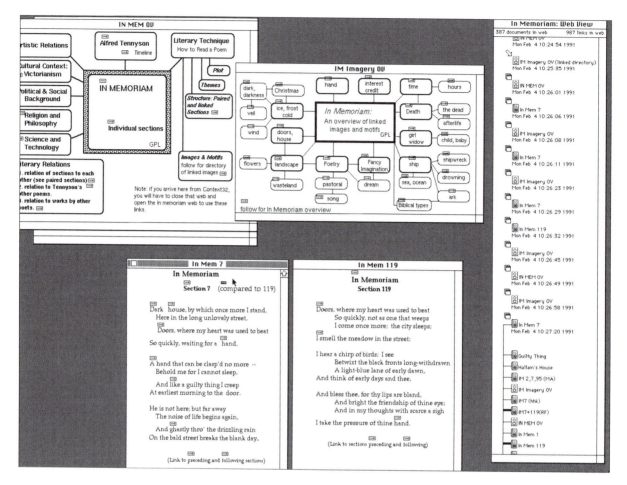

Figure 8. The Original Intermedia Version of *The In Memoriam Web*. In this snapshot of a typical screen during a session on Intermedia, the active document, *In Memoriam,* section 7 ("In Mem 7"), appears at the lower left center of the screen with a darkened strip across its top to indicate its status. Using the capacities of hypertext to navigate the poem easily, a reader has juxtaposed sections 119 and 7, which echo and complete each other. *In Memoriam* overview (IN MEM OV), which appears at the upper left, is a graphic document that serves as a sitemap: it organizes linked materials under generalized headings, such as "Cultural Context: Victorianism," or "Images and Motifs." The *In Memoriam* imagery overview ("IM Imagery OV"), a second visual index document, overlies the right border for the entire poem. On the right appears the Web View, which the system automatically creates for each document as the document becomes active either by being opened, or, if it is already open on the desktop, by being clicked on. In contrast to the hierarchically organized overviews the author creates, the Web View shows titled icons representing all documents connected electronically to the active document, here section 7 of the poem. Touching any link marker with the arrow-shaped cursor darkens the icons representing the documents linked to it; in this case, the reader has darkened the marker above the phrase "compared to 119" and thereby darkened icons representing both the text of section 7 and a student essay comparing it to section 119.

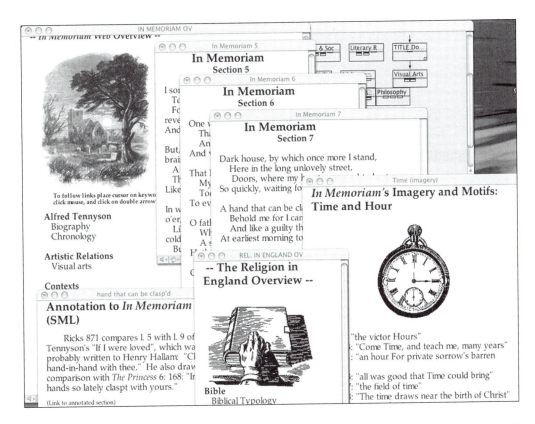

Figure 9. The Storyspace Version of *The In Memoriam Web*. Readers can make their way through this body of interlinked documents in a number of ways. One can proceed by following links from principal overviews, such as that for the entire web (*at left*), religion in England (*lower right*), or individual motifs—in this case that for time (*middle right*). One can also explore the folderlike structure of the Storyspace view (*upper right*), which can contain a dozen or more layers, or one can follow links from individual sections of the poem. This screen shot indicates how multiwindow hypertext systems, such as Storyspace, Intermedia, and Multicosm, enable authors to fix the location of windows, thereby permitting one to arrange the screen in ways that help orient the reader. Readers can easily move between parts of the poem and commentary on it.

Tennyson's radically experimental *In Memoriam* provides an exemplification of the truth of Benjamin's remark that "the history of every art form shows critical epochs in which a certain art form aspires to effects which could be fully obtained only with a changed technical standard, that is, to say, in a new art form" (*Illuminations*, 237). Another manifestation of this principle appears in Victorian word-painting, particularly in the hands of Ruskin and Tennyson, which anticipates in abundant detail the techniques of cinematography. Whereas word-painting anticipates a future medium (cinema) by using narrative to structure description, *In Memoriam* anticipates electronic hypertextuality precisely by challenging narrative and literary form based on it. Convinced that the thrust of elegiac narrative, which drives the reader and the mourner relentlessly from grief to consolation, falsified his own experiences, the poet constructed a poem of 131 fragments to communicate the ebb and flow of emotion, particularly the way the aftershocks of grief irrationally intrude long after the mourner has supposedly recovered.

Arthur Henry Hallam's death in 1833 forced Tennyson to question his faith in nature, God, and poetry. *In Memoriam* reveals that Tennyson, who found that brief lyrics best embodied the transitory emotions that buffeted him after his loss, rejected conventional elegy and narrative because both presented the reader with a too unified—and hence too simplified—version of the experiences of grief and acceptance. Creating an antilinear poetry of fragments, Tennyson leads the reader of *In Memoriam* from grief and despair through doubt to hope and faith; but at each step stubborn, contrary emotions intrude, and one encounters doubt in the midst of faith and pain in the midst of resolution. Instead of the elegiac plot of "Lycidas," "Adonais," and "Thyrsis," *In Memoriam* offers fragments interlaced by dozens of images and motifs and informed by an equal number of minor and major resolutions, the most famous of which is section 95's representation of Tennyson's climactic, if wonderfully ambiguous, mystical experience of contact with Hallam's spirit. In addition, individual sections, like 7 and 119 or 28, 78, and 104, variously resonate with one another.

The protohypertextuality of *In Memoriam* atomizes and disperses Tennyson the man. He is to be found nowhere, except possibly in the epilogue, which appears after and outside the poem itself. Tennyson, the real, once-existing man, with his actual beliefs and fears, cannot be extrapolated from within the poem's individual sections, for each presents Tennyson only at a particular moment. Traversing these individual sections, the reader experiences a somewhat idealized version of Tennyson's moments of grief and recovery. *In Memoriam* thus fulfills Paul Valéry's definition of poetry as a machine that reproduces an emotion. It also fulfills another of Benjamin's observations, one he makes in the course of contrasting painter and cameraman: "The painter maintains in his work a natural distance from reality, the cameraman penetrates deeply into its web. There is a tremendous difference between the pictures they obtain. That of the painter is a total one, that of the cameraman consists of multiple fragments which are assembled under a new law" (*Illuminations*, 233–34). Although speaking of a different information medium, Benjamin here captures some sense of the way hypertext, when compared to print, appears atomized; and in doing so, he also conveys one of the chief qualities of Tennyson's antilinear, multisequential poem.

The *In Memoriam* web attempts to capture the nonlinear organization of the poem by linking sections, such as 7 and 119, 2 and 39, or the Christmas poems, which echo across the poem to one another. More important, using the capacities of hypertext, the web permits the reader to trace from section to section several dozen leitmotifs that thread through the poem. Working with section 7, for example, readers who wish to move through the poem following a linear sequence can do so by using links to previous and succeeding sections, but they can also look up any word in a linked electronic dictionary or follow links to variant readings, critical commentary (including a comparison of this section and 119), and discussions of the poem's intertextual relations. Furthermore, activating indicated links near the words *dark, house, doors, hand,* and *guilty* produces a choice of several kinds of materials. Choosing *hand* instantly generates a menu that lists all the links to that word, and these include a graphic directory of *In Memoriam*'s major images, critical commentary on the image of the hand, and, most important, a concordance-like list of each use of the word in the poem and the phrase in which it appears; choosing any one item in the list produces the linked document, the graphic overview of imagery, a critical comment, or the full text of the section in which a particular use of hand appears.

Using the capacities of Intermedia and Storyspace to join an indefinite number of links to any passage (or block) of text, the reader moves through the poem along many different axes. Although, like the previously mentioned hypertext materials, the *In Memoriam* web contains reference materials and variant readings, its major difference appears in its use of link paths that permit the reader to organize the poem by means of its network of leitmotifs and echoing sections. In addition, this hypertext presentation of Tennyson's poem also contains a heavily linked graphic overview of the poem's literary relations—its intertextual relations, sources, analogues, confluences, and influences—that permits one to read the poem along axes provided by sets of links relating to the Bible and to works by thirty-eight other writers, chiefly poets, including Vergil, Horace, Dante, Chaucer, Shakespeare, and Milton as well as the Romantics and Victorians. Although Lanestedt, various students, and I created these links, they represent a form of objective links that could have been created automatically by a full-text search in systems such as Microcosm. Here, as in other respects, this web represents an adaptive form of hypertext.

In contrast to adapting texts whose printed versions already divide into sections analogous to lexias, one may, in the manner of Barthes's treatment of "Sarrasine" in *S/Z*, impose one's own divisions on a work. Obvious examples of possible projects of this sort include hypertext versions of either "Sarrasine" alone or of it and Barthes's *S/Z*. Stuart Moulthrop's version of *Forking Paths: An Interaction after Jorge Luis Borges* (1987) adapts Borges's "Forking Paths" in an electronic version that activates much of the work's potential for variation (see Moulthrop, "Reading from the Map"). Other fiction that obviously calls for translation into hypertext includes Julio Cortázar's *Hopscotch* and Robert Coover's "The Babysitter."

These instances of adaptive hypertext all exemplify forms of transition between textuality and hypertextuality. In addition, works originally conceived for hypertext already exist as well. These webs electronically link blocks of text, that is, lexias, to one another and to various graphic supplements, such as illustrations, maps, diagrams, and visual directories and overviews, some of which are foreign to print technology. In the future there will be more metatexts formed by linking individual sections of individual works, although the notion of an individual, discrete work becomes increasingly undermined and untenable within this form of information technology, as it already has within much contemporary critical theory. Such materials include hypertextual poetry and fiction, which I shall discuss later in this volume, and the already abundant World Wide Web equivalents of scholarly and critical work in print.

One of the first such works in this new medium—certainly the first on Intermedia—was Barry J. Fishman's "The Works of Graham Swift: A Hypertext Thesis," a 1989 Brown University honors thesis on the contemporary British novelist. Fishman's thesis takes the form of sixty-two lexias, of which fifty-five are text documents and seven diagrams or digitized photographs. The fifty-five text documents he created, which range from one-half to three single-space pages in length, include discussions of Swift's six published book-length works, the reviews each received, correspondence with the novelist, and essays on themes, techniques, and intertextual relations of both each individual book and Swift's entire oeuvre up to 1989. Although Fishman created his hypermedia corpus as a relatively self-contained set of documents, he linked his materials to several dozen documents already present on the system, including materials by faculty members in at least three different departments and comments by other students. Since Fishman created his web, it has grown as many other students added their own lexias, and it moved first to Storyspace and, more recently, to the World Wide Web, where it constitutes an important part of a web containing materials on recent Anglophone postcolonial and postimperial literature.

New Forms of Discursive Prose—Academic Writing and Weblogs

I have been describing new kinds of discursive prose, for at the very least hypertext enables new forms of the academic essay, book review, and thesis. More than a decade's work by thousands of scholars using the Internet has shown that these academic genres can take three basic forms. At their simplest, the author simply places a text without links into an HTML template that includes navigation links. As Peter Brusilovsky and Riccardo Rizzo have pointed out in a prize-winning conference paper, a great deal of current academic writing for the Web follows this model, which does not take advantage of the possibilities of hypertext.

In a second kind of hypertext prose, the author creates a document with links to documents on the same as well as on other websites. In essence this means, as I urge my students, that we must write with an awareness that we are writing in the presence of other texts. These other texts may support or contradict our argument, or some of them may serve as valuable annotations to it. For example, my review in *The Victorian Web* of Joseph Bizup's *Manufacturing Culture: Vindications of Early Victorian Industry* (2003) contains more than a dozen links to on-site materials about authors, novels, and historical events. In contrast, a review of Dale H. Porter's *The Thames Embankment: Environment, Technology, and Society in Victorian London* (1998) contains few links to existing lexias but more than a dozen to brief sample passages from Porter's book on topics including Oxford in 1850, the invention of urban green spaces, civil engineering as a profession, and Victorian wages for skilled and unskilled labor. In addition to making links to these brief lexias from the main text, I also appended a list of them plus a few others at the end of the review. Links to Porter's materials were also added to the sitemaps for science, technology, social history, and economics. Both author and publisher were delighted with this approach to reviewing because they believed, correctly, that it spread word about Porter's work in a particularly effective way.

The third kind of hypertext essay, as we have seen from Fishman's honors thesis, takes the form of a set of networked documents, created either to stand alone, as it largely was, or to take part in a larger web. Either way, an author wanting to conceive of an argument in terms of networked documents can write a concise essay to which she or he links a wide range of supporting evidence. Readers can then choose what areas they want to investigate in greater depth, and these auxiliary materials thereby become paratexts, easily accessible add-ons to the lexia one is currently reading.

The Weblog, or blog, as it is commonly known, is another new kind of discursive prose in digital form that makes us rethink a genre that originally arose when writing took the form of physical marks on physical surfaces. Blogging, the latest Internet craze, has major importance for anyone interested in hypertext because one form of it provides the first widely available means on the Web of allowing the active reader-author envisaged by Nelson, van Dam, and other pioneers. Blogs take the form of an online journal or diary most commonly written by a single person, and, like paper journals and diaries, they present the author's words in dated segments. Unlike their paper predecessors, they present entries in reverse chronological order. They can employ two different forms of hypertextuality. First, unlike discussion lists, all bloggers can link chronologically distant individual entries to each other, thereby "allowing readers to put events in context and get the whole story without the diarist having to explain again" (McNeill, 30). The second form of hypertextuality occurs only in those blogger systems that permit readers to comment on entries. Here's how it works: encountering a comment on my son's blog about the legality of China's revoking the patent for Viagra, I clicked on the word "Comment" and thereby opened a form into which I pasted my remarks plus a few sentences from Vincent Mosco's historical account of nineteenth-century American information piracy (which I use in chapter 8, below). Before I could submit it, the form containing my comment requested three bits of information: my e-mail address (required), my name ("real is appreciated"), and the URL of my website, if any (optional). Returning to the blog, I discovered that the zero next to "comment" had changed to "1." Clicking on the word "comment" opened a document containing what I had just submitted plus a space for other people to add their responses to my comment. There are more than a dozen kinds of blogging software, and many, including b2Evolution, Moveable Type, and Serendipity, have the Trackback feature that also allows bloggers to post links back to the site of anyone who commented on them.

Visually, blogs take many forms, but most have several columns, the widest dedicated to the dated entries and one or more others containing links to archives of older entries, personal information, associated blogs, and major topics of interest to the person who owns the site. Many contain images and even video, and most contain a personal statement or description of the site, which may be very brief, such as "i stash things here so i can find them again. sometimes other people come visit. or so the tracker implies." Some bloggers maintain two or more sites, one devoted to their academic or professional interests and another to their personal diary. Many users prefer reading blogs of friends and family members to e-mail, because they have no spam, and for that and other reasons they have become enormously popular. Many sophisticated bloggers use special software to subscribe to their favorite sites, thereby ensuring that they know when something new is posted on them. RSS and Atom feed represent the two main standards for such subscription tools. Whereas RSS sends the subscriber only the headline of a new blog, Atom feed adds a summary and includes its links as well. So-called feed readers obtain, organize, and display materials from large numbers of websites and Weblogs.

Blogs themselves can take as many forms and have as many principal organizing ideas as other forms of websites, but the majority of bloggers (and also the most intense ones) are highly skilled computer users whose professional activities demand technical information. According to www.techno-rati.com, which claims to have watched 3,145,522 Weblogs (entries) and tracked 456,140,934 links, Slashdot, a famous multiuser techie site, proved most popular, with 12, 904 blogs and 21,041 links. Fark, another popular blog that claims to have received over 350 million pageviews in 2003, offers a digest of the news. Each entry takes the form of a one-line summary that links to sources accompanied by little icons containing comments, such as "amusing," "cool," "obvious," "scary," and "stupid." An example of this last category on July 17, 2004: "Martha Stewart compares herself to Nelson Mandela," which links to its source, CNN. This item provoked ninety-two comments that stated almost every possible opinion about the criminal case.

NGOs and other organizations dealing with economic and political issues have blogs, as well as those concentrating on specific diseases, such as AIDS, asthma, tuberculosis, and cancer. Disease blogs take two very different forms, the first created by organizations that work toward the prevention and cure of a particular illness, the second written by people suffering from the disease; some of their intensely personal online diaries have acquired large followings. On a lighter note, many hobbies or leisure activities, such as model railroading and gardening, have blogs, though I found very few for hunting or flying; I did discover an Australian one, though, on fear of flying. Perhaps the most extreme personal-interest blogs involve sexual fetishes, including one by a nonsmoker who finds pictures of women smoking in public erotic.

Phil Gyford's translation of Samuel Pepys's famous seventeenth-century diary into a Weblog, which exemplifies the way ingenious people find unexpected uses for computer genres, creates a new form of participatory scholarship. As Gyford explained in an interview available in the online version of BBC News (World Edition), "I thought Pepys' diary could make a great Weblog. The published diary takes the form of nine hefty volumes—a daunting prospect. Reading it day by day on a website would be far more manageable, with the real-time aspect making it a more involving experience." As I am writing this on July 22, 2004, I find Pepys's diary entry for July 21, 1661, which has been up at least since the 19th, since a comment on the place name "Sturtloe" by Mark Ynys-Mon points out that it has changed to "Stirtloe," and a second submitted the next day by Vincente points out that "Sam on his nag could have had a nice ride down by the Ouse" and provides a link to a map of the area. Gyford has tried to ensure that most of the annotations are helpful by advising contributors: "Before posting an annotation please read the annotation guidelines. If your comment isn't directly relevant to this page, or is more conversational, try the discussion group." This blog, which contains approximately two hundred words, has thirteen in-text links, one by Gyford himself (a cross-reference to another *Diary* entry) and others leading to one or more readers' comments. The entry for July 21, 1661, dominates the screen, though if one scrolls down one can find entries back to the 13th. At the top of the window, Gyford has provided links to an introduction to the *Diary*, background information, archives, and a summary ("Read the Story So Far") for first-time readers. At the top right a search window appears and below it a column of links to seventeen categories of background information starting with art and literature, including food and drink, and ending with work and education. As this brief description makes clear, Gyford has not only made an appropriate Web translation of a classic text but he has also contributed importantly to the creation of a new form of public, collaborative online scholarship. Two interesting points: (1) Gyford's name does not appear on the main lexias of the blog, though if one explores "About this site," one can find it, and following a link to his personal site, one can learn about his fascinating career as an art student, professional model maker, system administrator, and web designer. (2) This elaborate scholarly project, which one expects that any Web-savvy undergraduate and graduate student will use, exists completely *outside* the Academy. Pause for a moment and think about the implications of that.

Many special-interest blogs, like some famous ones by AIDS and cancer victims, exemplify the Internet version of the personal journal or diary. Laurie McNeill's excellent article on the blog as personal diary (about which I learned from Adrian Miles's blog) points to "an unparalleled explosion of public life writing by private citizens. By March 2002, more than 800,000 blogs were registered on the Net; in July 2002 an average of 1.5 'Blogger blogs' were created *per minute* (blogger.com 6 Aug. 2002)" (32). When I checked two years later, some hosts of blogs boasted millions of users, estimates of the total ranging between two and eight million, though one commentator pointed out that only a quarter of people who begin blogs keep them going.

Googling the phrase "how many bloggers," I received the URLs for several sites with some of the information for which I was looking, but among the top entries appeared one from the blog of a young woman enumerating her sexual experiences (I hadn't meant that "how many"!). Her entry, which appeared as a separate lexia, contained links to another blog with similar material, and when I clicked on the link in the original blog labeled "Home," I found a site whose contents reminded me of the HBO television show, "Sex and the City"— more for the comedy, though, than the sex. Although the blogger identifies herself only as "Blaise K. ," she includes enough personal information, including photographs and the assertion that she is black and Jewish, that her anonymity doesn't seem very well protected. I assume the blogger intends the site for her friends, but Google mistakenly brought me there, as it may well bring her parents and employers. It is very difficult to maintain this kind of public privacy.

McNeill points out that such sites "often reinforce the stereotype of the diary as a genre for unbridled narcissism" because they assume that other people care about what bloggers have to say. That narcissism, McNeill admits, often turns out be justified, for some online diaries receive thousands of visitors and make their authors famous. They also place the author's remarks about private matters in a very public space. In fact, one of the most interesting effects of blogging lies in the way it unsettles our accustomed borders between the private and public spheres. "In their immediacy and accessibility, in their seemingly unmediated state, Web diaries blur the distinction between online and offline lives, 'virtual reality' and 'real life,' 'public' and 'private,' and most intriguingly for auto/biography studies, between the life and the text" (McNeill, 25). Those blogs that accept comments allow, McNeill claims, the "reader of an online diary" to participate actively

in constructing the text the diarist writes, and the identities he or she takes on in the narrative. Though active and even intimate, however, that participation remains virtual, disembodied. The confessor stays behind the "grille" of the Internet, allowing the diarist—and the reader—the illusion of anonymity necessary for "full" self-exposure. Janet Murray notes that "some people put things on their home page . . . that they have not told their closest friends. The enchantment of the computer creates for us a public space that also feels very private and intimate" (99) . . . For the online diarist, having readers means that the diarist has both joined and created communities, acts that inform the texts he or she will produce. (27, 32)

Many bloggers don't in fact allow comments, or else they screen them, and some intend their online diaries solely for a circle of friends and control access to them by using passwords. Nonetheless, once an entry goes online,

Internet search tools can bring it to the attention of Web surfers. The edges of a blog, like the borders of any document on the Internet, are porous and provisional at best. Most of the time when we consider the way digital media blur the borders of documents, we mean that links and search tools limit the power of authorship. In blogs we encounter a new prose genre that also unsettles our long-standing assumptions about public and private.

Problems with Terminology:

What Is the Object We Read, and

What Is a Text in Hypertext?

Writing about hypertext in a print medium immediately produces terminological problems much like those Barthes, Derrida, and others encountered when trying to describe a textuality neither instantiated by the physical object of the printed book nor limited to it. Since hypertext radically changes the experiences that *reading, writing,* and *text* signify, how, without misleading, can one employ these terms, so burdened with the assumptions of print technology, when referring to electronic materials? We still read *according to* print technology, and we still direct almost all of what we write toward print modes of publication, but we can already glimpse the first appearances of hypertextuality and begin to ascertain some aspects of its possible futures. Terms so implicated with print technology necessarily confuse unless handled with great care. Two examples will suffice.

An instance of the kind of problem we face appears when we try to decide what to call the object at which or with which one reads. The object with which one reads the production of print technology is, of course, the book, or smaller print-bearing forms, such as the newspaper or instruction sheet; for the sake of simplicity I shall refer to "book" as the most complex instance of printing technology. In our culture the term *book* can refer to three very different entities—the object itself, the text, or the instantiation of a particular technology. Calling the machine one uses to read hypertext an "electronic book," however, would be misleading, since the machine at which one reads (and writes, and carries out other operations, including sending and receiving mail) does not itself constitute a book, a text: it does not coincide either with the virtual text or with a physical embodiment of it.

Additional problems arise when one considers that hypertext involves a more active reader, one who not only chooses her reading paths but also has the opportunity (in true read-write systems) of reading as someone who creates text; that is, at any time the person reading can assume an authorial role and either attach links or add text to the text being read. Therefore, a term like *reader,* such as some computer systems employ for their electronic mailboxes or message spaces, does not seem appropriate either.[3]

One earlier solution was to call this reading-and-writing site a worksta-tion by analogy to the engineer's workstation, the term assigned to a relatively high-powered machine, often networked with others, that in the early 1990s had far more computing power, memory, and graphic capacities than the per-sonal computer. However, because *workstation* seems to suggest that such objects exist only in the workplace and find application only for gainful labor or employment, this choice of terminology also misleads. Nonetheless, I shall employ it occasionally, if only because it seems closer to what hypertext demands than any of the other terms thus far suggested. The problem with terminology arises, as has now become obvious, because the roles of reader and author change so much in hypermedia technology that our current vocabulary does not have much appropriate to offer.

Whatever one wishes to call the reading-and-writing site, one should think of the actual mechanism that one will use to work (and play) in hyper-text not as a free-standing machine, like today's personal computer. Rather, the "object one reads" must be seen as the entrance, the magic doorway, into the docuverse, since it is the individual reader's and writer's means of partic-ipating in—of being linked to—the world of linked hypermedia documents.

A similar terminological problem appears in what to do with the term *text,* which I have already employed so many times thus far in this study. More than any other term crucial to this discussion, *text* has ceased to inhabit a single world. Existing in two very different worlds, it gathers contradictory meanings to itself, and one must find some way of avoiding confusion when using it. Frequently, in trying to explain certain points of difference, I have found myself forced to blur old and new definitions or have discovered my-self using the old term in an essentially anachronistic sense. For example, in discussing that hypertext systems permit one to link a passage "in" the "text" to other passages "in" the "text" as well as to those "outside" it, one confronts precisely such anachronism. The kind of text that permits one to write, how-ever incorrectly, of insides and outsides belongs to print, whereas we are here considering a form of electronic virtual textuality for which these already sus-pect terms have become even more problematic and misleading. One solu-tion has been to use *text* as an anachronistic shorthand for the bracketed material in the following expression: "If one were to transfer a [complete printed] text (work), say, Milton's *Paradise Lost,* into electronic form, one could link passages within [what had been] the [original] text (Milton's poem) to each other; and one could also link passages to a wide range of materials out-side the original text to it." The problem is, of course, that as soon as one con-verts the printed text to an electronic one, it no longer possesses the same kind of textuality. In the following pages such references to text have to be understood, therefore, to mean "the electronic version of a printed text."

The question of what to call "text" in the medium of hypermedia leads directly to the question of what to include under that rubric in the first place. This question in turn immediately forces us to recognize that hypertext reconfigures the text in a fundamental way not immediately suggested by the fact of linking. Hypertextuality, like all digital textuality, inevitably includes a far higher percentage of nonverbal information than does print; the comparative ease with which such material can be appended encourages its inclusion. Hypertext, in other words, to some degree implements Derrida's call for a new form of hieroglyphic writing that can avoid some of the problems implicit and therefore inevitable in Western writing systems and their printed versions. Derrida argues for the inclusion of visual elements in writing as a means of escaping the constraints of linearity. Commenting on this thrust in Derrida's argument, Gregory Ulmer explains that grammatology thereby "confronts" four millennia during which anything in language that "resisted linearization was suppressed. Briefly stated, this suppression amounts to the denial of the pluridimensional character of symbolic thought originally present in the 'mythogram' (Leroi-Gourhan's term), or nonlinear writing (pictographic and rebus writing)" (*Applied Grammatology*, 8). Derrida, who asks for a new pictographic writing as a way out of logocentrism, has to some extent had his requests answered in hypertext. N. Katherine Hayles argues that digital text alone, even without links, emphasizes the visual, because "the computer restores and heightens the sense of word as image—an image drawn in a medium as fluid and changeable as water" (26).

Because hypertext systems link together passages of verbal text with images as easily as they link two or more passages of text, hypertext includes hypermedia, and I therefore use the two terms interchangeably. Moreover, since computing digitizes both alphanumeric symbols and images, electronic text in theory easily integrates the two. In practice, popular word-processing programs, such as Microsoft Word, have increasingly featured the capacity to include graphic materials in text documents, and, as we shall see, this capacity to insert still and moving images into alphanumeric text is one of the characterizing features of HTML. Linking, which permits an author to send the reader to an image from many different portions of the text, makes such integration of visual and verbal information even easier.

In addition to expanding the quantity and diversity of alphabetic and nonverbal information included in the text, computer text provides visual elements not found in printed work. Perhaps the most basic of these is the cursor, the blinking arrow, line, or other graphic element that represents the reader-author's presence in the text. The cursor, which the user moves either from the keypad by pressing arrow-marked keys or with devices like a mouse, rollerball, or trackpad, provides a moving intrusive image of the reader's presence in the text. Holding the mouse over a footnote number in Microsoft Word produces the text in the note in a pop-up window. The reader can also change the text by using the mouse to position the cursor between the letters in a word, say, between *t* and *h* in *the*. Pressing a button on the mouse inserts a vertical blinking line at this point; pressing the backspace or delete key removes the *t*. Typing will insert characters at this point. In a book one can always move one's finger or pencil across the printed page, but one's intrusion always remains physically separate from the text. One may make a mark on the page, but one's intrusion does not affect the text itself.

The cursor, which adds reader presence, activity, and movement, combines in most previous and extant hypertext systems with another graphic element, a symbol that indicates the presence of linked material.[4] The World Wide Web offers several kinds of changing cursors: the cursor changes from an arrow to a hand when positioned over a linked word or image, and commonly used Java scripts change the appearance of linked objects upon mouse-over—when, that is, the user positions the mouse over them. Yet other scripts produce drop-down menus of links. All these graphic devices remind readers that they are processing and manipulating a new kind of text, in which graphic elements play an important part.

Visual Elements in Print Text

This description of visual elements of all computer text reminds one that print also employs more visual information than people usually take into account: visual information is not limited, as one might at first think, merely to the obvious instances, such as illustrations, maps, diagrams, flow charts, or graphs.[5] Even printed text without explicitly visual supplementary materials already contains a good bit of visual information in addition to alphanumeric code. The visual components of writing and print technology include spacing between words, paragraphing, changes of type style and font size, formatting to indicate passages quoted from other works, assigning specific locations on the individual page or at the end of sections or of the entire document to indicate reference materials (foot notes and endnotes).

Despite the considerable presence of visual elements in print text, they tend to go unnoticed when contemporary writers contemplate the nature of text in an electronic age. Like other forms of change, the expansion of writing from a system of verbal language to one that centrally involves nonverbal information—visual information in the form of symbols and representational elements as well as other forms of information, including sound—has encountered stiff resistance, often from those from whom one is least likely to expect it, namely, from those who already employ computers for writing. Even those who advocate a change frequently find the experience of advocacy and change so tiring that they resist the next stage, even if it appears implicit in changes they have themselves advocated.

This resistance appears particularly clearly in the frequently encountered remark that writers should not concern themselves with typesetting or desktop publishing but ought to leave those activities to the printer. Academics and other writers, we are told, do not design well; and even if they did, the argument continues, such activities are a waste of their time. Such advice, which has recently become an injunction, should make us ask why. After all, when told that one should not avail oneself of some aspect or form of empowerment, particularly as a writer, one should ask why. What if someone told us: "Here is a pencil. Although it has a rubber apparatus at the opposite end from that with which you write, you should not use it. Real writers don't use it"? At the very least we should wonder why anyone had included such capacities to do something; experimenting with it would show that it erases; and very likely, given human curiosity and perversity, which may be the same thing in certain circumstances, we would be tempted to try it out. Thus a capacity would evolve into a guilty pleasure!

Anyone with the slightest interest in design who has even casually surveyed the output of commercial and university presses has noticed a high percentage of appallingly designed or obviously undesigned books. Despite the exemplary work of designers like P. J. Conkwright, Richard Eckersley, and Glen Burris, many presses continue to produce nasty-looking books with narrow margins and gutters, type too small or too coarse for a particular layout, and little sense of page design. Financial constraints are usually offered as the sole determinant of the situation, though good design does not have to produce a more costly final product, particularly in an age of computer typesetting. In several cases I am aware of, publishers have assigned book design to beginning manuscript editors who have had no training or experience in graphic design. As one who has been fortunate enough to have benefited from the efforts of first-rate, talented designers far more than I have suffered from those of poor ones, I make these observations not as a complaint but as a preparation for inquiring why authors are told they should not concern themselves with the visual appearance of their texts and why authors readily accept such instruction.

They do so in part because this injunction clearly involves matters of status and power. In particular, it involves a certain interpretation—that is, a social construction—of the idea of writer and writing. According to this conception, the writer's role and function is just to write. Writing, in turn, is conceived solely as a matter of recording (or creating) ideas by means of language. On the surface, such an approach seems neutral and obvious enough, and that in itself should warn one that it has been so naturalized as to include cultural assumptions that might be worth one's while to examine.

The injunction "just to write," which is based on this purely verbal conception of writing, obviously assumes the following: first, that only verbal information has value, at least for the writer as a writer and probably for the reader as reader;[6] second, that visual information has less value. Making use of such devalued or lesser-valued forms of information (or does visual material deserve the description "real information" at all?) in some way reduces the status of the writer, making him or her less of a real writer. This matter of status again raises its head when one considers another reason for the injunction "just to write," one tied more tightly to conceptions of division of labor, class, and status. In this view of things, it is thought that authors should not concern themselves with matters that belong to the printer. Although troubled by this exclusion, I accepted this argument until I learned that until recently (say, in the 1930s) authors routinely wandered around the typesetting shop at Oxford University Press while their books were being set and were permitted to render advice and judgment, something we are now told is none of our business, beneath us, and so on. The ostensible reason for instructing authors to refuse the power offered them by their writing implement also includes the idea that authors do not have the expertise, the sheer know-how to produce good design. Abundant papers by beginning undergraduates and beginning PC users, cluttered with dissonant typefaces and font sizes, used to be thrust forward to support this argument, one that we receive too readily without additional information. Now people point to ugly websites and blogs.

The fact that beginners in any field of endeavor do a fairly poor quality job at a new activity hardly argues forcefully for their abandoning that activity. If it did, we would similarly advise beginning students immediately to abandon their attempts at creative and discursive writing, at drawing and philosophy, and at mathematics and chemistry. One reason we do not offer such instructions is because we feel the skills involved in those endeavors are important—apparently in contrast to visual ones. Another reason, of course, is that teaching involves our livelihood and status. The question that arises, then, is why is visual information less important? The very fact that people experiment with visual elements of text on their computers shows the obvious pleasure they receive in manipulating visual effects. This pleasure suggests in turn that by forbidding the writer visual resources, we deny an apparently innocent source of pleasure, something that apparently must be cast aside if one is to be a true writer and a correct reader.

Much of our prejudice against the inclusion of visual information in text derives from print technology. Looking at the history of writing, one sees that it has a long connection with visual information, not least the origin of many alphabetic systems in hieroglyphics and other originally visual forms of writing. Medieval manuscripts present some sort of hypertext combination of font sizes, marginalia, illustrations, and visual embellishment, both in the form of calligraphy and that of pictorial additions.

This same prejudice against visual elements appears in recent supposedly authoritative guidelines for creating websites. Jakob Nielsen's *Designing Web Usability*, for example, advises web designers to avoid graphic elements, particularly for opening screens (homepages), because they unnecessarily consume both bandwidth and screen real estate. I certainly understand the reasons for such advice. Like many other users in the early days of commercial sites, I've waited many minutes for the opening screen of a national airline's site to download even though I had high-speed network access, finally giving up. Early web designers found themselves so understandably enthralled by elaborate graphics and animation that they cluttered sites with nonfunctional elements that consumed important resources. As the airline website shows, this approach proved disastrous for commercial applications at a time when most potential customers had slow Internet connections. Obviously, designers must balance ease of access against visual elements that encourage people to access the site in the first place, but avoiding graphic elements as a basic design principle doesn't make much sense for one obvious reason: images and other graphic elements are the single most important factor in the astonishing growth of the Word Wide Web. The invention of the image tag (<img src="picture.jpg">), which instructs the web browser to place a picture, icon, or other graphic element within text, made the World Wide Web immensely appealing, turning it into a medium rich with visual pleasures. The embed tag, which places QuicktimeVR, sound, and video in an HTML document, similarly converts the Web to a multimedia platform. Therefore, whether or not we believe it has an identifiable logic—a McLuhan-esque message in the medium—the Web certainly is significantly pictorial. Recommending that one should not use static or moving images in a medium popularized by their very presence therefore seems particularly bizarre.[7]

This blindness to the crucial visual components of textuality not only threatens to hinder our attempts to learn how to write in electronic space but has also markedly distorted our understanding of earlier forms of writing. In particular, our habits of assuming that alphanumeric—linguistic—text is the only text that counts has led to often bizarre distortions in scholarly editing. As Jerome J. McGann reminds us in *The Textual Condition,* "literary works typically secure their effects by other than purely linguistic means" (77), always deploying various visual devices to do so. Hence leaving such aspects of the text out of consideration—or omitting them from scholarly editions—drastically reconfigures individual works. "All poetry, even in its most traditional forms, asks the reader to decipher the text in spatial as well as linear terms. Stanzaic and generic forms, rhyme schemes, metrical orders: all of these deploy spatial functions in scripted texts, as their roots in oral poetry's 'visual' arts of memory should remind us" (113). One cannot translate such nonprint and even antiprint works like that of Blake and Dickinson into print without radically reconfiguring them, without creating essentially new texts, texts a large portion of whose resources have been excised. Although "textual and editorial theory has heretofore concerned itself almost exclusively with the linguistic codes," McGann urges, "the time has come, however, when we have to take greater theoretical account of the other coding network which operates at the documentary and bibliographical level of literary works" (78). Once again, as with the scholarly editing of medieval manuscripts and nineteenth-century books, digital word and digital image provide lenses through which we can examine the preconceptions—the blinders—of what Michael Joyce calls "the late age of print" (*Of Two Minds,* 111).

Animated Text

The essential visual components of all text find perhaps their fullest instantiation in the form of animated text—text that moves, even dances, on the computer screen, sweeping from one side to the other, appearing to move closer to readers or retreat away from them into a simulated distance. Text animation, which has become very popular in recent digital poetry, derives from the nature of computer text, which takes the form of code. Until the development of digital textuality, all writing necessarily took the form of physical marks on physical surfaces. With computers, writing, which had always been physical, now became a matter of codes—codes that could be changed, manipulated, and moved in entirely new ways. "Change the code, change the text" became the rule from which derive the advantages of so-called word processing (which is actually the composition, manipulation, and formatting of text in computer environments). The advantages of word processing over typewriters became so immediately obvious themselves in business and academia that dedicated word processors and then personal computers swiftly made typewriters obsolete. "Change the code, change the text" also produces the "styles" option in word-processing software, such as Microsoft Word, which permits a writer to create and deploy styles containing font, type size, and rules for various text entities (paragraph, inset quotation, bibliography, and so on). By simply highlighting a word, sentence, or paragraph, the user of such software can easily modify the appearance of text, whether it is intended to remain on screen or issue forth as a printout or as a typeset book.

The same text-as-code that permits word processing also permits moving words. In its simplest form, text animation simply involves moving the text on screen a line at a time, essentially dispensing the poem at a rate determined by the author. Kate Pullinger and Talan Memmot's elegant *Branded* (2003; Figure 10) functions in this way. Pearl Forss's *Authorship* (2000), which combines sound and text animation, exemplifies the use of this kind of animated text to create experimental discursive writing for e-space. First, to the accompaniment of a driving drumbeat, the words "What is" appear in white block letters against a black screen to which are quickly added in the red-orange words "an author?" The question mark then dances on screen, after which the sentence moves downward as words of Roland Barthes on authorship move on screen; these in turn are replaced by Forss's pronouncements about authorship; then in green appear the words "what matters who's speaking?"—a question immediately identified as having been asked by Beckett (whose name, in white, undulates on screen). Next, an image of a rose fills the entire screen, and on top of it appear many pink letters, which soon arrange themselves to state, "A rose by any other name would smell as sweet," an assertion immediately challenged by the question (in green) "or would it?" And this screen is rapidly obliterated by the appearance of images of theorists on authorship and covers of their books, all of which build to a collage. What I've described makes up the opening section or movement, several of which follow, each punctuated by the same assembling collage.

Such text animation, often accompanied by sound, appears more frequently in digital literary art than in discursive or informational projects. For example, several of the animated poems on the *Dotze Sentits: Poesia catalona d'avui* CD-ROM (1996), such as Josep Palau i Fabre's "La Noia" and Feliu For-

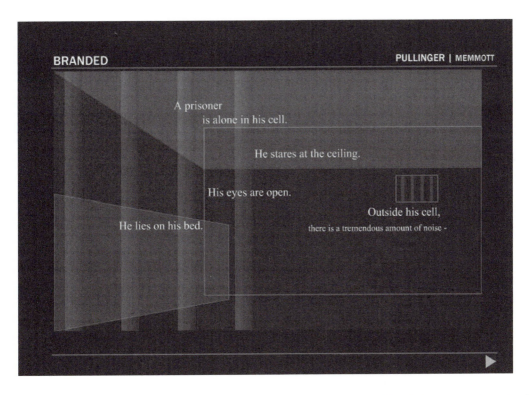

Figure 10. Animated Text. In Kate Pullinger and Talan Memmot's elegant animated poem, *Branded,* text animation takes the form of moving the text on screen a line at a time, dispensing the poem at a rate determined by the authors.

mosa's "Ell sort de seta l'aigua," accompany the sound of the poet's reading by moving words and phrases of different sizes and colors across the screen from top to bottom and from one edge of the screen to another; words pop in and out of existence, too, as the text performs itself.[8] More radical experimentation with animated text appears in Philadelpho Menezes and Wilton Azvedo's *Interpoesia: Poesia Hipmedia Interativa* (1998), in which elements (or fragments) of both spoken and written words react to the reader's manipulation of the computer mouse. Letters move, parts of words change color or disappear, and sounds become layered upon one another as the reader essentially performs the text using the sounds provided.

Moving text on screen, which has only become possible for most users with the advent of inexpensive computing power and broad bandwidth, has had an effect on digital literary arts almost as dramatic as that of word processing on academic institutions and the workplace. But are such projects hypertextual (and does it matter)?

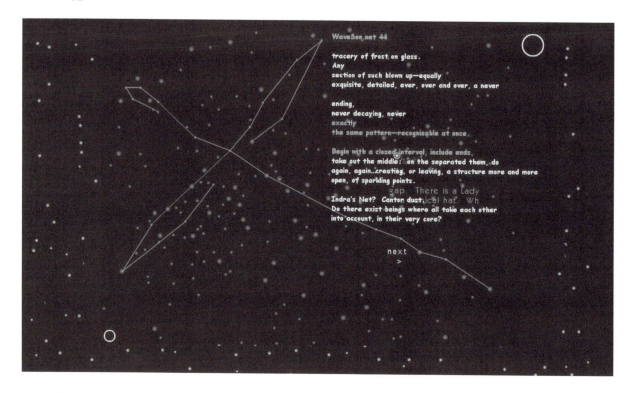

WaveSon.net 44

tracery of frost on glass.
Any
section of such blown up—equally
exquisite, detailed, ever, over and over, a never

ending,
never decaying, never
exactly
the same pattern—recognizable at once.

Begin with a closed interval, include ends,
take out the middle, on the separated them, do
again, again...creating, or leaving, a structure more and more
open, of sparkling points.

gap. There is a Lady
Indra's Net? Cantor dust ical hat. Wh
Do there exist beings where all take each other
into account, in their very core?

next
>

Figure 11. Stephanie Strickland's _Vniverse_. Both a constellation and poetic text appear against the starry sky as the reader manipulates the mouse.

In one important sense, these projects, like _Branded,_ appear essentially _anti_hypertextual. If one takes hypertext to be an information technology that demands readers take an active role, then these animated texts enforce the opposite tendency. In contrast to hypertext, they demand the reader assume a generally passive role as a member of an audience, rather than someone who has some say in what is to be read. They add, in other words, to the power of the author—or at least to the power of the text—and deny the possibility of a more empowered reader. Strickland's _Vniverse_ represents a comparatively rare example of text-animation hypermedia that strives to grant readers control; it is, however, quite unusual (Figure 11).

If one were to arrange print text, hypertext, video, and animated text along a spectrum, hypertext, perhaps surprisingly, would take its place closest to print. Reading written or printed text, one cannot change its order and progression, but because the text is fixed on the page, one can leave it, reading another text, taking notes, or simply organizing one's thoughts, and return to find the text where one left it, unchanged. The characteristic fixity of writing, therefore, endows the reader with the ability to process it asynchronously—that is, at the convenience of the reader.[9] Consider the difference of such fixed text from video and animated text: if one leaves the television set to answer the phone or welcome a guest, the program has moved on and one cannot retrieve it, unless, that is, one has a digital or analogue copy of it and can replay it. The very great difference in degree of audience control between video as seen on broadcast television and video viewed from storage media, such as videotape, DVD, or TiVo, suggests that they are experienced as different media. Still, since video, like cinema, is a temporal form—a technology that presents its information in necessary sequence—one generally has to follow long patches of the story or program in its original sequence to find one's place in an interrupted narrative. Animated text, in contrast, *entirely* controls the reader's access to information at the speed and at the time the author wishes. One could, it is true, replay the entire animated text, but the nature of the medium demands that the minimum chunk that can be examined takes the form of the entire sequence.

Another form of moving text appears in the timed links of Stuart Moulthrop's *Hegirascope,* links that dramatically affect the reader's relation to text. The reading experience produced by these timed links contrasts sharply with that possible with writing, print, and most hypertext. Since the text disappears at timed intervals outside the reader's control, the characteristic fixity of writing disappears as the document being read is replaced by another. Some of the replacements happen so quickly that this text enforces rapid reading, preventing any close reading, much less leisurely contemplation of it. Michael Joyce famously asserted that "hypertext is the revenge of text upon television" (*Of Two Minds,* 47, 111), by which I take him to mean that hypertext demands active readers in contrast to television's relatively passive audience. These examples of animated (or disappearing) text in contrast appear to be extensions of television and film to encompass and dominate text, or in Joyce's terms, the revenge of television (broadcast media) on hypertext. This is not necessarily a bad thing, any more than cinema is worse than print narrative. Animated text, like cinema and video, exists as an art form with its own criteria. It's just not hypertext.

Stretchtext

Not all animated alphanumeric text, it turns out, is nonhypertextual. In fact, Ted Nelson's stretchtext, which he advances as a complement to the by-now standard node-and-link form, produces a truly reader-activated form.[10] Except for researchers working with spatial hypertext, most students of hypermedia, like all users of the Web, work on the assumption that it must take the form of node and link. A good deal of theoretical and practical attention has appropriately been paid to the description, implementation, and categorization of linking. However, as Noah Wardop-Fruin has reminded us, Ted Nelson, who did not confine hypertext to the node-and-link form, also proposed stretchtext. According to Nelson's *Computer Lib/Dream Machines* (1974), "this form of hypertext is easy to use without getting lost . . . Gaps appear between phrases; new words and phrases pop into the gaps, an item at a time . . . The Stretchtext is stored as a text stream with extras, coded to pop in and pop out at the desired altitudes" (315).

Compare a reader's experience of stretchtext to that when reading on the Web. When one follows a link on the World Wide Web, one of two things happens: either the present text disappears and is replaced by a new one or the destination text opens in a new window. (On Windows machines, in which the newly opened document obscures the previous one because it appears on top of it, only an experienced user would know that one can move the most recently opened window out of the way. Macintosh machines follow a different paradigm, emphasizing a multiple-window presentation.) By and large, standard web browsers follow the replacement paradigm whereas other hypertext environments, such as Intermedia, Storyspace, and Microcosm, emphasize multiple windows. Stretchtext, which takes a different approach to hypertextuality, does what its name suggests and stretches or expands text when the reader activates a hot area.

For an example, let us look at a single sentence as it appears in a document based on passages from this book that I made using Nicholas Friesner's Web-based stretchtext. One first encounters the following:

Using hypertext, students of critical theory now have a laboratory with which to test its ideas. Most important, perhaps, an experience of reading hypertext or reading with hypertext greatly clarifies many of the most significant ideas of **critical theory.** As J. David Bolter points out in the course of explaining that hypertextuality embodies poststructuralist conceptions of the open text, "what is unnatural in print becomes natural in the electronic medium and will soon no longer need saying at all, because it can be shown." (www.cyberartsweb.org/cpace/ht/stretchtext/gpl2.html)

Clicking on "**critical theory**" produces "**critical theory.** In fact, some of the most exciting **student projects** take the form of testing, applying, or critiquing specific points of theory, including notions of the **author, text, and multivocality.**" Next, clicking on "**student projects**" in turn produces "**student projects** in **Intermedia**, Storyspace, html, and Flash, and published examples of hypermedia take the form of testing, applying, or critiquing specific points of theory, including notions of the **author, text,** and **multivocality**." Clicking on any of the four instances of bold text generates additional passages, the last three of which also contain a standard HTML link. Stretching "**text**," for example," produces "**text**—and Pearl Forss's *What Is an Author* acts as an experiment contrasting reader's reactions to moving text versus reader-centered hypertext." Clicking on the title of her project opens it in a new window.

A most important distinguishing characteristic of stretchtext follows from the manner in which it makes new text appear framed by the old: stretchtext does not fragment the text like other forms of hypermedia. Instead, it retains the text on the screen that provides a context to an anchor formed by a word or phrase even after it has been activated. Stretching the text provides a more immediate perceptual incorporation of the linked-to text with the text from which the link originates. In effect, *text* becomes *context* as new text is added; or rather, the previously present text remains while the new text appears and serves as its context. This conversion of text to context for other texts may be seen more abstractly in any textual medium, but stretchtext takes this notion quite literally.

The experience of using Friesner's Web-based implementation demonstrates that in certain situations stretchtext has an advantage over link-and-node hypertext; in other uses the link-and-node form works better. One strong advantage of stretchtext derives from the fact that hidden text is already present, though not visible, when the web browser loads the HTML file, and it therefore appears instantly when the text expands. The text also contracts instantly, thus providing two real advantages: first, because the newly appearing text appears in immediate physical proximity to the text one was reading before activating the stretchtext, the reader experiences none of the disorientation that may occur when following a link. Second, the very speed with which the stretchtext appears encourages readers to check stretchable areas to see if they in fact want the additional information on offer with the effect that readers feel they have more control over obtaining information.

Experience using stretchtext suggests that it provides a convenient means of obtaining definitions, brief explanations, and glossary-like annotations. In addition, a second or even third layer of stretchtext seems better suited than the replacement-window paradigm for more detailed information *directly related to the original anchor*. The one disadvantage the more clearly atomizing link-and-node hypertext does not have appears if one expands ancillary information—say, comments on Morris in an essay about Ruskin: reading a succession of increasingly more detailed stretchtext passages of the main topic can create reader disorientation when he or she returns to the passage's original form by shrinking the text. Here the occasionally criticized atomizing effect of link-and-note hypertext in fact proves a major advantage because when readers follow a link, they know they have moved to someplace new. The gap that always plays an essential role in linked hypermedia here has an orienting, rather than a disorienting, effect. One obvious way to take advantage of both forms of hypertext, of course, involves including links to exterior lexias at appropriate points within the stretched text ("appropriate" here meaning those places at which further expansion of the original text makes returning to the original contracted text confusing).

Friesner's Web-based stretchtext also works well with images and hence proves itself to be a form of hypermedia that provides authors with new options. Web-based hypermedia has three main ways of incorporating images: (1) placing images at the end of the link whether they appear alone or within text containing explanatory information, (2) placing images within a Javascript-created pop-up window usually smaller than the document it overlays, and (3) placing thumbnail images within a text, often at the right or left margin, which can link to larger images (a simple use of align = "left" or "right" plus hspace = "10" within the image tag provides an easy way to allow text to flow around an image, providing an aesthetically pleasing border that separates them). stretchtext offers a fourth way to handle images, and like its purely text-based form, it has some particularly effective applications. A lexia chiefly devoted to discussing one painting most effectively includes the image of it as a linked thumbnail within the text, but passing mentions of details, sources, or analogous paintings work better as stretchtext presentations, for they are nonintrusive, quickly viewed, and quickly closed and left behind. Thus, stretchtext-image presentation seems particularly well suited for introducing images to which one may wish to refer briefly. Since one can nest images as well as alphanumeric information in Friesner's stretchtext, it also provides a convenient way for the reader to access details of painting, earlier versions, and so on.

When I first read about stretchtext I envisioned it functioning vertically, as it does so effectively in Friesner's version; that is, I assumed the text would move apart above and below the stretching section, although I did not imagine the text would appear instantaneously. When Ian M. Lyons created TextStretcher as a Director demonstration of the concept, he experimented with several approaches to make words appear from within an already present text. In one version the text moves horizontally, and as new words arrive, they push the old text to the right. In a second version the stretchable text divides vertically, leaving an empty area. In one of these implementations of stretchtext, the words fill in from left to right until the space disappears; in another that Lyons considers especially useful for poetry, the first word of the new text appears in the center of the vacant area, and words then flow out from either side. If speed is calibrated with effective stops and starts, word-by-word presentation may give the impression of speech inscribed on the fly. Letter-by-letter presentation may then give the impression of the text's being typed out by either a human or an artificial source, depending on speed changes.

In linear stretch mode, activating the stretchtext anchor makes the sentence or phrase in which it appears divide, producing an empty space that new words begin to fill, moving from left to right. Once one begins animating the presentation of new text, a temporal delay occurs. As Lyons points out, this delay can perceptually create the impression of a certain amount of resistance the medium itself presents to the embodiment of the text. Just how much resistance is met in turn depends on the relative speed with which words and letters appear on the screen. With linear presentation (here presumed to be in the direction of normal reading), the potential for a different reading experience multiplies, depending on whether one presents text one word or one letter at a time. (Lyons chose to present one word at a time.)

In the expansion-from-center mode, Lyons explains,

Expansion and contraction move in two opposite directions at once. Again, if properly calibrated, the speed at which text appears to allow for reading at both ends simultaneously on the fly significantly heightens the feeling of contextual dependency. With expansion from the center of the new lexia, text moving backward works to complete the newly created hanging fragment that precedes the new insertion; text moving forward (in the normal direction of reading) aims to give reverse justification for the waiting text that now marks the latter half of the newly old, contextualizing material. Moreover, the cognitive experience of learning to read in multiple directions at once so as to realign severed context is at first rather challenging. What is exciting about this approach is that, from a cognitive standpoint, it creates an alternative reading procedure, one that while predictably difficult remains surprisingly possible. (Private communication)

This form of stretchtext, which Lyons created for writing poetry, obviously draws attention to the experience of text itself, intentionally preventing the reader from reading *through* the text, from too readily taking the text as transparent. With informational hypertext, however, one does not wish to foreground the linguistic aspect of the medium, and one therefore needs devices that enable reading.

Like all hypertext environments, TextStretcher needs a means of indicating to the reader the presence of an anchor, here defined as a span of text or other information that, when the proper protocol is followed, activates the hypertext functionality; in node-and-link hypertext such activation involves following a link; in stretchtext it involves inserting nested or hidden text within the present text. Different link-and-node systems have employed different ways of indicating hot text: Intermedia used static indications of links—a horizontal arrow within a rectangle at the end of a text span; the World Wide Web uses both static and dynamic means—the standard blue underlined text that authors and users can customize and the changing appearance of the user's mouse from an arrow to a hand when positioned over an anchor.

In creating TextStretcher, Lyons chose two simple symbols: (1) a vertical line within parentheses to indicate an anchor and (2) a hyphen between parentheses to indicate that one can contract the stretched text to produce its earlier state. When one clicks on an icon representing hot text in all versions of TextStretcher, new words appear, and depending on the icon, clicking again can either make the newly arrived text disappear or expand it further with new information. Friesner's Web-based stretchtext uses a boldface font to indicate an expandable anchor and a colored unboldface font to indicate contractible text. Some experimenting with his version have used phrases, such as "[Show/Hide Additional Content]" and icons, and others working with stretchtext combine a three- or four-inch gap followed by "More" on a colored rectangle. None of these approaches seems entirely satisfactory, and one possible solution might involve revealing the locations of stretchable text by mouse-overs. Stretchtext complements note-and-link hypermedia so valuably that solving these interface issues warrants considerable attention.

The Dispersed Text

At the same time that the individual hypertext lexia has looser, or less determining, bonds to other lexias from the same work (to use a terminology that now threatens to become obsolete), it also associates itself with text created by other authors. In fact, it associates with whatever text links to it, thereby dissolving notions of the intellectual separation of one text from others, just as some chemicals destroy the cell membrane of an organism. Destroying the cell membrane destroys the cell; it kills. In contrast, similarly destroying now-conventional notions of textual separation may destroy certain attitudes associated with text, but it will not necessarily destroy text. It will, however, reconfigure it and our expectations of it.

Another related effect of electronic linking is that it disperses "the" text into other texts. As an individual lexia loses its physical and intellectual separation from others when linked electronically to them, it also finds itself dispersed into them. The necessary contextualization and intertextuality produced by situating individual reading units within a network of easily navigable pathways weaves texts, including those by different authors and those in nonverbal media, tightly together. One effect is to weaken and perhaps destroy any sense of textual uniqueness.

Such notions are hardly novel to contemporary literary theory, but here as in so many other cases hypertext creates an almost literal reification or embodiment of a principle that had seemed particularly abstract and difficult when read from the vantage point of print. Since much of the appeal, even charm, of these theoretical insights lies in their difficulty and even preciousness, this more literal presentation promises to disturb theoreticians, in part, of course, because it greatly disturbs status and power relations within their field of expertise.

Hypertextual Translation of

Scribal Culture

Hypertext fragments, disperses, or atomizes text in two related ways. First, by removing the linearity of print, it frees the individual passages from one ordering principle—sequence—and threatens to transform the text into chaos. Second, hypertext destroys the notion of a fixed unitary text. Considering the "entire" text in relation to its component parts produces the first form of fragmentation; considering it in relation to its variant readings and versions produces the second.

Loss of a belief in unitary textuality could produce many changes in Western culture, many of them quite costly, when judged from the vantage point of our present print-based attitudes. Not all these changes are necessarily costly or damaging, however, particularly to the world of scholarship, where this conceptual change would permit us to redress some of the distortions of naturalizing print culture. Accustomed to the standard scholarly edition of canonical texts, we conventionally suppress the fact that such twentieth-century print versions of works originally created within a manuscript culture are bizarrely fictional idealizations that produce a vastly changed experience of text. To begin with, the printed scholarly edition of Plato, Vergil, or Augustine provides a text far easier to negotiate and decipher than any available to these authors' contemporary readers. They encountered texts so different from ours that even to suggest that we share common experiences of reading misleads.

Contemporary readers of Plato, Vergil, or Augustine processed texts without interword spacing, capitalization, or punctuation. Had you read these last two sentences fifteen hundred years ago, they would have taken the following form:

theyencounteredtextssodifferentfromoursthatateventosuggestthatwesharecommon experiencesofreadingmisleadscontemporaryreadersofplatovergiloraugustine processedtextswithoutinterwordspacingcapitalizationorpunctuationhadyouread theselasttwosentencesfifteenhundredyearsagotheywouldtakethefollowingform

Such unbroken streams of alphabetic characters made even phonetic literacy a matter of great skill. Since deciphering such texts heavily favored reading aloud, almost all readers experienced texts not only as an occasion for strenuous acts of code breaking but also as a kind of public performance.

The very fact that the text we would have read fifteen hundred years ago appeared in a manuscript form also implies that to read it in the first place we would first have had to gain access to a rare, even unique object—assuming, that is, that we could have discovered the existence of the manuscript and made an inconvenient, expensive, and often dangerous trip to see it. Having gained access to this manuscript, we would also have approached it much differently from the way we today approach the everyday encounter with a printed book. We would probably have taken the encounter as a rare, privileged opportunity, and we would also have approached the experience of reading this unique object with a very different set of assumptions than would a modern scholar. As Elizabeth Eisenstein has shown, the first role of the scholar in a manuscript culture was simply to preserve the text, which doubly threatened to degrade with each reading: each time someone physically handled the fragile object, it reduced its longevity, and each time someone copied the manuscript to preserve and transmit its text, the copyist inevitably introduced textual drift.

Thus, even without taking into account the alien presence of pagination, indices, references, title pages, and other devices of book technology, the encounter with and subsequent reading of a manuscript constituted a very different set of experiences than those which we take for granted. Equally important, whereas the importance of scholarly editions lies precisely in their appearance in comparatively large numbers, each manuscript of our texts by Plato, Vergil, or Augustine existed as a unique object. We do not know which particular version of a text by these authors any reader encountered. Presenting the history and relation of texts created within a manuscript culture in terms of the unitary text of modern scholarship certainly fictionalizes—and falsifies—their intertextual relations.

Modern scholarly editions and manuscripts combine both uniqueness and multiplicity, but they do so in different ways. A modern edition of Plato, Vergil, or Augustine begins by assuming the existence of a unique, unitary text, but it is produced in the first place because it can disseminate that text in a number of identical copies. In contrast, each ancient or medieval manuscript, which embodies only one of many potential variations of "a text," exists as a unique object. A new conception of text is needed by scholars trying to determine not some probably mythical and certainly long-lost master text but the ways individual readers actually encountered Plato, Vergil, or Augustine in a manuscript culture. In fact, we must abandon the notion of a unitary text and replace it with conceptions of a dispersed text. We must do, in other words, what some art historians working with analogous medieval problems have done—take the conception of a unique type embodied in a single object and replace it with a conception of a type as a complex set of variants. For example, trying to determine the thematic, iconological, and compositional antecedents of early-fourteenth-century ivory Madonnas, Robert Suckale and other recent students of the Court Style have abandoned linear derivations and the notion of a unitary type. Instead, they emphasize that sculptors chose among several sets of fundamental forms or "groundplans" as points of departure. Some sort of change in basic attitudes toward the creations of manuscript culture seems necessary.

The capacity of hypertext to link all the versions or variants of a particular text might offer a means of somewhat redressing the balance between uniqueness and variation in preprint texts. Of course, even in hypertext presentations, both modern printing conventions and scholarly apparatus will still infringe on attempts to recreate the experience of encountering these texts, and nothing can restore the uniqueness and corollary aura of the individual manuscript. Nonetheless, as the work of Peter Robinson shows, hypertext offers the possibility of presenting a text as a dispersed field of variants and not as a falsely unitary entity. High-resolution screens and other technological capacities also increasingly permit a means of presenting all the individual manuscripts. The Bodleian Library, Oxford, has already put online detailed, large-scale images of some of its most precious illuminated manuscripts. An acquaintance with hypertext systems might by itself sufficiently change assumptions about textuality to free students of preprint texts from some of their biases.

A Third Convergence: Hypertext and Theories of Scholarly Editing

All forms of hypertext, even the most rudimentary, change our conceptions of text and textuality. The dispersed textuality characteristic of this information technology therefore calls into question some of the most basic assumptions about the nature of text and scholarly textual editing. The appearance of the digital word has the major cultural effect of permitting us, for the first time in centuries, easily to perceive the degree to which we have become so accustomed to the qualities and cultural effects of the book that we unconsciously transfer them to the productions of oral and manuscript cultures. We so tend to take print and print-based culture for granted that, as the jargon has it, we have "naturalized" the book by assuming that habits of mind and manners of working associated with it have naturally and inevitably always existed. Eisenstein, McLuhan, Kernan, and other students of the cultural implications of print technology have demonstrated the ways in which the printed book formed and informed our intellectual history. They point out, for example, that a great part of these cultural effects derive from book technology's creation of multiple copies of essentially the same text. Multiple copies of a fixed text in turn produce scholarship and education as we know it by permitting readers in different times and places to consult and refer to the "same" text. Historians of print technology also point out that economic factors associated with book production led to the development of both copyright and related notions of creativity and originality. My reason for once again going over this familiar ground lies in the fact that all these factors combine to make a single, singular unitary text an almost unspoken cultural ideal. They provide, in other words, the cultural model and justification for scholarly textual editing as we have known it.

It is particularly ironic or simple poetic justice—take your pick—that digital technology so calls into question the assumptions of print-associated editorial theory that it forces us to reconceive editing texts originally produced for print as well as those created within earlier information regimes. Print technology's emphasis on the unitary text prompted the notion of a single perfect version of all texts at precisely the cultural moment that the presence of multiple print editions undercut that emphasis—something not much recognized, if at all, until the arrival of digitality. As the work of James Thorpe, George Bornstein, Jerome J. McGann, and others has urged, any publication during an author's lifetime that in some manner received his or her approval—if only to the extent that the author later chose not to correct changes made by an editor or printer—is an authentic edition. Looking at the works of authors such as Ruskin and Yeats, who radically rewrote and rearranged their texts throughout their careers, one recognizes that the traditional scholarly edition generally makes extremely difficult reconstructing the version someone read at a particular date. Indeed, from one point of view it may radically distort our experience of an individual volume of poems by the very fact that it enforces an especially static frozen model on what turns out to have been a continually shifting and changing entity.

This new conception of a more fluid, dispersed text, possibly truer than conventional editions, raises the issue if one can have a scholarly edition at all, or if we must settle for what McGann terms an archive ("Complete Writings")—essentially a collection of textual fragments (or versions) from which we assemble, or have the computer assemble, any particular version that suits a certain reading strategy or scholarly question, such as "What version of *Modern Painters,* Volume 1, did William Morris read at a particular date and how did the text he read differ from what American Ruskinians read?"

One does not encounter many of these issues when producing print editions because matters of scale and economy decide or foreclose them in advance. In general, physical and economic limitations shape the nature of annotations one attaches to a print edition just as they shape the basic conception of that edition. So what can we expect to happen when these limitations disappear? Or, to phrase the question differently, what advantages and disadvantages, what new problems and new advantages, will we encounter with the digital word?

Hypertext, Scholarly Annotation, and the Electronic Scholarly Edition

One answer lies in what hypertext does to the concept of annotation. As I argue at length in the following chapter, this new information technology reconfigures not only our experience of textuality but also our conceptions of the author's relation to that text, for it inevitably produces several forms of asynchronous collaboration, the first, limited one inevitably appearing when readers choose their own ways through a branching text. A second form appears only in a fully networked hypertext environment that permits readers to add links to texts they encounter. In such environments, which are exemplified by the World Wide Web, the editor, like the author, inevitably loses a certain amount of power and control. Or, as one of my friends who created the first website for a major computer company pointed out, "If you want to play this game, you have to give up control of your own text." Although one could envision a situation in which any reader could comment on another editor's text, a far more interesting one arises when successive editors or commentators add to what in the print environment would be an existing edition. In fact, one can envisage a situation in which readers might ultimately encounter a range of annotations.

An example taken from my recent experience with having students create an annotated version—read "edition"—of Carlyle's "Hudson's Statue" on the World Wide Web illuminates some of the issues here. I intended the assignment in part to introduce undergraduates to various electronic resources available at my university, including the online versions of the *Oxford English Dictionary* and *Encyclopaedia Britannica*. I wished to habituate them to using electronic reference tools accessible outside the physical precincts of the library both to acquaint them with these new tools and also to encourage students to move from them to those presently available only in print form. For this project students chose terms or phrases ranging from British political history ("Lord Ellenborough" and "People's League") to religion and myth ("Vishnu," "Vedas," "Loki"). They then defined or described the items chosen and then briefly explained Carlyle's allusion and, where known, his uses of these items in other writings.

This simple undergraduate assignment immediately raised issues crucial to the electronic scholarly edition. First of all, the absence of limitations on scale—or to be more accurate, the absence of the same limitations on scale one encounters with physical editions—permits much longer, more substantial notes than might seem suitable in a print edition. To some extent a hypertext environment always reconfigures the relative status of main text and subsidiary annotation. It also makes much longer notes possible. Electronic linking makes information in a note easily available, and therefore these more substantial notes conveniently link to many more places both inside and outside the particular text under consideration than would be either possible or conveniently usable in a print edition. Taking our present example of "Hudson's Statue," for instance, we see that historical materials on, say, democratic movements like Chartism and the People's International League, can shift positions in relation to the annotated text: unlike a print environment, an electronic one permits perceiving the relation of such materials in opposite manners. The historical materials can appear as annotations to the Carlyle text, or conversely "Hudson's Statue" can appear—be experienced as—an annotation to the historical materials. Both in other words exist in a networked textual field in which their relationship depends solely on the reader's need and purpose.

Such recognitions of what happens to the scholarly text in wide-area-networked environments, such as those created by World Wide Web and

HyperG, only complicates matters by forcing us to confront the question, "What becomes of the concept and practice of scholarly annotation?" Clearly, linking by itself isn't enough, and neither is text retrieval. At first glance, it might seem that one could solve many issues of scholarly annotation in an electronic environment by using sophisticated text retrieval. In the case of my student-created annotated edition of "Hudson's Statue," one could just provide instructions to use the available search tools, though this do-it-yourself approach would probably appeal only to the already-experienced researcher. Our textual experiment quickly turned up another, more basic problem when several bright, hard-working neophytes wrote elegant notes containing accurate, clearly attributed information that nonetheless referred to the wrong person, in two cases providing material about figures from the Renaissance rather than about the far-lesser-known nineteenth-century figures to whom Carlyle referred. What this simple-minded example suggests, of course, is nothing more radical than that for the foreseeable future scholarship will always be needed, or to phrase my point in terms relevant to the present inquiry, one cannot automate textual annotation. Text retrieval, however valuable, by itself can't do it all.

Fine, but what about hypertext? The problem, after all, with information retrieval lies in the fact that active readers might obtain either nonsignificant information or information whose value they might not be able to determine. Hypertext, in contrast, can provide editorially approved connections in the form of links, which can move from a passage in the so-called main text—here "Hudson's Statue"—to other passages in the same text, explanatory materials relevant to it, and so on. Therefore, assuming that one had permission to create links to the various online resources, such as the *OED*, one could do so. If one did not have such permission, one could easily download copies of the materials from them, choose relevant sections, and put them back online within a web to which one had access; this second procedure is in essence the one many students choose to follow. Although providing slightly more convenience to the reader than the text-retrieval do-it-yourself model, this model still confronts the reader with problems in the form of passages (or notes) longer than he or she may wish to read.

One solution lies in creating multilevel or linked progressive annotation. Looking at the valuable, if overly long, essay one student had written on Carlyle and Hindu deities, I realized that a better way of proceeding lay in taking the brief concluding section on Carlyle's satiric use of these materials and making that the first text or lexia the reader encounters; the first mention of, say, Vedas or Vishnu, in that lexia was then linked to the longer essays, thereby providing conveniently accessible information on demand but not before it was required.

I have approached these questions about scholarly editions through the apparently unrelated matters of a student assignment and educational materials because they remind us that in anything like a fully linked electronic environment, all texts have variable applications and purposes. One consequence appears in the variable forms that annotation and editorial apparatus will almost certainly have to take: since everyone from the advanced scholar down to the beginning student or reader outside the setting of an educational institution might be able to read such texts, they will require various layers or levels of annotation, something particularly necessary when the ultimate linked text is not a scholarly note but another literary text.

Thus far I have written only as if the linked material in the hypertext scholarly edition consists of textual apparatus, explanatory comment, and contextualization, but by now it should have become obvious that many of those comments inevitably lead to other so-called primary texts. Thus, in our putative edition of "Hudson's Statue" one cannot only link it to reference works, such as the *OED*, the *Britannica*, (and possibly in the future) to the *Dictionary of National Biography*, but also to entire linguistic corpora and to other texts by the same author, including working drafts, letters, and other publications. Why stop there? Even in the relatively flat, primitive version of hypertext offered by the present World Wide Web of the Carlylean text demands links to works on which he draws, such as Jonathan Swift's *Tale of a Tub*, and those that draw on him, such as Ruskin's "Traffic," whose satiric image of the Goddess-of-Getting-on (or Britannia of the Market) derives rather obviously from Carlyle's ruminations on the never-completed statue of a stock swindler. Finally, one cannot restrict the text field to literary works, and "Hudson's Statue" inevitably links not only to the Bible and contemporary guides to its interpretation but also to a wide range of primary materials, including parliamentary documents and contemporary newspapers, to which Carlyle's text obviously relates.

Once again, though, linking, which reconfigures our experience and expectations of the text, is not enough, for the scholarly editor must decide *how* to link various texts. The need for some form of intermediary lexias again seems obvious, the first, say, briefly pointing to a proposed connection between two texts, the next in sequence providing a summary of complex relations (the outline, in fact, of what might in the print environment have been a scholarly article or even book), the third an overview of relevant comparisons, and the last the actual full text of the other author. At each stage (or lexia), the reader should have the power not only to return to the so-called main text of "Hudson's Statue" but also to reach these linked materials out of sequence. Vannevar Bush, who invented the general notion of hypertext, thought that chains or trails of links might themselves constitute a new form of scholarly writing, and annotations in the form of such guided tours might conceivably become part of the future scholarly edition. We can be certain, however, that as constraints of scale lessen, increasing amounts of material will be summoned to illuminate individual texts and new forms of multiple annotation will develop as a way of turning availability into accessibility.

Hypertext and the Problem of Text Structure

The fact that a single lexia can function very differently within large networked hypertexts raises fundamental questions about the applicability of Standard Generalized Markup Language (SGML) and its heir, Extended Markup Language (XML), to electronic scholarly editions, which increasingly appear in vast electronic information spaces rather than in the stand-alone versions we see today in CD-ROMs, such as Peter Robinson's Chaucer project and Anne McDermott's edition of Johnson's *Dictionary*. The relation of markup languages and hypertext appears particularly crucial to scholarly editing since so many large projects depend on SGML, XML, and their more specific scholarly forms specified by the Text Encoding Initiative (TEI).

One of the fundamental strengths of XML, of course, lies in its creation of a single electronic text that can lend itself to many forms of both print and electronic presentation. Looking at medieval *scripta continua* above, we encountered text without any markup, not even spaces between words. Later manuscript and print text contains presentational markup—that is, the encoding takes the form of specific formatting decisions; one indicates a paragraph by skipping, say, an extra line and indenting five or seven spaces. Although perfectly suited to physical texts, such forms of encoding appear particularly inefficient and even harmful in electronic environments, since they prevent easy transference and manipulation of texts. So-called procedural markup characterizes handwritten and printed text; to indicate a paragraph, authors and scribes, as we have seen, follow a certain procedure, such as that described above. Electronic text works better when one creates a generalized markup that simply indicates the presence of a text entity, such as a paragraph, that is then defined in another place.

Once all aspects of any particular text have been indicated with the correct SGML and XML tags, the text appears in a wonderfully generalized, potentially multiplicitous form. For example, after one has tagged (or "marked up") each instance of a text element, such as titles at the beginning of each chapter, by placing them between a particular set of tags—say, <chaptertitle> and </chaptertitle>—one can easily configure such text elements differently in different versions of a text. Thus, if printed on my university's mainframe printer, which permits only a single proportional font, chapter titles appear bolded in the larger of two available sizes. If printed with a typesetting device, however, the same chapter titles automatically appear in a very different font and size, say, 30-point Helvetica. If presented electronically, moreover, chapter titles can appear in a color different from that of the main text; in the DynaText translation of the version of *Hypertext,* for example, they appear in green whereas the main text appears in black. My first point here is that once one has created such a generalized text, one can adapt it to different publication modes with a single instruction that indicates the specific appearance of all labeled text elements. My second point here is that such tagged text records its own abstract structure.

In "What Is Text Really?," their pioneering essay on SGML, Stephen J. DeRose, David G. Durand, Elli Mylonas, and Allen H. Renear argue that text consists of hierarchically organized context objects, such as sentences, paragraphs, sections, and chapters. Do hypertext and markup languages, therefore, conceive of text in fundamentally opposed ways? At first glance, this seems to be the case, since hypertext produces nonhierarchical text structures whereas SGML and XML record hierarchical book structures. The question arises, To what extent do such visions of markup languages and hypertext conflict? After all, SGML and XML fundamentally assert book structure. But do they assert a single essential structure, however reconfigurable? Hypertext subverts hierarchy in text and in so doing might seem to subvert markup languages and call into question their basic usefulness. In electronic space, as we have already observed, an individual lexia may inhabit, or contribute to, several text structures simultaneously. At first consideration this fact might appear to suggest that markup languages fundamentally oppose hypertext, but such is hardly the case.

Once again, Ted Nelson provides assistance, for it is he who pointed out that the problem with classification systems lies in the fact not that they are bad but that different people—and the same person at different times—require different ones. One of the great strengths of hypertext, after all, lies in its ability to provide access to materials regardless of how they are classified and (hence) how and where they are stored. From the Nelsonian point of view, hypertext does not so much violate classifications as supplement them, making up for their inevitable shortcomings. From the point of view of one considering either the relation of hypertext to markup languages or the hypertextualization of them, the problem becomes one of finding some way to encode or signal multiple structures or multiple classifications of structure. If a scholarly annotation and main text can exchange roles, status, and nature, then one needs a device that permits a SGML- or XML-marked lexia to present a different appearance, if so required, on being entered or opened from different locations.

Returning to our examples from "Hudson's Statue," we realize that readers starting from Carlyle's text will experience linked materials on Chartism and the People's League as annotations to it, but readers starting with primary or secondary materials concerning these political movements will experience "Hudson's Statue" as an annotation to them. When discussing writing for electronic space in chapter 5, I suggest ways in which both software designers and individual authors have to assist readers. For the moment I shall point out only that one such means of orienting and hence empowering readers takes the form of clearly indicating the permeable borders of the provisional text to which any lexia belongs. Using such orientation rhetoric might require that materials by Carlyle have a different appearance from those of conceivably related materials, such as lexias about the English Revolution of the 1640s and Victorian political movements. In such a case, one needs a way of configuring the text according to the means from which it is accessed. This textual polymorphism in turn suggests that in such environments text is alive, changing, kinetic, open-ended in a new way.

Argumentation, Organization, and Rhetoric

Electronic linking, which gives the reader a far more active role than is possible with books, has certain major effects. Considered from the vantage point of a literature intertwined with book technology, these effects appear harmful and dangerous, as indeed they must be to a cultural hegemony based, as ours is, on a different technology of cultural memory. In particular, the numerating linear rhetoric of "first, second, third" so well suited to print will continue to appear within individual blocks of text but cannot be used to structure arguments in a medium that encourages readers to choose different paths rather than follow a linear one. The shift away from linearization might seem a major change, and it is, but we should remind ourselves that it is not an abandonment of the natural.

"The structuring of books," Tom McArthur reminds us, "is anything but 'natural'—indeed, it is thoroughly unnatural and took all of 4,000 years to bring about. The achievement of the Scholastics, pre-eminently among the world's scribal elites, was to conventionalize the themes, plot and shapes of books in a truly rigorous way, as they also structured syllabuses, scripture and debate" (69). Their conventions of book structure, however, changed fundamentally with the advent of the printing press, which encouraged alphabetic ordering, a procedure that had never before caught on. Why?

One reason must certainly be that people had already become accustomed over too many centuries to thematically ordered material. Such material bore a close resemblance to the "normal" organization of written work: . . . Alphabetization may also have been offensive to the global Scholastic view of things. It must have seemed a perverse, disjointed and ultimately meaningless way of ordering material to men who were interested in neat frames for containing all knowledge. Certainly, alphabetization poses problems of fragmentation that may be less immediately obvious with word lists but can become serious when dealing with subject lists. (76–77)

McArthur's salutary remarks, which remind us how we always naturalize the social constructions of our world, also suggest that from one point of view, the Scholastics', the movement from manuscript to print and then to hypertext appears one of increasing fragmentation. As long as a thematic or other culturally coherent means of ordering is available to the reader, the fragmentation of the hypertext document does not imply the kind of entropy that such fragmentation would have in the world of print. Capacities such as fulltext searching, automatic linking, agents, and conceptual filtering potentially have the power to retain the benefits of hypertextuality while insulating the reader from the ill effects of abandoning linearity.

Beginnings in the Open Text

The concepts (and experiences) of beginning and ending imply linearity. What happens to them in a form of textuality not governed chiefly by linearity? If we assume that hypertextuality possesses multiple sequences rather than that it has an entire absence of linearity and sequence, then one answer to this query must be that it provides multiple beginnings and endings rather than single ones. Theorists whose model of hypertext is the Word Wide Web might disagree with this claim. Marie-Laure Ryan, for example, asserts that "every hypertext has a fixed entry point—there must be an address to reach first before the system of links can be activated" (226). Although many web and other hyperfictions seem to support that statement, examples in various hypertext environments show that such is not the case. Joyce's *afternoon,* which was published in the Page Reader format, does have a fixed starting point, but other works in Storyspace do not. Like most webfictions, *Patchwork Girl* has an opening screen, a drawing of Frankenstein's female monster that is equivalent to the frontispiece in a book, and it is followed by a second screen, which simultaneously serves as a title page and sitemap, presenting the reader with five points at which to begin reading: "a graveyard," "a journal," "a quilt," "a story," and "broken accents." Much, of course, depends on what Ryan intends by "system of links"; if one means the narrative, then her claim is not always true. If she means "that point of the hypertext that one sees first," then the claim is true but only in a trivial sense, because in those Storyspace hyperfictions that do not have an opening screen but present the reader with the software's graphic representation of folders and documents, the reader can choose any starting point in these spatial hypertexts. *Adam's Bookstore,* for example, which arranges its lexias in a circular pattern, invites readers to begin at any point. (Observation suggests most readers begin at the top center or top right.)

Similarly, even if we concentrate on webfiction—and Ryan's use of "address" suggests that she is thinking only about the Web, since a URL is an address—we encounter two ways in which readers do not enter narratives at a fixed point. First, search engines can guide readers to any lexia within a hyperfiction; authors of course do not intend such an apparently random starting point, but once they place their work on the Internet, they allow it to happen. (That is the reason I tell my students writing both hyperfiction and hypertext essays to be prepared for readers who "fall in through the living-room ceiling rather than entering through the front door," and therefore at least consider including navigation and orientation devices that will give readers some idea of where they have landed—and perhaps encourage them to keep reading.) Second, webfictions can also open, like *Patchwork Girl,* with a first screen that provides the reader with multiple beginnings. The opening screen, then, is no more the beginning of the narrative than are the title pages of *Jane Eyre* or *Waterland.*

Drawing on Edward W. Said's work on origins and openings, one can suggest that, in contrast to print, hypertext offers at least two different kinds of beginnings. The first concerns the individual lexia, the second a gathering of them into a metatext. Whenever one has a body of hypertext materials that stands alone—either because it occupies an entire system or because it exists, however transiently, within a frame, the reader has to begin reading at some point, and for the reader that point is a beginning. Writing of print, Said explains that "a work's beginning is, practically speaking, the main entrance to what it offers" (3). But what happens when a work offers many "main" entrances—in fact, offers as many entrances as there are linked passages by means of which one can arrive at the individual lexia (which, from one perspective, becomes equivalent to a work)? Said provides materials for an answer when he argues that a "'beginning' is designated in order to indicate, clarify, or define a *later* time, place, or action. In short, the designation of a beginning generally involves also the designation of a consequent *intention*" (5). In Said's terms, therefore, even atomized text can make a beginning when the link site, or point of departure, assumes the role of the beginning of a chain or path. According to Said, "we see that the beginning is the first point (in time, space, or action) of an accomplishment or process that has duration and meaning. *The beginning, then, is the first step in the intentional production of meaning*" (5).

Said's quasi-hypertextual definition of a beginning here suggests that "in retrospect we can regard a beginning as the point at which, in a given work, the writer departs from all other works; a beginning immediately establishes relationships with works already existing, relationships of either continuity or antagonism or some mixture of both" (3).

Endings in the Open Text

If hypertext makes determining the beginning of a text difficult because it both changes our conception of text and permits readers to "begin" at many different points, it similarly changes the sense of an ending. Readers in read-write systems cannot only choose different points of ending, they can also continue to add to the text, to extend it, to make it more than it was when they began to read. As Nelson, one of the originators of hypertext, points out: "There is no Final Word. There can be no final version, no last thought. There is always a new view, a new idea, a reinterpretation. And literature, which we propose to electronify, is a system for preserving continuity in the face of this fact . . . Remember the analogy between text and water. Water flows freely, ice does not. The free-flowing, live documents on the network are subject to constant new use and linkage, and those new links continually become interactively available. Any detached copy someone keeps is frozen and dead, lacking access to the new linkage" (*Computer Lib*, 2/61, 48). Here, as in several other ways, Bakhtin's conception of textuality anticipates hypertext. Caryl Emerson, his translator and editor, explains that "for Bakhtin 'the whole' is not a finished entity; it is always a relationship . . . Thus, the whole can never be finalized and set aside; when a whole is realized, it is by definition already open to change" (*Problems*, xxxix).

Hypertext blurs the end boundaries of the metatext, and conventional notions of completion and a finished product do not apply to hypertext, whose essential novelty makes difficult defining and describing it in older terms, since they derive from another educational and information technology and have hidden assumptions inappropriate to hypertext. Particularly inapplicable are the related notions of completion and a finished product. As Derrida recognizes, a form of textuality that goes beyond print "forces us to extend . . . the dominant notion of a 'text,'" so that it "is henceforth no longer a finished corpus of writing, some content enclosed in a book or its margins but a differential network, a fabric of traces referring endlessly to something other than itself, to other differential traces" ("Living On," 83–84).

Hypertextual materials, which by definition are open-ended, expandable, and incomplete, call such notions into question. If one put a work conventionally considered complete, such as *Ulysses,* into a hypertext format, it would immediately become "incomplete." Electronic linking, which emphasizes making connections, immediately expands a text by providing large numbers of points to which other texts can attach themselves. The fixity and physical isolation of book technology, which permits standardization and relatively easy reproduction, necessarily closes off such possibilities. Hypertext opens them up.

Boundaries of the Open Text

Hypertext redefines not only beginnings and endings of the text but also its borders—its sides, as it were. Hypertext thus provides us with a means to escape what Gérard Genette terms a "sort of idolatry, which is no less serious, and today more dangerous" than idealization of the author, "namely, the fetishism of the work—conceived of as a closed, complete, absolute object" (*Figures,* 147). When one moves from physical to virtual text, and from print to hypertext, boundaries blur— a blurring that Derrida works so hard to achieve in his print publications— and one therefore no longer can rely on conceptions or assumptions of inside and out. As Derrida explains, "To keep the outside out . . . is the inaugural gesture of 'logic' itself, of good 'sense' insofar as it accords with the self-identity of *that which is:* being is what it is, the outside is outside and the inside inside. Writing must thus return to being what it *should never have ceased to be:* an accessory, an accident, an excess" (*Dissemination,* 128). Without linearity and sharp bounds between inside and out, between absence and presence, and between self and other, philosophy will change. Working within the world of print, Derrida presciently argues, using Platonic texts as an example, that "the textual chain we must set back in place is thus no longer simply 'internal' to Plato's lexicon. But in going beyond the bounds of that lexicon, we are less interested in breaking through certain limits, with or without cause, than in putting in doubt the right to posit such limits in the first place. In a word, we do not believe that there exists, in all rigor, a Platonic text, closed upon itself, complete with its inside and outside" (130).

Derrida furthermore explains, with a fine combination of patience and wit, that in noticing that texts really do not have insides and outsides, one does not reduce them to so much mush: "Not that one must then consider that it [the text] is leaking on all sides and can be drowned confusedly in the undifferentiated generality of the element. Rather, provided the articulations are rigorously and prudently recognized, one should simply be able to untangle the hidden forces of attraction linking a present word with an absent word in the text of Plato" (130).

Another sign of Derrida's awareness of the limitations and confinements of contemporary attitudes, which arise in association with the technology of the printed book, is his protohypertextual approach to textuality and meaning, an approach that remains skeptical of "a fundamental or totalizing principle," since it recognizes that "the classical system's 'outside' can no longer take the form of the sort of extra-text which would arrest the concatenation of writing" (5).

Hypertext both on the Internet and in its read-write forms thus creates an open, open-bordered text, a text that cannot shut out other texts and therefore embodies the Derridean text that blurs "all those boundaries that form the running border of what used to be called a text, of what we once thought this word could identify, i.e., the supposed end and beginning of a work, the unity of the corpus, the title, the margins, the signatures, the referential realm outside the frame, and so forth." Hypertext therefore undergoes what Derrida describes as "a sort of overrun [*débordement*] that spoils all these boundaries and divisions" ("Living On," 83). Anyone who believes Derrida is here being overly dramatic should consider the power of the open hypermedia systems discussed in chapter 1 to add links to someone else's Web document.

In hypertext systems, links within and without a text—intratextual and intertextual connections between points of text (including images)—become equivalent, thus bringing texts closer together and blurring the boundaries among them (Figure 12). Consider the case of intertextual links in Milton. Milton's various descriptions of himself as prophet or inspired poet in *Paradise Lost* and his citations of Genesis 3:15 provide obvious examples. Extratextual and intratextual links, in contrast, are exemplified by links between a particular passage in which Milton mentions prophecy and his other writings in prose or poetry that make similar or obviously relevant points, as well as biblical texts, commentaries throughout the ages, comparable or contrasting

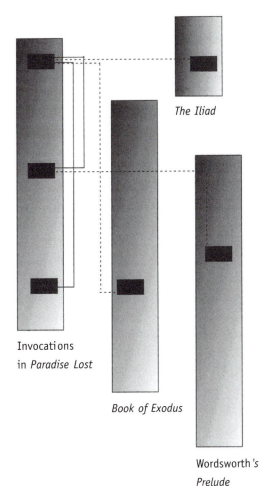

The Iliad

Invocations
in *Paradise Lost*

Book of Exodus

Wordsworth's
Prelude

The Borderless Text

Linking changes the experience of text and authorship by rendering the borders of all text permeable:

By reifying allusions, echoes, references, and so on, linking

(1) makes them material,

(2) draws individual texts experientially closer together.

Consider, for example, a hypertext presentation (or "edition") of Milton's *Paradise Lost*.

Figure 12. The Borderless Electronic Text

poetic statements by others, and scholarly comment. Similarly, Miltonic citations of the biblical text about the heel of man crushing the serpent's head and being in turn bruised by the serpent obviously link to the biblical passage and its traditional interpretations as well as to other literary allusions and scholarly comment on all these subjects. Hypertext linking simply allows one to speed up the usual process of making connections while providing a means of graphing such transactions, if one can apply the word *simply* to such a radically transformative procedure.

The speed with which one can move between passages and points in sets of texts changes both the way we read and the way we write, just as the high-speed number-crunching computing changed various scientific fields by making possible investigations that before had required too much time or risk. One change comes from the fact that linking permits the reader to move with equal facility between points within a text and those outside it. Once one can move with equal facility between, say, the opening section of *Paradise Lost* and a passage in Book 12 thousands of lines "away," and between that opening section and a particular anterior French text or modern scholarly comment, then, in an important sense, the discreteness of texts, which print culture creates, has radically changed and possibly disappeared. One may argue that, in fact, all the hypertext linking of such texts does is embody the way one actually experiences texts in the act of reading; but if so, the act of reading has in some way gotten much closer to the electronic embodiment of text and in so doing has begun to change its nature.

These observations about hypertext suggest that computers bring us much closer to a culture some of whose qualities have more in common with that of preliterate man than even Walter J. Ong has been willing to admit. In *Orality and Literacy* he argues that computers have brought us into what he terms an age of "secondary orality" that has "striking resemblances" to the primary, preliterate orality "in its participatory mystique, its fostering of a communal sense, its concentration on the present moment, and even its use of formulas" (136). Nonetheless, although Ong finds interesting parallels between a computer culture and a purely oral one, he mistakenly insists: "The sequential processing and spatializing of the word, initiated by writing and raised to a new order of intensity by print, is further intensified by the computer, which maximizes commitment of the word to space and to (electronic) local motion and optimizes analytic sequentiality by making it virtually instantaneous" (136). In fact, hypertext systems, which insert every text into a web of relations, produce a very different effect, for they allow non- or multisequential reading and thinking.

One major effect of such nonsequential reading, the weakening of the boundaries of the text, can be thought of either as correcting the artificial isolation of a text from its contexts or as violating one of the chief qualities of the book. According to Ong, writing and printing produce the effect of discrete, self-contained utterance:

By isolating thought on a written surface, detached from any interlocutor, making utterance in this sense autonomous and indifferent to attack, writing presents utterance and thought as uninvolved with all else, somehow self-contained, complete. Print in the same way situates utterance and thought on a surface disengaged from everything else, but it also goes farther in suggesting self-containment. (132)

We have already observed the way in which hypertext suggests integration rather than self-containment. Another possible result of such hypertext may also be disconcerting. As Ong points out, books, unlike their authors, cannot really be challenged:

The author might be challenged if only he or she could be reached, but the author cannot be reached in any book. There is no way to refute a text. After absolutely total and devastating refutation, it says exactly the same thing as before. This is one reason why "the book says" is popularly tantamount to "it is true." It is also one reason why books have been burnt. A text stating what the whole world knows is false will state falsehood forever, so long as the text exists. (79)

The question arises, however, If hypertext situates texts in a field of other texts, can any individual work that has been addressed by another still speak so forcefully? One can imagine hypertext presentations of books (or the equivalent) in which the reader can call up all the reviews and comments on that book, which would then inevitably exist as part of a complex dialogue rather than as the embodiment of a voice or thought that speaks unceasingly. Hypertext, which links one block of text to myriad others, destroys that physical isolation of the text, just as it also destroys the attitudes created by that isolation. Because hypertext systems permit a reader both to annotate an individual text and to link it to other, perhaps contradictory, texts, it destroys one of the most basic characteristics of the printed text—its separation and univocality. Whenever one places a text within a network of other texts, one forces it to exist as part of a complex dialogue. Hypertext linking, which tends to change the roles of author and reader, also changes the limits of the individual text.

Electronic linking radically changes the experience of a text by changing its spatial and temporal relation to other texts. Reading a hypertext version of Dickens's *Great Expectations* or Eliot's *Wasteland,* for example, one follows links to predecessor texts, variant readings, criticism, and so on. Following an electronic link to an image of, say, the desert or a wasteland in a poem by Tennyson, Browning, or Swinburne takes no more time than following one from a passage earlier in the poem to one near its end. Therefore, readers experience these other, earlier texts outside *The Wasteland* and the passage inside the work as existing equally distant from the first passage. Hypertext thereby blurs the distinction between what is inside and what is outside a text. It also makes all the texts connected to a block of text collaborate with that text.

The Status of the Text, Status in the Text

Alvin Kernan claims that "Benjamin's general theory of the demystification of art through numerous reproductions explains precisely what happened when in the eighteenth century the printing press, with its logic of multiplicity, stripped the classical texts of the old literary order of their aura" (152), and it seems likely that hypertext will extend this process of demystification even further. Kernan convincingly argues that by Pope's time a "flood of books, in its accumulation both of different texts and identical copies of the same texts, threatened to obscure the few idealized classics, both ancient and modern, of polite letters, and to weaken their aura by making printed copies of them" (153). Any information medium that encourages rapid dissemination of texts and easy access to them will increasingly demystify individual texts. But hypertext has a second potentially demystifying effect: by making the borders of the text (now conceived as the individual lexia) permeable, it removes some of its independence and uniqueness.

Kernan further adds that "since printed books were for the most part in the vernacular, they further desacralized letters by expanding its canon from a group of venerable texts written in ancient languages known only to an elite to include a body of contemporary writing in the natural language understood by all who read" (153–54). Will electronic Web versions of the Bible accompanied by commentaries, concordances, and dictionaries, like *Nave's Topical Bible* (http://bible.christiansunite.com/Naves_Topical_Bible/), which seem to be essentially democratizing, similarly desacralize the scriptures? They have the potential to do so in two ways. First, by making some of the scholar's procedures easily available to almost any reader, this electronic Bible might demystify a text that possesses a talismanic power for many in its intended audience.

Second, and more fundamental, the very fact that this hypertext Bible enforces the presence of multiple versions potentially undercuts belief in the possibility of a unique, unitary text. Certainly, the precedent of Victorian loss of belief in the doctrine of verbal inspiration of the scriptures suggests that hypertext could have a potentially parallel effect (Landow, *Victorian Types*, 54–56). In Victorian England the widescale abandonment of belief that every word of the Bible was divinely inspired, even in its English translation, followed from a variety of causes, including influence of German higher criticism, independent British applications of rational approaches by those like

Bishop Colenso, and the discoveries of geology, philology, and (later) biology. The discovery, for instance, that Hebrew did not possess the uniqueness as a language that some believers, particularly Evangelicals, long assumed it did, eroded faith, in large part because believers became aware of unexpected multiplicity where they had assumed only unity. The discovery of multiple manuscripts of scripture had parallel effects. Hypertext, which emphasizes multiplicity, may cause similar crises in belief.

Although the fundamental drive of the printed page is a linear, straight-ahead thrust that captures readers and forces them to read along if they are to read at all, specialized forms of text have developed that use secondary codes to present information difficult or impossible to include in linear text. The footnote or endnote, which is one of the prime ways that books create an additional space, requires some code, such as a superscript number or one within parentheses, that signals readers to stop reading what is conventionally termed the *main* text or the body of the text and begin reading some peripheral or appended patch of text that hangs off that part of the main text.

In both scholarly editing and scholarly prose such divisions of text partake of fixed hierarchies of status and power. The smaller size type that presents footnote and endnote text, like the placement of that text away from the normal center of the reader's attention, makes clear that such language is subsidiary, dependent, less important. In scholarly editing, such typographic and other encoding makes clear that the editor's efforts, no matter how lavish or long suffering, are obviously less important than the words being edited, for these appear in the main text. That's why Barthes's *S/Z*, whose organization makes the reader encounter its many notes before coming to the text on which they comment, is both such a reconfiguration of conventional scholarly editions and an effective parody of them. In scholarly and critical discourse that employs annotation, these conventions also establish the importance of the dominant argument in opposition to the author's sources, scholarly allies, and opponents, and even the work of fiction or poetry on which the critical text focuses.

One experiences hypertext annotation of a text very differently. In the first place, electronic linking immediately destroys the simple binary opposition of text and note that founds the status relations that inhabit the printed book. Following a link can bring the reader to a later portion of the text or to a text to which the first one alludes. It may also lead to other works by the same author, or to a range of critical commentary, textual variants, and the like. The assignment of text and annotation to what Tom Wolfe calls different "statu-spheres" therefore becomes very difficult, and in a fully networked environment such text hierarchies tend to collapse.

Hypertext linking situates the present text at the center of the textual universe, thus creating a new kind of hierarchy, in which the power of the center dominates that of the infinite periphery. But because in hypertext that center is always a transient, decenterable virtual center—one created, in other words, only by one's act of reading that particular text—it never tyrannizes other aspects of the network in the way a printed text does.

Barthes, well aware of the political constraints of a text that makes a reader read in a particular way, himself manipulates the political relations of text in interesting ways. The entire procedure or construction of *S/Z,* for example, serves as a commentary on the political relationships among portions of the standard scholarly text, the problem of hierarchy. Barthes playfully creates his own version of complex footnote systems. Like Derrida in *Glas,* he creates a work or metatext that the reader accustomed to reading books finds either abrasively different or, on rare occasion, a wittily powerful commentary on the way books work—that is, on the way they force readers to see relationships between sections and thereby endow certain assemblages of words with power and value because they appear in certain formats rather than others.

Barthes, in other words, comments on the footnote, and all of *S/Z* turns out to be a criticism of the power relations between portions of text. In a footnote or endnote, we recall, that portion of the text conventionally known as the main text has a value for both reader and writer that surpasses any of its supplementary portions, which include notes, prefaces, dedications, and so on, most of which take the form of apparatuses designed to aid information retrieval. These devices, almost all of which derive directly from print technology, can function only when one has fixed, repeatable, physically isolated texts. They have great advantages and permit certain kinds of reading: one need not, for example, memorize the location of a particular passage if one has system features such as chapter titles, tables of contents, and indices. So the reference device has enormous value as a means of reader orientation, navigation, and information retrieval.

It comes at certain costs, costs that, like most paid by the reader of text, have become so much a part of our experiences of reading that we do not notice them at all. Barthes makes us notice them. Barthes, like most late-twentieth-century critical theorists, is at his best seeing the invisible, breathing on it in hopes that the condensate will illuminate the shadows of what others have long missed and taken to be not there. What, then, does the footnote imply, and how does Barthes manipulate or avoid it? Combined with the physical isolation of each text, the division between main text and footnote establishes the primary importance of main text in its relation to other texts even when thinking about the subject instantly reveals that such a relationship cannot in fact exist.

Take our scholarly article, the kind of articles we academics all write. One wishes to write an article on some aspect of the Nausicaa section of Joyce's *Ulysses,* a text that by even the crudest quantitative measures appears to be more important, more powerful than our note identifying, say, one of the sources of Gerty McDowell's phrasing from a contemporary women's magazine. Joyce's novel, for example, exists in more copies than our article can or will and it therefore has an enormously larger readership and reputation—all problematic notions, I admit, all relying on certain ideologies; and yet most of us, I expect, will accede to them for they are the values by which we work. Ostensibly, that is. Even deconstructionists privilege the text, the great work.

Once, however, one begins to write one's article, the conventions of print quickly call those assumptions into question, since anything in the main text is clearly more important than anything outside it. The physically isolated discrete text is very discreet indeed, for as Ong makes clear, it hides obvious connections of indebtedness and qualification. When one introduces other authors into the text, they appear as attenuated, often highly distorted shadows of themselves. Part of this is necessary, since one cannot, after all, reproduce an entire article or book by another author in one's own. Part of this attenuation comes from authorial inaccuracy, slovenliness, or outright dishonesty. Nonetheless, such attenuation is part of the message of print, an implication one cannot avoid, or at least one cannot avoid since the advent of hypertext, which by providing an alternative textual mode reveals differences that turn out to be, no longer, inevitabilities and invisibilities.

In print when I provide the page number of an indicated or cited passage from Joyce, or even include that passage in text or note, that passage—that occasion for my article—clearly exists in a subsidiary, comparatively minor position in relation to my words, which appear, after all, in the so-called main text. What would happen, though, if one wrote one's article in hypertext? Assuming one worked in a fully implemented hypertextual environment, one would begin by calling up Joyce's novel and, on one side of the video screen, opening the passage or passages involved. Next, one would write one's comment, but where one would usually cite Joyce, one now does so in a very different way. Now one creates an electronic link between one's own text and one or more sections of the Joycean text. At the same time one also links one's text to other aspects of one's own text, texts by others, and earlier texts by oneself. Several things have happened, things that violate our expectations. First, attaching my commentary to a passage from Joyce makes it exist in a far different, far less powerful, relation to Joyce, the so-called original text, than it would in the world of physically isolated texts. Second, as soon as one attaches more than one text block or lexia to a single anchor (or block, or link marker), one destroys all possibility of the bipartite hierarchy of footnote and main text. In hypertext, the main text is that which one is presently reading. So one has a double revaluation: with the dissolution of this hierarchy, any attached text gains an importance it might not have had before.

In Bakhtin's terms, the scholarly article, which quotes or cites statements by others—"some for refutation and others for confirmation and supplementation—is one instance of a dialogic interrelationship among directly signifying discourses within the limits of a single context . . . This is not a clash of two ultimate semantic authorities, but rather an objectified (plotted) clash of two represented positions, subordinated wholly to the higher, ultimate authority of the author. The monologic context, under these circumstances, is neither broken nor weakened" (*Problems,* 188). Trying to evade the constraints, the logic, of print scholarship, Bakhtin himself takes an approach to quoting other authors more characteristic of hypertext or postbook technology than that of the book. According to his editor and translator, Emerson, when Bakhtin quotes other critics, "he does so at length, and lets each voice sound fully. He understands that the frame is always in the power of the framer, and that there is an outrageous privilege in the power to cite others. Thus Bakhtin's footnotes rarely serve to narrow down debate by discrediting totally, or (on the other hand) by conferring exclusive authority. They might identify, expand, illustrate, but they do not pull rank on the body of the text—and thus more in the nature of a marginal gloss than an authoritative footnote" (xxxvii).

Derrida also comments on the status relations that cut and divide texts, but unlike Barthes, he concerns himself with oppositions between preface and main text and main text and other texts. Recognizing that varying levels of status accrue to different portions of a text, Derrida examines the way each takes on associations with power or importance. In discussing Hegel's introduction to the *Logic,* Derrida points out, for example, that the preface must be distinguished from the introduction. They do not have the same function, or even the same dignity, in Hegel's eyes (*Dissemination,* 17). Derrida's new textuality, or true textuality (which I have continually likened to hypertextuality), represents "an entirely other typology where the outlines of the preface and the 'main' text are blurred" (39).

Hypertext and Decentrality: The Philosophical Grounding

One tends to think of text from within the position of the lexia under consideration. Accustomed to reading pages of print on paper, one tends to conceive of text from the vantage point of the reader experiencing that page or passage, and that portion of text assumes a centrality. Hypertext, however, makes such assumptions of centrality fundamentally problematic. In contrast, the linked text, the annotation, exists as the *other* text, and it leads to a conception (and experience) of text as Other.

In hypertext this annotation, or commentary, or appended text can be any linked text, and therefore the position of any lexia in hypertext resembles that of the Victorian sage. For like the sage, say, Carlyle, Thoreau, or Ruskin, the lexia stands outside, off center, and challenges. In other words, hypertext, like the sage, thrives on marginality. From that essential marginality to which he stakes his claim by his skillful, aggressive use of pronouns to oppose his interests and views to those of the reader, he defines his discursive position or vantage point.

Hypertext similarly emphasizes that the marginal has as much to offer as does the central, in part because hypertext refuses to grant centrality to anything, to any lexia, for more than the time a gaze rests on it. In hypertext, centrality, like beauty and relevance, resides in the mind of the beholder. Like Andy Warhol's modern person's fifteen minutes of fame, centrality in hypertext only exists as a matter of evanescence. As one might expect from an information medium that changes our relations to data, thoughts, and selves so dramatically, that evanescence of this (ever-migrating) centrality is merely a given—that's the way things are—rather than an occasion for complaint or satire. It is simply the condition under which—or within which—we think, communicate, or record these thoughts and communications in the hypertextual docuverse.

This hypertextual dissolution of centrality, which makes the medium such a potentially democratic one, also makes it a model of society of conversations in which no one conversation, no one discipline or ideology, dominates or founds the others. It is thus the instantiation of what Richard Rorty terms *edifying philosophy,* the point of which "is to keep the conversation going rather than to find objective truth." It is a form of philosophy

having sense only as a protest against attempts to close off conversation by proposals for universal commensuration through the hypostatization of some privileged set of descriptions. The danger which edifying discourse tries to avert is that some given vocabulary, some way in which people might come to think of themselves, will deceive them into thinking that from now on all discourse could be, or should be, normal discourse. The resulting freezing-over of culture would be, in the eyes of edifying philosophers, the dehumanization of human beings. (377)

Hypertext, which has a built-in bias against "hypostatization" and probably against privileged descriptions as well, therefore embodies the approach to philosophy that Rorty urges. The basic experience of text, information, and control, which moves the boundary of power away from the author in the direction of the reader, models such a postmodern, antihierarchical medium of information, text, philosophy, and society.

4

Reconfiguring the Author

Erosion of the Self

Like contemporary critical theory, hypertext reconfigures—rewrites—the author in several obvious ways. First of all, the figure of the hypertext author approaches, even if it does not merge with, that of the reader; the functions of reader and writer become more deeply entwined with each other than ever before.[1] This transformation and near merging of roles is but the latest stage in the convergence of what had once been two very different activities. Although today we assume that anyone who reads can also write, historians of reading point out that for millennia many people capable of reading could not even sign their own names. Today when we consider reading and writing, we probably think of them as serial processes or as procedures carried out intermittently by the same person: first one reads, then one writes, and then one reads some more. Hypertext, which creates an active, even intrusive reader, carries this convergence of activities one step closer to completion; but in so doing, it infringes on the power of the writer, removing some of it and granting it to the reader. These shifts in the relations of author and reader do not, however, imply that hypertext automatically makes readers into authors or co-authors—except, that is, in hypertext environments that give readers the ability to add links and texts to what they read.[2]

One clear sign of such transference of authorial power appears in the reader's abilities to choose his or her way through the metatext, to annotate text written by others, and to create links between documents written by others. Read-write hypertext like Intermedia or Weblogs that accept comments do not permit the active reader to change the text produced by another person, but it does narrow the phenomenological distance that separates individual documents from one another in the worlds of print and manu-script. In reducing the autonomy of the text, hypertext reduces the autonomy of the author. In the words of Michael Heim, "as the authoritativeness of text diminishes, so too does the recognition of the private self of the creative author" (*Electric Language,* 221). Granted, much of that so-called autonomy had been illusory and existed as little more than the difficulty that readers had in perceiving connections between documents. Nonetheless, hypertext—which I am here taking as the convergence of poststructuralist conceptions of textuality and electronic embodiments of it—does do away with certain aspects of the authoritativeness and autonomy of the text, and in so doing it does reconceive the figure and function of authorship. One powerful instance of the way hypermedia environments diminish the author's control over his or her own text appears in the way so-called open systems permit readers to insert links into a lexia written by someone else. Portal Maximizer, for example, permits overlaying one author's Web documents with another author's links, although the original document remains unchanged.[3]

William R. Paulson, who examines literature from the vantage point of information theory, arrives at much the same position when he argues that "to characterize texts as artificially and imperfectly autonomous is not to eliminate the role of the author but to deny the reader's or critic's submission to any instance of authority. This perspective leaves room neither for author-ial mastery of a communicative object nor for the authority of a textual coherence so complete that the reader's (infinite) task would be merely to receive its rich and multilayered meaning." Beginning from the position of information theory, Paulson finds that in "literary communication," as in all communication, "there is an irreducible element of noise," and therefore "the reader's task does not end with reception, for reception is inherently flawed. What literature solicits of the reader is not simply receptive but the active, independent, autonomous construction of meaning" (139). Finding no reason to exile the author from the text, Paulson nonetheless ends up by assigning to the reader a small portion of the power that, in earlier views, had been the prerogative of the writer.

Hypertext and contemporary theory reconceive the author in a second way. As we shall observe when we examine the notion of collaborative writ-ing, both agree in configuring the author of the text as a text. As Barthes explains in his famous exposition of the idea, "this 'I' which approaches the text is already itself a plurality of other texts, of codes which are infinite" (*S/Z,* 10). Barthes's point, which should seem both familiar and unexceptional to anyone who has encountered Joyce's weaving of Gerty McDowell out of the texts of her class and culture, appears much clearer and more obvious from the vantage point of intertextuality. In this case, as in others at which we have already looked, contemporary theory proposes and hypertext disposes; or, to be less theologically aphoristic, hypertext embodies many of the ideas and attitudes proposed by Barthes, Derrida, Foucault, and others.

One of the most important of these ideas involves treating the self of author and reader not simply as (print) text but as a hypertext. For all these authors the self takes the form of a decentered (or centerless) network of codes that, on another level, also serves as a node within another centerless network. Jean-François Lyotard, for example, rejects nineteenth-century Romantic paradigms of an islanded self in favor of a model of the self as a node in an information network: "A self does not amount to much," he assures us with fashionable nonchalance, "but no self is an island; each exists in a fabric of relations that is now more complex and mobile than ever before. Young or old, man or woman, rich or poor, a person is always located at 'nodal points' of specific communication circuits, however tiny these may be. Or better: one is always located at a post through which various kinds of messages pass" (*Postmodern Condition,* 15). Lyotard's analogy becomes even stronger if one realizes that by "post" he most likely means the modern European post office, which is a telecommunications center containing telephones and other networked devices.

Some theorists find the idea of participating in a network to be demeaning and depressing, particularly since contemporary conceptions of textuality deemphasize autonomy in favor of participation. Before succumbing to posthumanist depression, however, one should place Foucault's statements about "the author's disappearance" in the context of recent discussions of machine intelligence (Foucault, "What Is an Author?" 119). According to Heinz Pagels, machines capable of complex intellectual processing will "put an end to much discussion about the mind-body problem, because it will be very hard not to attribute a conscious mind to them without failing to do so for more human beings. Gradually the popular view will become that consciousness is simply 'what happens' when electronic components are put together the right way" (92). Pagels's thoughts on the eventual electronic solution to the mind-body problem recall Foucault's discussion of "the singular relationship that holds between an author and a text [as] the manner in which a text apparently points to this figure who is outside and precedes it" ("What Is an Author?" 115). This point of view makes apparent that literature generates precisely such appearance of a self, and that, moreover, we have long read a self "out" of texts as evidence that a unified self exists "behind" or "within" or "implicit in" it. The problem for anyone who yearns to retain older conceptions of authorship or the author function lies in the fact that radical changes in textuality produce radical changes in the author figure derived from that textuality. Lack of textual autonomy, like lack of textual centeredness, immediately reverberates through conceptions of authorship as well. Similarly, the unboundedness of the new textuality disperses the author as well. Foucault opens this side of the question when he raises what, in another context, might be a standard problem in a graduate course on the methodology of scholarship:

If we wish to publish the complete works of Nietzsche, for example, where do we draw the line? Certainly, everything must be published, but can we agree on what "everything" means? We will, of course, include everything that Nietzsche himself published, along with the drafts of his works, his plans for aphorisms, his marginalia; notations and corrections. But what if, in a notebook filled with aphorisms, we find a reference, a reminder of an appointment, an address, or a laundry bill, should this be included in his works? Why not? . . . If some have found it convenient to bypass the individuality of the writer or his status as an author to concentrate on a work, they have failed to appreciate the equally problematic nature of the word "work" and the unity it designates. (119)

Within the context of Foucault's discussion of "the author's disappearance" (119), the illimitable plenitude of Nietzsche's oeuvre demonstrates that there's more than one way to kill an author. One can destroy (what we mean by) the author, which includes the notion of sole authorship, by removing the autonomy of text. One can also achieve the same end by decentering text or by transforming text into a network. Finally, one can remove limits on textuality, permitting it to expand, until Nietzsche, the edifying philosopher, becomes equally the author of *The Gay Science* and laundry lists and other such trivia—as indeed he was. Such illimitable plenitude has truly "transformed" the author, or at least the older conception of him, into "a victim of his own writing" (117).

Fears about the death of the author, whether in complaint or celebration, derive from Claude Lévi-Strauss, whose mythological works demonstrated for a generation of critics that works of powerful imagination take form without an author. In *The Raw and the Cooked* (1964), for example, where he showed, "not how men think in myths, but how myths operate in men's minds without their being aware of the fact," he also suggests "it would perhaps be better to go still further and, disregarding the thinking subject completely, proceed as if the thinking process were taking place in the myths, in the reflection upon themselves and their interrelation" (12).[4] Lévi-Strauss's presentation of mythological thought as a complex system of transformations without a center turns it into a networked text—not surprising, since the network serves as one of the main paradigms of synchronous structure.[5] Edward Said claims that the "two principal forces that have eroded the authority of the human subject in contemporary reflection are, on the one hand, the host of problems that arise in defining the subject's authenticity and, on the other, the development of disciplines like linguistics and ethnology that dramatize the subject's anomalous and unprivileged, even untenable, position in thought" (293). One may add to this observation that these disciplines' network paradigms also contribute importantly to this sense of the attenuated, depleted, eroding, or even vanishing subject.

Some authors, such as Said and Heim, derive the erosion of the thinking subject directly from electronic information technology. Said, for example, claims it is quite possible to argue "that the proliferation of information (and what is still more remarkable, a proliferation of the hardware for disseminating and preserving this information) has hopelessly diminished the role apparently played by the individual" (51).[6] Michael Heim, who believes loss of authorial power to be implicit in all electronic text, complains: "Fragments, reused material, the trails and intricate pathways of 'hypertext,' as Ted Nelson terms it, all these advance the disintegration of the centering voice of contemplative thought. The arbitrariness and availability of database searching decreases the felt sense of an authorial control over what is written" (*Electric Language,* 220). A database search, in other words, permits the active reader to enter the author's text at any point and not at the point the author chose as the beginning. Of course, as long as we have had indices, scholarly readers have dipped into specialist publications before or (shame!) instead of reading them through from beginning to end. In fact, studies of the way specialists read periodicals in their areas of expertise confirm that the linear model of reading is often little more than a pious fiction for many expert readers (McKnight, Richardson, and Dillon, "Journal Articles").

Although Heim here mentions hypertext in relation to the erosion of authorial prerogative, the chief problem, he argues elsewhere, lies in the way "digital writing turns the private solitude of reflective reading and writing into a public network where the personal symbolic framework needed for original authorship is threatened by linkage with the total textuality of human expressions" (*Electric Language,* 215). Unlike most writers on hypertext, he finds participation in a network a matter for worry rather than celebration, but he describes the same world they do, though with a strange combination of prophecy and myopia. Heim, who sees this loss of authorial control in terms of a corollary loss of privacy, argues that "anyone writing on a fully equipped computer is, in a sense, directly linked with the totality of symbolic expressions—more so and essentially so than in any previous writing element" (215). Pointing out that word processing redefines the related notions of publishing, making public, and privacy, Heim argues that anyone who writes with a word processor cannot escape the electronic network: "Digital writing, because it consists of electronic signals, puts one willy-nilly on a network where everything is constantly published. Privacy becomes an increasingly fragile notion. Word processing manifests a world in which the public itself and its publicity have become omnivorous; to make public has therefore a different meaning than ever before" (215). Although in 1987 Heim much exaggerated the loss of privacy inherent in writing with word-processing software per se, he turns out, as Weblog diaries prove, to have been prescient. When he wrote, most people did not in fact do most of their writing on networks, but the Internet changes everything: e-mail and personal blogs blur the boundaries between public and private.[7] Although Heim may possibly overstate the case for universal loss of privacy—the results are not in yet—he has accurately presented both some implications of hypertext for writers and the reactions against them by the print author accustomed to the fiction of the autonomous text.

The third form of reconfiguration of self and author shared by theory and hypertext concerns the decentered self, an obvious corollary to the network paradigm. As Said points out, major contemporary theorists reject "the human subject as grounding center for human knowledge. Derrida, Foucault, and Deleuze . . . have spoken of contemporary knowledge (*savoir*) as decentered; Deleuze's formulation is that knowledge, insofar as it is intelligible, is apprehensible in terms of *nomadic centers,* provisional structures that are never permanent, always straying from one set of information to another" (376). These three contemporary thinkers advance a conceptualization of thought best understood, like their views of text, in an electronic, virtual, hypertextual environment.

Before mourning too readily for this vanished or much diminished self, we would do well to remind ourselves that, although Western thought long held such notions of the unitary self in a privileged position, texts from Homer to Freud have steadily argued the contrary position. Divine or demonic possession, inspiration, humors, moods, dreams, the unconscious—all these devices that serve to explain how human beings act better, worse, or just different from their usual behavior argue against the unitary conception of the self so central to moral, criminal, and copyright law. The editor of the Soncino edition of the Hebrew Bible reminds us that

> Balaam's personality is an old enigma, which has baffled the skill of commentators . . . He is represented in Scripture as at the same time heathen sorcerer, true Prophet, and the perverter who suggested a peculiarly abhorrent means of bringing about the ruin of Israel. Because of these fundamental contradictions in character, Bible Critics assume, that the Scriptural account of Balaam is a combination of two or three varying traditions belonging to different periods . . . Such a view betrays a slight knowledge of the fearful complexity of the mind and soul of man. It is only in the realm of Fable that men and women display, as it were in a single flash of light, some one aspect of human nature. It is otherwise in real life. (668)

Given such long observed multiplicities of the self, we are forced to realize that notions of the unitary author or self cannot authenticate the unity of a text.[8] The instance of Balaam also reminds us that we have access to him only in Scriptures and that it is the biblical text, after all, which figures the unwilling prophet as a fractured self.

How the Print Author Differs from the Hypertext Author

Authors who have experienced writing within a hypertext environment often encounter certain predictable frustrations when returning to write for the linear world of print. Such frustrations derive from repeated recognitions that effective argument requires closing off connections and abandoning lines of investigation that hypertextuality would have made available. Here are two examples of what I mean. Near the opening of this chapter, in the midst of discussing the importance of Lévi-Strauss to recent discussions of authorship, I made the following statement: "Lévi-Strauss's presentation of mythological thought as a complex system of transformations without a center turns it into a networked text—not surprising, since the network serves as one of the main paradigms of synchronous structure"; and to this text I appended a note, pointing out that in *The Scope of Anthropology* "Lévi-Strauss also employs this model for societies as a whole: 'Our society, a particular instance in a much vaster family of societies, depends, like all others, for its coherence and its very existence on a network—grown infinitely unstable and complicated among us—of ties between consanguineal families.'" At this point in the main text, I had originally planned to place Foucault's remark that "we can easily imagine a culture where discourse would circulate without any need for an author" ("What Is an Author?" 138), and to this remark I had considered adding the observation that, yes, we can easily "imagine" such a culture, but we do not have to do so, since Lévi-Strauss's mythographic works have provided abundant examples of it. Although the diachronic relationship between these two influential thinkers seemed worthy of notice, I could not add the passage from Foucault and my comment because it disturbed my planned line of argument, which next required Said's relation of ethnology and linguistics to the erosion of "the authority of the human subject" in contemporary thought. I did not want to veer off in yet another direction. I then considered putting this observation in note 7, but again, it also seemed out of place there.

Had I written this chapter within a hypertext environment, the need to maintain a linear thrust would not have required this kind of choice. It would have required choices, but not this kind, and I could have linked two or more passages to this point in the main text, thereby creating multiple contexts both for my argument and for the quoted passage that served as my point of departure. I am not urging, of course, that in its print form this chapter has lost something of major importance because I could not easily append multiple connections without confusing the reader. (Had my abandoned remark seemed important enough to my overall argument, I could have managed to include it in several obvious ways, such as adding another paragraph or rewriting the main text to provide a point from which to hang another note.) No, I make this point to remind us that, as Derrida emphasizes, the linear habits of thought associated with print technology often influence us to think in particular ways that require narrowness, decontextualization, and intellectual attenuation, if not downright impoverishment. Linear argument, in other words, forces one to cut off a quoted passage from other, apparently irrelevant contexts that in fact contribute to its meaning. The linearity of print also provides the passage with an illusory center whose force is intensified by such selection.

A second example points to another kind of exclusion associated with linear writing. During the course of composing the first three chapters of this book, several passages, such as Barthes's description of the writerly text and Derrida's exposition of borders, boundaries, and *débordement,* forced themselves into the line of argument and hence deserved inclusion seven or eight times. One can repeatedly refer to a particular passage, of course, by combining full quotation, selections, and skillful paraphrase, but in general the writer can concentrate on a quoted section of text in this manner only when it serves as the center, or one of the centers, of the argument. If I wished to write a chapter or an entire book about Derridean *débordement,* I could return repeatedly to it in different contexts, thereby revealing its richness of implication. But that is not the book I wished to write in 1991, or wish to write now, nor is that the argument I wish to pursue here, and so I suppress that text and argument, which henceforth exist only in potentia. After careful consideration, I decide which of the many places in the text would most benefit from introduction of the quotation and then at the appropriate moment, I trundle it forward. As a result, I necessarily close off all but a few of its obvious points of connection.

As an experienced writer accustomed to making such choices, I realize that selection is one of the principles of effective argument. But why does one have to write texts in this way? If I were writing a hypertext version of this text—and the versions would exist so differently that one has to place quotation marks around "version" and "text," and probably "I" as well—I would not have to choose to write a single text. I could, instead, produce one that contained a plurality of ways through it. For example, after preparing the reader for Derrida's discussion of *débordement*, I could then link my preparatory remarks either to the passage itself or to the entire text of "Living On," and I could provide temporary markings that would indicate the beginning and end of the passage I wished to emphasize. At the same time, my hypertext would link the same passage to other points in my argument. How would I go about creating such links?

To answer this question, let me return to my first and simpler example, which involved linking passages from Lévi-Strauss's *Scope of Anthropology* and Foucault's "What Is an Author?" to a remark about the anthropologist's use of the network model. Let's look at how one makes a link in three different hypermedia environments, Intermedia, Storyspace, and HTML (for the Web). Unlike creating links in HTML, linking in Microcosm, Storyspace, and Intermedia follows the now common cut-and-paste paradigm found in word processors, graphics editors, and spread sheets. Using the mouse or other pointing device, one places the cursor immediately before the first letter of the first word in the passage in question, the sign of which is that the text appears highlighted—that is, it appears within a black rectangle, and the black type against a white background now appears in reverse video, white lettering against a black background. With the text highlighted, one moves the mouse until the point of the arrow-shaped cursor covers any part of the word "Intermedia" that appears in a horizontal list of words at the top of the screen ("File," "Edit," "Intermedia," and so on). Holding down the mouse button, one draws the cursor down, thereby producing the Intermedia menu, which contains choices. Placing the pointer over "Start Link," one releases the mouse button, proceeds to the second text, and carries out the same operation until one opens the Intermedia menu, at which point one chooses "Complete Link."

The system then produces a panel containing places to type any desired labels for the linked passages; it automatically adds the title of the entire text, and the writer can describe the linked passage within that text. For example, if I created a link between the hypermedia equivalent of my text for the previous section of this chapter and a passage in *The Scope of Anthropology*, Intermedia would automatically add the title of that text, "The Erosion of the Author," to which I would add a phrase, say, "Lévi-Strauss & myth as network." At the other end of the link, the system would furnish "Claude Lévi-Strauss, *The Scope of Anthropology*," and I would add something like "Lévi-Strauss & society as network." When a reader activates the link marker in the main text, the new entry appears as an option: "Claude Lévi-Strauss, *The Scope of Anthropology* : (Lévi-Strauss & society as network)." Storyspace linking involves a roughly similar, if simpler, procedure: to link from a phrase to another document, one highlights the phrase, moves the cursor to a palette containing an arrow, clicks on it, and then clicks on the other document, at which point a panel appears in which one can place a description. To make a link in HTML (which only permits one link per anchor), one has to type something like <a href = "../../levistrauss/1.html" target= "blank"> how men think in myths</a>, or use a handy html editor like BBEdit or Dreamweaver, which would add the HTML tag (<a href =></a>) after I typed in the information between quotation marks (../../levistrauss/1.html).

In Storyspace, Intermedia, and similar programs, linking the second text, the passage from Foucault, follows the identical procedure with the single exception that one no longer has to provide a label for the lexia in the main text, since it already has one. In HTML one has to sacrifice this second link or find another appropriate phrase to which one could add a link.

If instead of linking these two brief passages of quotation, documentation, and commentary, I created a more complex document set, focused on Derridean *débordement*, one would follow the same procedure to create links. In addition, one would also create kinds of documents not found in printed text, some of which would be primarily visual or hieroglyphic. One, for example, might take the form of a concept map showing, among other things, uses of the term *débordement* in "Living On," other works by Derrida in which it appears, and its relation to a range of contexts and disciplines from cartography and histology to etymology and French military history. Current hypermedia systems, including popular World Wide Web browsers, permit linking to interactive video, music, and animation as well as dictionaries, text, time lines, and static graphics. In the future these links will take more dynamic forms, and following them will animate some procedure, say, a search through a French thesaurus, or a reader-determined tracking of *débordement* created after I had completed my document would automatically become available.

My brief description of how I would go about producing this text were I writing it in something like a complete hypertext environment might trouble some readers because it suggests that I have sacrificed a certain amount of authorial control, ceding some of it to the reader. The act of writing has also changed to some extent. Electronic hypertext and contemporary discussions of critical theory, particularly those of the poststructuralists, display many points of convergence, but one point on which they differ is tone. Whereas most writings on theory, with the notable exception of Derrida, are models of scholarly solemnity, records of disillusionment and brave sacrifice of humanistic positions, writers on hypertext are downright celebratory. Whereas terms like *death, vanish, loss,* and expressions of depletion and impoverishment color critical theory, the vocabulary of freedom, energy, and empowerment marks writings on hypertextuality. One reason for these different tones may lie in the different intellectual traditions, national and disciplinary, from which they spring. A more important reason, I propose, is that critical theorists, as I have tried to show, continually confront the limitation—indeed, what they somewhat prematurely take to be the exhaustion—of the culture of print. They write from an awareness of limitation and shortcoming, and from a moody nostalgia, often before the fact, at the losses their disillusionment has brought and will bring. Writers on hypertext, in contrast, glory in possibility, excited by the future of textuality, knowledge, and writing. Another way of putting this opposing tone and mood is that most writers on critical theory, however brilliantly they may theorize a much-desired new textuality, nonetheless write from within daily experience of the old and only of the old. Many writers on hypertext, on the other hand, have already had some experience of hypertext systems, and they therefore write from a different experiential vantage point. Most poststructuralists write from within the twilight of a wished-for coming day; most writers of hypertext write about many of the same things from within the dawn.

Virtual Presence

Many features of hypermedia derive from its creating the virtual presence of all the authors who contribute to its materials. Computer scientists draw on optics for an analogy when they speak of "virtual machines" created by an operating system that provides individual users sharing a system with the sense of working on their own individual machines. In the first chapter, when discussing electronic textuality,

I pointed to another kind of "virtual" existence, the virtual text: all texts that one encounters on the computer screen are virtual, rather than real. In a similar manner, the reader experiences the virtual presence of other contributors.

Such virtual presence is of course a characteristic of all technology of cultural memory based on writing and symbol systems. Since we all manipulate cultural codes—particularly language but also mathematics and other symbols—in slightly different ways, each record of an utterance conveys a sense of the one who makes that utterance. Hypermedia differs from print technology, however, in several crucial ways that amplify this notion of virtual presence. Because the essential connectivity of hypermedia removes the physical isolation of individual texts characteristic of print technology, the presence of individual authors becomes both more available and more important. The characteristic flexibility of this reader-centered information technology means, quite simply, that writers have a much greater presence in the system, as potential contributors and collaborative participants but also as readers who choose their own paths through the materials.

Collaborative Writing, Collaborative Authorship

The virtual presence of other texts and other authors contributes importantly to the radical reconception of authorship, authorial property, and collaboration associated with hypertext. Within a hypertext environment all writing becomes collaborative writing, doubly so. The first element of collaboration appears when one compares the roles of writer and reader, since the active reader necessarily collaborates with the author in producing the particular version of the text she or he reads by the choices she or he makes—a fact much more obvious in very large hypertexts than in smaller hyperfictions. The second aspect of collaboration appears when one compares the writer with other writers—that is, the author who is writing now with the virtual presence of all writers "on the system" who wrote then but whose writings are still present.

The word *collaboration*, which derives from the Latin for *working* plus that for *with* or *together*, conveys the suggestion, among others, of working side by side on the same endeavor. Most people's conceptions of collaborative work take the form of two or more scientists, songwriters, or the like continually conferring as they pursue a project in the same place at the same time. I have worked on an essay with a fellow scholar in this manner. One of us would a type a sentence, at which point the other would approve, qualify, or rewrite it, and then we would proceed to the next sentence. Far more common a form of collaboration, I suspect, is that second mode described as "versioning," in which one worker produces a draft that another person then edits by modifying and adding. The first and the second forms of collaborative authorship tend to blur, but the distinguishing factor here is that versioning takes place out of the presence of the other collaborator and at a later time.

Both of these models require considerable ability to work productively with other people, and evidence suggests that many people either do not have such ability or do not enjoy putting it into practice. In fact, according to those who have carried out experiments in collaborative work, a third form proves more common than the first two—the assembly-line or segmentation model of working together, according to which individual workers divide the overall task and work entirely independently. This last mode is the form that most people engaged in collaborative work choose when they work on projects ranging from programming to art exhibitions.

Networked hypertext systems like the World Wide Web, Hyper-G, Sepia, and Intermedia offer a fourth model of collaborative work that combines aspects of the previous ones. By emphasizing the presence of other texts and their cooperative interaction, networked hypertext makes all additions to a system simultaneously a matter of versioning and of the assembly-line model. Once ensconced within a network of electronic links, a document no longer exists by itself. It always exists in relation to other documents in a way that a book or printed document never does and never can. From this crucial shift in the way texts exist in relation to others derive two principles that, in turn, produce this fourth form of collaboration: first, any document placed on any networked system that supports electronically linked materials potentially exists in collaboration with any and all other documents on that system; second, any document electronically linked to any other document collaborates with it.

According to the *American Heritage Dictionary of the English Language*, the verb *to collaborate* can mean either "to work together, especially in a joint intellectual effort," or "to cooperate treasonably, as with an enemy occupying one's country." The combination of labor, political power, and aggressiveness that appears in this dictionary definition well indicates some of the problems that arise when one discusses collaborative work. On the one hand, the notion of collaboration embraces notions of working together with others, of forming a community of action. This meaning recognizes, as it were, that we all exist within social groups, and it obviously places value on contributions to that group. On the other hand, collaboration also includes a deep suspicion of working with others, something aesthetically as well as emotionally engrained since the advent of Romanticism, which exalts the idea of individual effort to such a degree that it, like copyright law, often fails to recognize, or even suppresses, the fact that artists and writers work collaboratively with texts created by others.

Most of our intellectual endeavors involve collaboration, but we do not always recognize that fact for two reasons. The rules of our intellectual culture, particularly those that define intellectual property and authorship, do not encourage such recognitions, and furthermore, information technology from Gutenberg to the present—the technology of the book—systematically hinders full recognition of collaborative authorship.

Throughout the past century the physical and biological sciences have increasingly conceived of scientific research, authorship, and publication as group endeavors. The conditions of scientific research, according to which many research projects require the cooperating services of a number of specialists in the same or (often) different fields, bear some resemblances to the medieval guild system in which apprentices, journeymen, and masters all worked on a single complex project. Nonetheless, "collaborations differ depending on whether the substance of the research involves a theoretical science, such as mathematics, or an empirical science, such as biology or psychology. The former are characterized by collaborations among equals, with little division of labor, whereas the latter are characterized by more explicit exchange of services, and more substantial division of labor" (Galegher, Egido, and Kraut, 151). The financing of scientific research, which supports the individual project, the institution at which it is carried out, and the costs of educating new members of the discipline all nurture such group endeavors and consequent conceptions of group authorship.[9]

In general, the scientific disciplines rely on an inclusive conception of authorship: anyone who has made a major contribution to finding particular results, occasionally including specialized technicians and those who develop techniques necessary to carry out a course of research, can appear as authors of scientific papers, and similarly, those in whose laboratories a project is carried out may receive authorial credit if an individual project and the publication of its results depend intimately on their general research. In the course of a graduate student's research for a dissertation, he or she may receive continual advice and evaluation. When the student's project bears fruit and appears in the form of one or more publications, the advisor's name often appears as co-author.

Not so in the humanities, where graduate student research is supported largely by teaching assistantships and not, as in the sciences, by research funding. Although an advisor of a student in English or art history often acts in ways closely paralleling the advisor of the student in physics, chemistry, or biology, explicit acknowledgments of cooperative work rarely appear. Even when a senior scholar provides the student with a fairly precise research project, continual guidance, and access to crucial materials that the senior scholar has discovered or assembled, the student does not include the advisor as co-author.

The marked differences between conceptions of authorship in the sciences and the humanities demonstrate the validity of Michel Foucault's observation that "the 'author-function' is tied to the legal and institutional systems that circumscribe, determine, and articulate the realm of discourses; it does not operate in a uniform manner in all discourses, at all times, and in any given culture it is not defined by the spontaneous attribution of a text to its creator, but through a series of precise and complex procedures; it does not refer, purely and simply, to an actual individual" ("What Is an Author?" 131). One reason for the different conceptions of authorship and authorial property in the humanities and the sciences lies in the different conditions of funding and the different discipline-politics that result.

Another corollary reason is that the humanistic disciplines, which traditionally apply historical approaches to the areas they study, consider their own assumptions about authorship, authorial ownership, creativity, and originality to be eternal verities.[10] In particular, literary studies and literary institutions, such as departments of English, which still bathe themselves in the afterglow of Romanticism, uncritically inflate Romantic notions of creativity and originality to the point of absurdity. An example comes readily to hand from the preface of Lisa Ede and Andrea Lunsford's recent study of collaborative writing, the production of which they discovered to have involved "acts of subversion and of liberatory significance": "We began collaborating in spite of concerned warnings of friends and colleagues, including those of Edward P. J. Corbett, the person in whose honor we first wrote collaboratively. We knew that our collaboration represented a challenge to traditional research conventions in the humanities. Andrea's colleagues (at the University of British Columbia) said so when they declined to consider any of her coauthored or coedited works as part of a review for promotion" (ix–x).

Ede and Lunsford, whose interest in their subject grew out of the "difference between our personal experience as coauthors and the responses of many of our friends and colleagues" (5), set the issue of collaborative writing within the contexts of actual practice in the worlds of business and academia, the history of theories of creative individualism and copyright in recent Western culture, contemporary and feminist analyses of many of these other contexts. They produce a wide range of evidence in convincingly arguing that "the pervasive commonsense assumption that writing is inherently and necessarily a solitary, individual act" (5) supports a traditional patriarchal construction of authorship and authority. After arguing against "univocal psychological theories of the self" (132) and associated notions of an isolated individualism, Ede and Lunsford call for a more Bakhtinian reconception of the self and for what they term a *dialogic,* rather than a hierarchical, mode of collaboration.

I shall return to their ideas when I discuss the role of hypertext in collaborative learning, but now I wish to point out that as scholars from McLuhan and Eisenstein to Ede and Lunsford have long argued, book technology and the attitudes it supports are the institutions most responsible for maintaining exaggerated notions of authorial individuality, uniqueness, and ownership that often drastically falsify the conception of original contributions in the humanities and convey distorted pictures of research. The sciences take a relatively expansive, inclusive view of authorship and consequently of text ownership.[11] The humanities take a far more restricted view that emphasizes individuality, separation, and uniqueness—often creating a vastly distorted view of the connection of a particular text to those that preceded it. Neither view possesses an obvious rightness. Each is obviously a social construction, and each has on occasion proved to distort actual conditions of intellectual work carried out in a particular field.

Whatever the political, economic, and other discipline-specific factors that maintain the conception of noncooperative authorship in the humanities, print technology has also contributed to the sense of a separate, unique text that is the product—and hence the property—of one person, the author. Hypertext changes all this, in large part because it does away with the isolation of the individual text that characterizes the book. As McLuhan and other students of the cultural influence of print technology have pointed out, modern conceptions of intellectual property derive both from the organization and financing of book production and from the uniformity and fixity of text that characterizes the printed book. J. David Bolter explains that book technology itself created new conceptions of authorship and publication:

Because printing a book is a costly and laborious task, few readers have the opportunity to become published authors. An author is a person whose words are faithfully copied and sent round the literary world, whereas readers are merely the audience for those words. The distinction meant less in the age of manuscripts, when "publication" was less of an event and when the reader's own notes and glosses had the same status as the text itself. Any reader could decide to cross over and become an author:

one simply sat down and wrote a treatise or put one's notes in a form for others to read. Once the treatise was written, there was no difference between it and the works of other "published" writers, except that the more famous works existed in more copies. (*Writing Space*, 148–49)

Printing a book requires a considerable expenditure of capital and labor, and the need to protect that investment contributes to notions of intellectual property. But these notions would not be possible in the first place without the physically separate, fixed text of the printed book. Just as the need to finance printing of books led to a search for the large audiences that in turn stimulated the ultimate triumph of the vernacular and fixed spelling, so, too, the fixed nature of the individual text made possible the idea that each author produces something unique and identifiable as property.

The needs of the marketplace, as least as they are conceived by editors and publishing houses, reinforce all the worst effects of these conceptions of authorship in both academic and popular books. Alleen Pace Nilsen reports that Nancy Mitford and her husband wrote the best-selling *The High Cost of Death* together, but only her name appears because the publisher urged that multiple authors would cut sales. Another common solution involves resorting to a pseudonym: Perri O'Shaunessy is the pen-name of Pam and Mary O'Shaunessy, created when their editors would not permit a double-author byline, and John Case, "author" of *The Genesis Code,* "is really husband-and-wife team Jim and Carolyn Hougan."[12] In another case, to make a book more marketable a publisher replaced the chief editor of a major psychiatric textbook with the name of a prestigious contributor who had not edited the volume at all (cited by Ede and Lunsford, 3–4). I am sure everyone has examples of such distortion of authorial practice by what a publisher believes to be good business. I have mine: a number of years ago after an exercise in collaborative work and writing with three graduate students produced a publishable manuscript, we decided by mutual agreement on the ordering of authors' names on the title page. By the time the volume appeared, the three former graduate students all held teaching positions, and its appearance, one expects, might have helped them professionally. Unfortunately, the publisher insisted on including only the first editor's name in all notices, advertisements, and catalogues. Such an action, of course, does not have so serious an effect as removing the editors' names from the title page, but it certainly discriminates unfairly between the first two editors, who did equal amounts of work, and it certainly conveys a strong message to beginning humanists about the culturally assigned value of cooperation and collaboration.

Even though print technology is not entirely or even largely responsible for current attitudes in the humanities toward authorship and collaboration, a shift to hypertext systems would change them by emphasizing elements of collaboration. As Tora K. Bikson and J. D. Eveland point out in relation to other, nonhumanities work, "The electronic environment is a rich context in which doing work and sharing work becomes virtually indistinguishable" (286). If we can make ourselves aware of the new possibilities created by these changes, we can at the very least take advantage of the characteristic qualities of this new form of information technology.

One relevant characteristic quality of networked hypertext systems is that they produce a sense of authorship, authorial property, and creativity that differs markedly from those associated with book technology. Hypertext changes our sense of authorship and creativity (or originality) by moving away from the constrictions of page-bound technology. In so doing, it promises to have an effect on cultural and intellectual disciplines as important as those produced by earlier shifts in the technology of cultural memory that followed the invention of writing and printing (see Bolter, McLuhan, and Eisenstein).

Examples of Collaboration

in Hypertext

Collaborative work in hypertext takes many forms, one of the most interesting of which illustrates the principle that one almost inevitably works collaboratively whenever creating documents on a multiauthor hypertext system. I discovered the inevitably collaborative nature of hypermedia authorship in the old Intermedia days. While linking materials to the overview (or sitemap) for Graham Swift's *Waterland* (1983), I observed Nicole Yankelovich, project coordinator of the Intermedia project at IRIS, working on materials for a course in arms control and disarmament offered by Richard Smoke of Brown University's Center for Foreign Policy Development. Those materials, which were created by someone from a discipline very different from mine for a very different kind of course, filled a major gap in a project I was working on. Although my co-authors and I had created materials about technology, including graphic and text documents on canals and railroads, to attach to the science and technology section of the *Waterland* overview, we did not have the expertise to create parallel documents about nuclear technology and the antinuclear movement, two subjects that play a significant part in Swift's novel. Creating a brief introduction to the subject of *Waterland* and nuclear disarmament, I linked it first to the science and technology section in the *Waterland* overview and then to the time line that the nuclear arms course materials employ as a directory file. A brief document and a few links enable students in the introductory survey of English literature to explore the materials created for a course in another discipline. Similarly, students from that course could now encounter materials showing the effects on contemporary fiction of the concerns covered in their political science course. Hypertext thus allows and encourages collaborative work, and at the same time it encourages interdisciplinary approaches by making materials created by specialists in different disciplines work together—collaborate.

This kind of collaboration-by-link occurs all the time on the World Wide Web. Each time a student or faculty member from another institution has one of their documents added to *The Victorian Web*—say, on characterization or race, class, and gender in *Jane Eyre*—they automatically join in a discussion on these topics. Similarly, Phil Gyford's translation of Pepys's *Diaries* into a Weblog, at which we looked in the previous chapter, exemplifies yet another approach to collaboration on the Web.

The important point here is that hypermedia linking automatically produces collaboration. Looking at the way the arms control materials joined to those supporting the four English courses, one encounters a typical example of how the connectivity that characterizes hypertext transforms independently produced documents into collaborative ones and authors working alone into collaborative authors. When one considers the arms control materials from the point of view of their originator, they exist as part of a discrete body of materials. When one considers them from the vantage point of a reader, their status changes: as soon as they appear within a hypertext environment, these and all other documents then exist as part of a larger system and in relation therefore to other materials on that system. By forming electronic pathways between blocks of texts, links actualize the potential relations between them. Just as hypertext as an educational medium transforms the teacher from a leader into a kind of coach or companion, hypertext as a writing medium metamorphoses the author into an editor or developer. Hypermedia, like cinema and video or opera, is a team production.

5 — Reconfiguring Writing

Since writing hypermedia successfully involves finding ways to prevent readers from becoming confused and discouraged when they encounter text in e-space, let us examine this notion of disorientation before considering some of the methods used to prevent it. Crucial as disorientation might seem to discussions of hypertext authoring, this term remains unexamined and inadequately defined. Such a claim might appear particularly odd because writers on the subject since Jeff Conklin have apparently provided fairly precise statements of what they mean by what Conklin himself termed the *disorientation problem*. According to his initial statement of the issue, disorientation seems to inhere in the medium itself: "Along with the power of being able to organize information much more complexly comes the problem of having to know (1) where you are in the network and (2) how to get to some other place that you know (or think) exists in the network. I call this the *disorientation problem*. Of course, one has a disorientation in traditional linear text documents, but in a linear text the reader has only two options: He can search for the desired text earlier in the text or later in the text" (38). Kenneth Utting and Nicole Yankelovich, who similarly point out that "hypermedia . . . has the potential to dramatically confuse and confound readers, writers, teachers, and learners," quote Conklin's definition of disorientation as "the tendency to lose one's sense of location and direction in a nonlinear document" (58), and in their example of three aspects of disorientation, they mention "confusion about where to go or, having decided on a destination, how to get there," and also disorientation in the sense of not knowing "the boundaries of the information space" (61) one is exploring.

Three points here demand notice. First, the concept of disorientation relates closely to the tendency to use spatial, geographical, and travel metaphors to describe the way users experience hypertext. Such uses are obviously appropriate to dictionary definitions of *disorient*. Neither *The American Heritage Dictionary* nor *Webster's Collegiate Dictionary* defines *disorientation*, but according to *The American Heritage Dictionary*, to disorient is "to cause to lose one's sense of direction or location, as by removing from a familiar environment," and *Webster's* offers three definitions of *disorient*: (1) "to cause to lose one's bearings: displace from normal position or relationship"; (2) "to cause to lose the sense of time, place, or identity"; and (3) "to confuse."

In general, authors writing about hypertext seem to mean *confuse* and specifically *lose bearings* when they use the term, and this usage derives from commonplace application of spatial metaphors to describe the reader's behavior in a hypertext environment. Thus, in "The Art of Navigating through Hypertext," Jakob Nielsen points out in the usual formulation that "one of the major usability problems with hypertext is the user's risk of disorientation while navigating the information space. For example, our studies showed that 56 percent of the readers of a document written in one of the most popular commercial hypertext systems agreed fully or partially with the statement *I was often confused about where I was*" (298). Nielsen believes that "true hypertext should also make users *feel* that they can move freely through the information according to their own needs" (298).

Second, as Conklin and others writing in this field state the problem of disorientation, it obviously concerns the design of the information technology alone. In other words, the related concepts of disorientation and confusion appear, in their terms, to have nothing to do with the materials, the content, on the hypertext system. Nonetheless, we all know that readers often experience confusion and disorientation simply because they fail to grasp the logic or even meaning of a particular argument. Even if the works of Kant, Einstein, and Heidegger were to appear on the finest hypertext and information retrieval system in the world, they would still disorient many readers. Although Conklin and other students of hypertext have not naively or incompletely defined what they mean by disorientation, their restriction of this term to system-generated disorientation in practice does not take into account a large portion of the actual reading experience—and its implications for hypertext authors. The issue has a bearing on a third point about the notion of disorientation.

Third, disorientation, as these comments make clear, is conceived by these authors as crippling and disenabling, as something, in other words, that blocks completion of a task one has set for oneself or that has been set for one by others. Disorientation, furthermore, is presented as such a massive, monolithic problem that these authors pay little or no attention to how people actually cope with this experience. Is it, in fact, crippling, and do users of hypertext systems simply give up or fail in whatever tasks they have engaged themselves when they meet disorientation? As we shall see, expert users of hypertext do not always find the experience of disorientation to be particularly stressful, much less paralyzing.

The role of disorientation in literature suggests some reasons why this might be the case. Readers of literature in fact often describe the experience here presented as disorientation as pleasurable, even exciting, and some forms of literature, particularly those that emphasize either allegory or stylistic and narrative experimentation, rely on disorienting the reader as a primary effect. Although the kind of pleasurable disorientation that one finds in Dante's *Divine Comedy,* Browning's *Ring and the Book,* and Eliot's *Wasteland* derives from what we have termed the content and not from the information technology that presents it, this effect has one important parallel to that encountered in some forms of hypertext: in each case the neophyte or inexperienced reader finds unpleasantly confusing materials that more expert ones find a source of pleasure.

The Concept of Disorientation in the Humanities

The reasons for the radically different ways people in the humanities and technological disciplines regard disorientation become particularly clear in three areas—aesthetic theories of disorientation, conceptions of modernism and postmodernism as cultural movements, and the related conceptions of hypertext fiction.

The classic statement of the positive value of cognitive and other disorientation in aesthetic works appears in Morse Peckham's *Man's Rage for Chaos: Biology, Behavior, and the Arts* (1967), which argues that "art offers not order but the opportunity to experience more disorder than any other human artifact, and . . . artistic experience, therefore, is characterized . . . by disorientation" (41). According to him, "the artist's role is to create occasions for disorientation, and of the perceiver's role to experience it. The distinguishing mark of the perceiver's transaction with the work of art is discontinuity of experience, not continuity; disorder, not order; emotional disturbance, not emotional catharsis, even though some works have a cadential close" (254). Human beings so "passionately" want "a predictable and ordered world" that "only in protected situations, characterized by high walls of psychic insulation," can they permit themselves to perceive the gap between "expectancy or set or orientation, and the data . . . interaction with the environment actually produces. . . Art offers precisely this kind of experience" (313).

Peckham argues finally that art is "an adaptational mechanism" that reinforces our ability to survive:

Art is rehearsal for those real situations in which it is vital for our survival to endure cognitive tension, to refuse the comforts of validation by affective congruence when such validation is inappropriate because too vital interests are at stake; art is the reinforcement of the capacity to endure disorientation so that a real and signifi-cant problem can emerge. Art is the exposure to the tensions and problems of a false world so that man can endure exposing himself to the tensions and problems of a real world. (314)

Peckham's positive views of aesthetic disorientation, which seem to grow out of the arts and literature of modernism, clearly present it as a matter of free-dom and human development.

Students of literature and the arts have long emphasized the role of dis-orientation in both modernism and postmodernism. Like the works of the cubists, expressionists, and other movements of twentieth-century art, James Joyce's *Ulysses,* T. S. Eliot's *Wasteland,* and William Faulkner's *Sound and the Fury*—to cite three classics of literary modernism—all make disorientation a central aesthetic experience. Similarly, as recent writers on postmodernist fiction point out, it is characterized by a range of qualities that produce cog-nitive disorientation: "contradiction, discontinuity, randomness," "intractable epistemological uncertainty," and "cognitive estrangement" (McHale, 7, 11, 59). These attitudes, which students of the past century's culture almost univer-sally view positively, appear throughout discussions of hypertext fiction as well. Robert Coover, for example, makes quite clear the relations between dis-orientation, hypertext, and the traditions of the avant garde when he describes the way hypertext fiction promises to fulfill the liberating functions of the experimental tradition in fiction.[1] He also emphasizes the effect on writers of this disorienting freedom. Discussing the conservatism of writing students, he claims that

getting them to consider trying out alternative or innovative forms is harder than talk-ing them into chastity as a life-style. But suddenly, confronted with hyperspace, they have no choice: all the comforting structures have been erased. It's improvise or go home. Some frantically rebuild those old structures, some just get lost and drift out of sight, most leap in fearlessly without even asking how deep it is (*infinitely* deep), admitting, even as they continue to paddle for dear life, that this new arena is indeed an exciting, provocative, if frequently frustrating medium for the creation of new narratives, a potentially revolutionary space, empowered, exactly as advertised, to transform the very art of fiction. ("End of Books," 24)

Michael Joyce describes potentially disorienting qualities of hypertext fiction in terms that praise the necessary activism required of readers: "Constructive hypertexts require a capability to act: to create, to change and to recover particular encounters within the developing body of knowledge. These encounters . . . are maintained as versions, i.e., trails, paths, webs, notebooks, etc.; but they are versions of what they are becoming, a structure for what does not yet exist" (*Of Two Minds,* 42). In much the same vein Stuart Moulthrop, like Coover, relates the experience of encountering gaps and disorientation that characterize the reader's experience in hypertext as potentially liberating. "In a world where the 'global variables' of power and knowledge tend to orient themselves toward singular, hegemonic world orders, it becomes increasingly difficult to jump outside 'the system.' And as Thomas Pynchon reminds us: 'Living inside the System is like riding across the country in a bus driven by a maniac bent on suicide' (*Gravity's Rainbow,* 412)" ("Beyond the Electronic Book," 76). Given the fact that many humanities users of hypertext, like those specifically concerned with hypertext fiction, associate the general experience of disorientation with avant-garde, liberating, and culturally approved aesthetic experience, it should be no surprise that they treat the issue of disorientation far differently than do almost all who consider it in the technical disciplines.

The Love of Possibilities

In experiments that Paul Kahn and I conducted in 1991 experienced student-users of hypertext showed a love of browsing and of the serendipity it occasions very much at odds with by-now conventional attitudes toward disorientation in hypertext. For example, one user explained that by "accidentally clicking" on a particular link he found that he had made a "delightful detour," since it led to an answer to one of the problems in information retrieval. "Although I guess this mistake has an analog in book technology, it would be the improbable act of being in the wrong section of the library, the wrong row of books, the wrong shelf, picking up the wrong book, and opening up magically to the correct page."

Some of these responses were disconcertingly unexpected and for that reason potentially quite valuable to anyone considering the design, implementation, and educational application of hypertext. In two cases very experienced programmers had more difficulties with certain aspects of information-retrieval tasks than did comparative neophytes. It would appear that their expectations of systems and retrieval mechanisms served to hinder rather than to assist their explorations. Accustomed to using full-text search mechanisms in other kinds of computer systems, one of these students spent some fifteen minutes searching for one in Intermedia—the version used did not have the system's later search tools—and then gave up on the assignment, assuming that no other methods of locating the information existed.

In contrast, a relatively unsophisticated user solved the first problem of locating works by one scholar in a matter of moments. As he explained: "I found these references by opening the Critics Quoted Document in the Bibliographical Folder in the Dickens Folder . . . Total Time: 6 min." Another similarly responded, "I answered the first question of the assignment using [the Intermedia folder system]. Since the folders were labeled well, I found it quick and easy to first find the 'Bibliography' folder, and then open the 'Critics Quoted' document. There I found the names of the three authors in the question. Since the web was already engaged, I could activate the link markers and see all the destination documents connected to a particular author (if in fact the web was well linked). Thus, I approached the web from an odd angle, from the actual document folders, but it was the one which I felt to be the easiest and quickest for this question. This same information could be found in the Bibliography Overview . . . If I had never come across the Victorianism Overview, for whatever reason, I might never have come across the sought after bibliographic information. But I did find the information, outside the system's (few) attempts at organization. I felt so comfortable with the sight of the Macintosh document icon that I felt there was no 'violation,' as Intermedia depicts no structure to be violated—documents seem to be either autonomous or within the web" (MF).

This student's narrative forcefully restates the truism that people who want to find information will find it as much by what they know about that information as by system features alone. In other words, orientation by content seems able to solve potential problems of disorientation caused by the system design considered in isolation. In this case, some experienced users of computer tended to conceive this task as a means of testing system capacities whereas those who were content experts, or who *took the approach of a content expert,* conceived the task in terms that made the desired information the center of the task.

One important lesson for both designers of hypertext software and those who teach or write with hypertext appears in the problems encountered by the students with more computer skills. In relying too heavily on system features, they implicitly made the assumption that the system, rather than the author, does most of the work. In doing so, they tended to ignore the stylistic and other author-created devices that made the search quick and easy for a majority of users.

We should also note that a preference for browsing up to and including the sense of "disorientation" can create disconcerting results for hypertext designers, despite the fact that hypertext theorists often praise this approach to wandering through a database. For example, one user criticized one of the systems precisely because it proved "more difficult to become disoriented in the good way that Intermedia and Storyspace tend to facilitate. I found that links continually brought me back to crossroads or overviews, rather than to other documents. For this reason I felt less like an active reader. Orientation devices such as these explained and categorized links rather than allowing me to make my own connections and categories" (AM). To those who find disorientation a negative quality, these comments might seem puzzling, because apparently negative qualities here come in for praise. In fact, this student specifically mentions "the good way" Intermedia and Storyspace create a sense of disorientation, which she takes to be a condition that empowers hypertext users because it places them in an active role—one particularly appropriate to this new information medium.

The reactions of these student-evaluators suggest six points about reader disorientation. First, although it represents a potentially significant problem in some systems, a priori concerns about it may well arise from lack of experience with hypertext systems, specifically from attempting to apply reading and information-retrieval protocols appropriate to book technology to this new medium.

Second, what one reader experiences as disorientation, another may find pleasurable.

Third, disorientation has quite different connotations in the writings of those based in technological as opposed to literary disciplines. The technologically based conception of disorientation relates to a conception of education essentially limited to factual information. Literary or humanistic assumptions about disorientation seem related to a conception of education in which students learn to deal with complex matters of interpretation.

Fourth, disorientation—let me emphasize this point yet again—arises both in the normal act of reading difficult material *and* in poorly designed systems. Knowledge of content, as some of our evaluators demonstrate, has to be considered as part of any solution to issues of system-generated or system-permitted disorientation.

Fifth, since for the foreseeable future, book and electronic technologies will exist together, in some applications supplementing, in others competing with, each other, designers of hypertext systems will continue to find themselves in a terribly difficult situation. Systems they design will almost certainly encounter a heterogeneous pool of users, some still trying to read according to the rules of books, others, increasingly sophisticated in electronic media, who find the specific qualities of hypertext reading and exploration, including occasional "disorientation," as pleasurable, desirable qualities.

Sixth—and most important—writing, as much as system design, as much as software, prevents the less pleasant forms of disorientation. We must therefore develop a rhetoric and stylistics of hypertext writing.

The Rhetoric and Stylistics of Writing for E-Space; or, How Should We Write Hypertext?

Linking, by itself, is not enough. The hypermedia author cannot realize the enormous potential of the medium simply by linking one passage or image to others. The act of connecting one text to another fails to achieve all the expected benefits of hypermedia and can even alienate the user. On the briefest consideration, such a recognition will hardly surprise, since authors of print essays, poems, narratives, and books do not expect to *write* merely by stringing together sentences and paragraphs without the assistance of stylistic devices and rhetorical conventions. If to communicate effectively, hypermedia authors must employ devices suited to their medium, two questions arise. First, what are the defining characteristics or qualities of hypertext as reading and writing medium? Second, to what extent do they depend on specific hardware and software? What effect, for example, does the presence or absence of color, size of one's monitor, and the speed of one's computer have on reading hypertext?

Then there are questions less immediately derived from the hardware. Assuming that writing at the level of phrase, sentence, and paragraph will not change in some fundamental way—and this, I admit, may be too large an assumption to make at this stage—what new forms of organization, rhetoric, and structure must we develop to communicate effectively in electronic space? In other words, if hypertext demands a new rhetoric and a new stylistics, of what do they consist, and how, if at all, do they relate to issues like system speed and the like?

To begin, let us look once again at the nature of the medium. I have just written that "hypertext changes the way texts exist and the way we read them," and in earlier chapters we have observed many examples of such difference from chirographic and print textuality. Whether or not it is true that the digital word produces a secondary or new kind of orality, many of the devices required by hypertext appear in oral speech, just as they do in its written versions or dialects. Many of these devices to which I wish to direct our attention fall into a single category: they announce a change of direction and often also provide some indication of what that new direction will be. For example, words and phrases like in *contrast, nevertheless,* and *on the other hand* give advance notice to listeners and readers of something, say, an instance or assertion, is coming contrary to what has come before. *For example* announces a category shift as the discourse switches, most likely, from general or abstract statement to proposed instances of it. Causal or temporal terms, such as *because* or *after,* similarly ready listeners for changes of intellectual direction. In both print and oral communication, they are means, in other words, of preparing us for breaks in a linear stream of language. One must take care in using this term *linear* since, as we have already seen when looking at hypertext narrative, all experience of listening or reading in whatever medium is in an important sense linear, unidirectional. Thus, although readers—or, to be precise, *readings*—take different paths through *afternoon, Patchwork Girl,* or *Quibbling,* each path—each experience of reading—takes the form of a sequence. It is the text that is multisequential, not a particular reading path through it. I emphasize this obvious point because the problem of preparing for change of direction (and openings and closings are also such changes) has been with us since the beginnings of human language.

Since hypertext and hypermedia are chiefly defined by the link, a writing device that offers potential changes of direction, the rhetoric and stylistics of this new information technology generally involve such change—potential or actual change of place, relation, or direction. Before determining which techniques best accommodate such change, we must realize that, together, they attempt to answer several related questions: First, what must one do to orient readers and help them read efficiently and with pleasure? Second, how can one help readers retrace the steps in their reading path? Third, how can one inform those reading a document where the links in that document lead? Finally, how can one assist readers who have just entered a new document to feel at home there?

Drawing on the analogy of travel, we can say that the first problem concerns *orientation* information necessary for finding one's place within a body of interlinked texts. The second concerns *navigation* information necessary for making one's way through the materials. The third concerns *exit* or *departure* information and the fourth *arrival* or *entrance* information. In each case, creators of hypermedia materials must decide what readers need to know at either end of a hypermedia link in order to make use of what they find there. The general issue here is one of interpretation. More specifically, to enable visitors to this new kind of text to read it pleasurably, comfortably, and efficiently, how much interpretation must the designer-author attach to the system as a whole, to link pathways, and to documents at the end of links?

Unfortunately, no analogy maps reality with complete accuracy. Navigation, the art of controlling the course of a plane or ship, presupposes a spatial world, but one does not *entirely* experience hypertext as such. In navigation, we remember, one must determine one's spatial position in relation to landmarks or astral locations and then decide on a means of moving toward one's goal, which lies out of sight at some spatial distance. Because it takes time to move across the separating distance, one also experiences that distance as time: one's ship lies so many nautical miles and therefore so many days and hours from one's goal. The reader, however, does not experience hypertext in this way. The reader of *Paradise Lost,* for example, experiences as equally close the linked parts of Homer and Vergil to which the poem's opening section allude and linked lines on the next page or in the next book (see Figure 12). Because hypertext linking takes relatively the same amount of time to traverse, all linked texts are experienced as lying at the same "distance" from the point of departure. Thus, whereas navigation presupposes that one finds oneself at the center of a spatial world in which desired items lie at varying distances from one's own location, hypertext (and other forms of addressable, digital textuality) presupposes an experiential world in which the goal is always potentially but one jump or link away.

General Observations. Hypermedia as a medium conveys the strong impression that its links signify coherent, purposeful, and above all *useful* relationships. From which follows that the very existence of links conditions the reader to expect purposeful, important relationships between linked materials. One of the presuppositions in hypertext, particularly when applied to educational uses, is that linking materials encourages habits of relational thinking in the reader. Such intrinsic hypermedia emphasis on interconnectedness (or connectivity) provides a powerful means of teaching sophisticated critical thinking, particularly that which builds on multicausal analyses and relating different kinds of data. But since hypermedia systems predispose users to expect significant relationships among lexias, those that disappoint these expectations tend to appear particularly incoherent and without significance.

When users follow links and encounter materials that do not appear to possess a significant relation to the document from which the link pathway originated, they feel confused and resentful. In reading materials on the Web, the delays encountered by users tend to exaggerate this effect, thus providing another reason for avoiding time-consuming graphic or other elements whenever possible if one wishes to include an audience without high-speed access to the Internet.

System-Generated Means of Reader Orientation. Devices of orientation permit readers (1) to determine their present location, (2) to have some idea of that location's relation to other materials, (3) to return to their starting point, and (4) to explore materials not directly linked to those in which they presently find themselves.

The graphic presentation of information embodied in the useful, if limited, desktop metaphor proves an especially effective means of reader orientation in the systems that use them, but World Wide Web browsers, in which the risks of disorientation are particularly grave, do not. Of course, the "Show Location" window in IE, Firefox, Safari, Netscape, and other HTML browsers does provide the exact address of a lexia, and as I write today, following a link to one student's essay on *Patchwork Girl*, say, Lars Hubrich's "Stitched Identity," would produce the following information in the location window: http://www.cyberspaceweb.org/ht/pg/lhpatch.html. Such information not only appears in a form daunting to most readers, it fails to be very helpful on two counts: first, the need to create economically brief directory names often renders the file incomprehensible to all but the person who maintains a website and, second, it provides very little information about the relation of this particular lexia to the information space it inhabits.

One way of providing the benefits of graphic presentation of a folder structure takes the form of Java applets, such as those created by Dynamic Diagrams for IBM's website (Figure 13), which generate an animated three-dimensional image of an individual lexia's location within a file structure. Where such software solutions are not available, authors have employed two solutions. One involves organizing an entire site according to what is essentially a folder structure and then making that organization clear. Thus, Susan Farrell's *Art-Crimes,* a site containing graffiti from around the world, presents its information in terms of country, city, and additional collections for each city. Of course, this beautiful site, which provides a visual archive, has little intrinsically hypertextual about it and therefore cannot serve as an example for other kinds of webs.

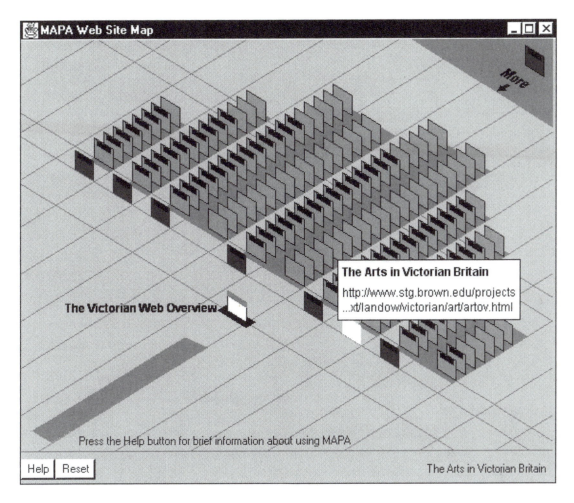

Figure 13. A View of *The Victorian Web*. This map was created by Dynamic Diagrams MAPA from the vantage point of this web's homepage (main sitemap or overview). In this screen shot, a user has activated a pop-up window displaying the URL and title of "The Arts in Victorian Britain," a second-level overview. By using a computer mouse to move the cursor farther away from the top level, users can also learn the titles of lexias linked to this and other overviews. Double clicking on the icon for any lexia opens it. (Copyright 1996 Dynamic Diagrams. Used by permission.)

Keeping (the) Track: Where've I Been, What Did I Read? In addition to helping readers discern their general location within an information space at any moment, systems also have to provide both some means of informing them from whence they came and a means of allowing them to return. As one of its functions, the Roadmap in Storyspace provides a sequential list of lexias that one has visited and provides a slightly cumbersome means of returning to any one. Web browsers also have means of providing current reading history: mousing down on the "Go" or "History" menu at the top of the screen produces a chronologically ordered list of lexias one has opened.[2] Current versions of webviewers, including Internet Explorer and Safari, solve this problem by retaining lists of the web documents one has visited; Explorer permits the user to specify how many sites should be retained.

Another valuable orientation tool takes the form of permanent bookmarks in World Wide Web browsers, a feature anticipated by HyperCard-based systems like Voyager Expanded Book, Keyboard, and Toolbook. Such a bookmark function permits readers to record places to which they might like to return at some future time, and when designing large websites, most of whose links remain internal, authors can advise readers who contemplate following links to materials offsite that they might first wish to use their viewer's bookmark facility, thus making return easier, particularly in a complex session. One activates a bookmark simply by choosing it from a list available at the top of the screen. Although these devices play an important role in allowing readers to customize their own experiences of the webs they read, they do not compensate for the absence of long-term reading histories in very large, complex corpora, such as one finds on the Web.

Dynamic and Static Tables of Contents. One often encounters the table of contents, a device directly transferred from book design, in hypertext documents, often to very good effect. Readers of materials on the Web will have frequently encountered it since a good many homepages and title screens consist essentially of linked tables of contents. I have used it myself, particularly when creating hypertext versions of print materials, a subject I shall discuss at greater length in a separate section below. The World Wide Web version of the opening chapter of the first version of this book, for instance, employs two such contents screens, one for the entire volume and a second for the first chapter, the only one available on the web. Although such a table of contents provides a familiar, often effective means of presenting a work's organization, in its static form it often overemphasizes the element of the electronic book to the detriment of its hypertextuality.

Electronic Book Technologies' DynaText, which features a dynamic, automatically generated contents screen, offers a much more powerful version of this device. DynaText uses text in the form of SGML, a far richer, more powerful older relation of the Web's HTML. Since SGML requires that one begin and end every chapter title, section heading, and all other text structures with specific tags (markup), DynaText employs this information to produce an automatically generated contents screen, which authors and designers can arrange to appear a particular place on the screen. In *Hypertext in Hypertext*, the electronic version of this book's first version, this contents section appeared to the left of the main text.

This electronic table of contents differs in several ways from the static versions one encounters in the printed book and on the Web. First, clicking on an icon near the title of a chapter immediately causes the section headings next level down to appear, and clicking on them in turn displays subheadings, and so on. Essentially, this dynamic table of contents acts very much like Nelsonian stretchtext. Since I had added additional subdivisions to almost every section better to suit reading in an electronic environment, this feature permitted *Hypertext in Hypertext* to display both the book's original organization and the added elements as readers needed them. The second point at which DynaText's dynamic contents screen differs from static ones is that clicking on any section immediately brings up the relevant section in the text window immediately to the right of the section title (Figure 14). Finally, because the designers of this system combined this feature with its full-text search engine, the results of a search appear in the contents screen as well as in the text itself. Searching for "Derrida," one learns that this name appears seventy-four times in the entire book, forty-three times in the first chapter, and five times in the first section of that chapter. This dynamic listing proves particularly valuable when a DynaText web is configured as an electronic book, for then following a link from one point in the text to another causes the destination text to replace the departure one. The system works so quickly—nearly instantaneously—that without the contents listing at the side, readers become disoriented.

Tables of contents, whether static or dynamic, certainly have their uses, particularly when hypertextualizing material originally conceived for print presentation. Linked static tables are already common in HTML, but one can also create some of the effects of the DynaText form by using HTML frames, placing the contents at the left and text at the right.

Suppose You Could Have Everything?—The Intermedia Web View and Some Partial Analogues. The most important Intermedia feature that current systems, particularly web browsers, lack is its system-generated dynamic tracking map, whose basic idea evolved through three stages. The first, the Global Tracking Map, provided graphic information about all links and documents in a particular body of linked documents. Clicking twice on the icon for a particular hypertext corpus, such as *Context32, Nuclear Arms,* or *Biology,* simultaneously activated that hypertext web—that is, opened it—and generated a

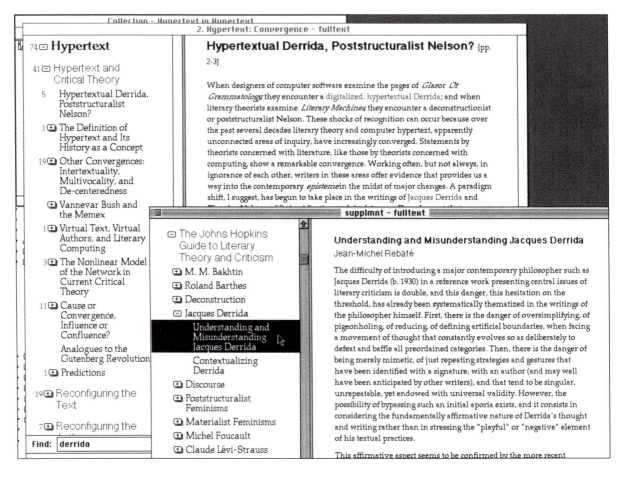

Figure 14. The Dynamic Table of Contents in Electronic Book Technologies' DynaText. This system, which combines the features of an electronic book with hypertext linking, automatically generates a reconfigurable, linked table of contents from the SGML codes used to mark elements of a text, such as chapter and section titles. In this example from *Hypertext in Hypertext,* mousing down on the plus signs to the left of items in the table of contents immediately displays titles of subsections. Clicking on the subsection title immediately brings up the relevant section in the right-hand panel. The table of contents also reinforced DynaText's full-text retrieval functions: in this case, after a reader has typed in "derrida" (the search tool is not case-sensitive), DynaText both highlights all occurrences of the word throughout the text (*top center*) and lists the number of occurrences next to each chapter and section heading (*left*). Having observed that "Derrida" appears five times in the book's opening section, the reader has moved that section into view; noticing that "Derrida" appears in red ink, the sign of a link, the reader has then clicked once on that link and opened a second DynaText "book" at Jean-Michel Rabaté's "Understanding and Misunderstanding Derrida" from *The Johns Hopkins Guide to Literary Theory and Criticism.*

document in which icons representing each document in the web were joined by lines representing all links between documents. This Global Tracking Map, which functioned only during early stages of Intermedia's development, immediately demonstrated that such a device was virtually useless for all but the smallest document sets or webs. (Although pictures of it have appeared in articles on hypertext, the Global Tracking Map was never used educationally and was never part of any released version of Intermedia.) It is worth noting this failed approach because, according to the computer science literature, it seems to reappear again and again as a solution to orientation problems for the Web.

IRIS next developed the Local Tracking Map, a dynamic hypergraph whose icons represented the destinations of all the links in the current document. Upon opening a new lexia or activating a previously opened one, this graphic navigational tool morphed, informing readers where links in the new lexia would bring them. This dynamic hypergraph, which did much to prevent disorientation, became even more useful in its third and final version, the Web View, with the addition of two more features: a graphic representation of the reader's history and transformation of the icons into links. Double clicking on any icon in the Web View opened the document it represented, thereby adding another way of making one's way through webs. (For illustrations, see Utting and Yankelovich.)

Although this feature succeeded well in orienting the reader, it worked even better when combined with author-generated concept maps, such as the overviews (sitemaps) I have employed on systems including Intermedia, Interleaf World View, Storyspace, Microcosm, MacWeb, and the World Wide Web. One basic form of these overviews surrounds a single concept (Victorianism, Darwinism, Gender Matters) or entity (Gaskell's *North and South*, Dickens) by a series of others (literary relations, cultural context, economic background), to each of which many documents link. Whereas the Web View presented all documents attached to the entire overview, the overview has a hierarchical organization but does not reveal the nature or number of documents linked to each block. Intermedia provided two ways of obtaining this information—a menu that following links from a particular link marker activated the Web View (see Figure 8). Clicking on a particular link and thus activating it darkened all the links attached to that block in the Web View. Thus, working together, individual documents and the Web View continually informed the reader what information was one jump away from the current text. This combination of materials generated by authors and system features well exemplifies the way hypertext authors employ what are essentially stylistic and rhetorical devices to supplement system design and work synergistically with it.

These features of this no-longer available system solved the basic problem of orienting readers. Unfortunately, current World Wide Web browsers are very disorienting because they provide no overall view of materials and neither do they indicate to readers where links will take them. The use of sitemaps, HTML documents that list or graphically display destinations of links, have greatly contributed to web usability, and many sites now include them (see Kahn and Krzysztof, *Mapping Websites*).

Various research and commercially available systems have had partial analogues to the Intermedia Web View. One research version of the University of Southampton Microcosm system, for example, had something very like the local tracking map, but it was not implemented in the released version, and Storyspace, a commercially available system, has its Roadmap, which has many of the Web View's functions (Figure 15). Like the Web View, the Roadmap records one's reading path, shows linked lexias, and permits one to open them; unlike the Web View, the Storyspace device also lists all links coming into the current lexia. Unfortunately, the Roadmap, which takes the form of a menu containing scrollable lists, lacks the Web View's dynamic quality, for it does not run continuously and has to be opened from a menu or by means of a key combination for each individual document.

Intermedia's dynamic hypergraph proved so valuable as a means of orientation and navigation that I hope someone will develop an equivalent application either as part of widely used World Wide Web viewers or as an add-on that will function with them. Certain steps have already been taken in that direction. The University of Heidelburg's Hyper-Tree, for example, offers graphic representations of the file structure of individual servers, but, unfortunately, like the first Intermedia attempt to graph links, it provides too much information, thus rendering it of little practical use. The Touchgraph Google Browser (Figure 16),which is more selective, draws on search results from Google to map what it takes to be the most popular connections between the lexia (or entire website) whose URL the user provides and other lexias within and without the site, producing results quite different than a sitemap or overview. The Touchgraph for *The Victorian Web's* "Religion in England" sitemap (Figure 16) has only a few of the dozens of the documents on religion, such as "High Church: Tractarianism" and "The Broad Church Party," but it omits perhaps the most important, heavily linked document on religion in *The Victorian Web*, the essay on the Church of England. Interestingly, it reveals a close connection between religion and British art and, rather unexpectedly,

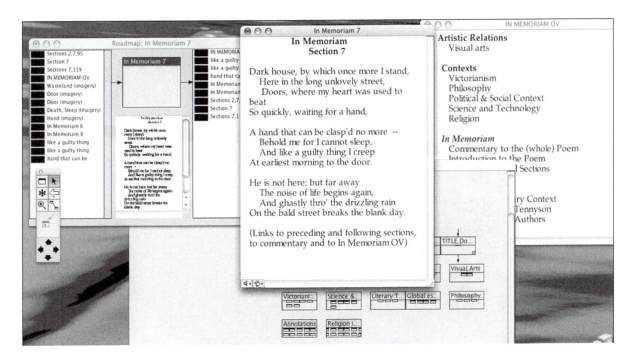

Figure 15. The Storyspace Roadmap Feature. The Roadmap (*upper left*) represents a static analogue to the Intermedia Web View. At the top center of the Roadmap appears one's reading history and, immediately below it, the first few lines of the currently active lexia. Like the Web View, the Roadmap informs readers of possible destination lexias and permits readers to open them directly, but unlike the Intermedia tool, which displayed only destination lexias, the Roadmap displays all links in and out of the current lexia. Unfortunately, whereas the Web View always remained in sight and automatically reconfigured itself as each new document opened, the Roadmap appears only on demand and has to open each document separately.

Lewis Carroll has a major presence. This diagram also contains off-site lexias, and whereas the *Artcyclopedia* essay on D. G. Rossetti might be expected, since the Pre-Raphaelites appear repeatedly, I'm intrigued by the outlying "Virtual Tour of Brasenose College." A similar Touchgraph for Charlotte Brontë, which is much denser, includes a substantial number of obvious external sites on the Brontës, Derbyshire, and other women writers, but I am mystified by the presence (occupying the diagram's entire upper-left quadrant) of a concentration of materials on the Harlem Renaissance, including the Red Hot Jazz Archive. The Touchgraph for *The Victorian Web*, which is too dense to reproduce effectively, makes many obvious connections to external sites and quite a few unexpected ones, too. The very fact that so many unexpected connections appear in these diagrams makes them quite fascinating,

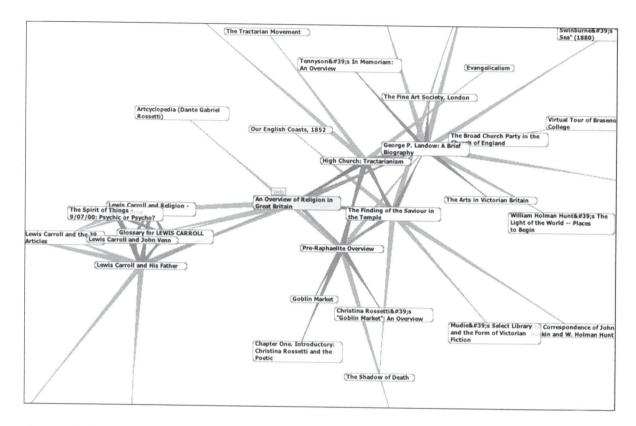

Figure 16. The Touchgraph Google Browser. This software draws upon search results from Google to map the most popular connections between an individual lexia or entire website whose URL the user provides and other lexias within and without the site. Its results, which differ markedly from an author-created sitemap or overview, often reveal unexpected relationships among websites.

and I can see a value in either including a few screenshots to show readers some interesting connections or linking to the Touchgraph site, so they can explore for themselves. The Touchgraph approach shows, however, the shortcomings for our purposes of mapping a website according to popularity (most visited and linked-to lexias): the resulting diagram omits one of the most valuable characteristics of hypermedia—its capacity to support individual, even idiosyncratic, approaches to information.

Author-Created Orientation Devices: Overviews. As the Web View and Roadmap show, readers need effectively organized preview functions—what Mark Bernstein terms "airlocks"—that show them what lies one jump away. In the next section I shall suggest stylistic, rhetorical techniques that hypertext authors can employ in the absence of such software tools. Some hypertext systems like Microcosm, Storyspace, and Intermedia provide several means of helping orient the reader; others provide little built-in assistance to solving basic problems of orientation. But whatever system authors employ, they should use overview and gateway documents, which are devices entirely under their own control. Overviews or sitemaps, which can take many forms, are author-created (as opposed to system-generated) documents that serve as directories to aid in navigating the materials. Overviews assist readers to gain convenient access to all the materials in many documents or to a broad topic that cuts across several disciplines.

Overviews or sitemaps take six forms, the first of which is a graphic concept map that suggests that various ideas relate to some central phenomenon or impinge on it. This center, the subject of the overview, can be an author (Tennyson, Darwin), chronological or period term (eighteenth century, Victorian), idea or movement (realism, feminism), or other concept (biblical typology, cyborg). The implied and often reinforced message of such arrangements is simply that any idea that the reader makes the center of his or her investigations exists situated within a field of other phenomena, which may or may not relate to it causally. Such graphic presentation of materials depicts one informing idea or hidden agenda of hypermedia materials, namely, that one proceeds in understanding any particular phenomenon by relating it to other contexts.

These kinds of overview lexias, which I have used since the first days of Intermedia, have particular value for the World Wide Web, which tends toward a flattened form of hypertext. Their emphasis on multiple approaches simultaneously provides a way of breaking out of the implied page format that confines the Web and also of creating what Paul Kahn has termed a "crossroads document," a point to which the reader can return repeatedly and before departing in new directions. The various websites I maintain use different kinds of overviews. *The Victorian Web* surrounds a central image by a range of related topics. In that for Elizabeth Gaskell's *North and South,* for example, a linked icon for political and social context appears at the top center, and immediately beneath comes those for biography, other works by the same author, Victorianism, and women's lives (Figure 17). The icons for literary relations and visual arts flank the image representing the novel. In the line below are five icons representing aspects of technique—setting, symbolism, characterization, narration, and genre; centered beneath them appears that for religion and philosophy.

Although one could use a single image map for such an overview on a website, using separate icons has some advantages, the first of which is that by using the "Alt" option in HTML that permits one to include a text label,

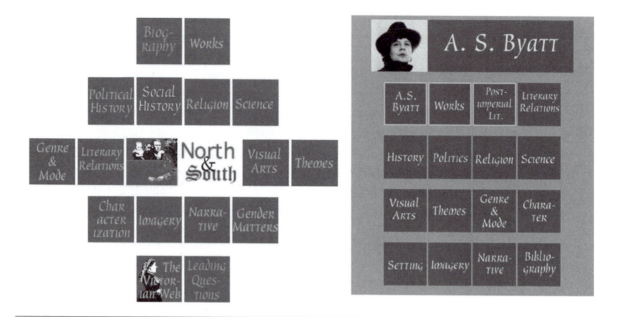

Figure 17. Two World Wide Web Overviews. These examples show two different approaches to creating overview lexias for the World Wide Web. That on the left, the overview for Elizabeth Gaskell's *North and South* in *The Victorian Web,* represents the latest version of the Intermedia-style overviews, which emphasize that readers can approach a subject from multiple points of view. The A. S. Byatt overview, in contrast, presents a similarly nonhierarchical approach to organizing information, by arranging its linked headings in a series of horizontal rows. This approach to creating overviews with HTML (text) documents has several advantages over image maps: (1) this text-based overview loads (opens) faster than server-based image maps; (2) since Internet Explorer, Safari, Netscape, and other web browsers retain images in a cache, building different overviews with the same elements creates documents that load very quickly; (3) such overviews are easily modified by adding or subtracting individual icon-and-link combinations; (4) these overviews have the advantage of employing the same files for both overviews and footer icons, thus reducing storage space and access time.

these kinds of overviews will work with old-fashioned browsers that do not have graphics capacities—an important consideration when portions of one's intended audience may not have the kind of computer access or equipment needed to handle large images. Moreover, one may create standard templates for all the overviews in a particular web, thus producing a kind of visual consistency, and yet one can easily modify appropriate elements. In the Gaskell overview, for instance, "Works" replaces "Other Works," and in those for other texts other icons appear, including those for themes, bibliography, related World Wide Web materials, and so on.

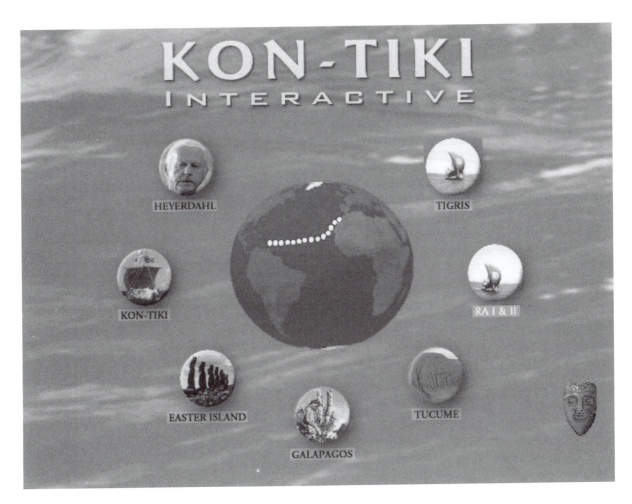

Figure 18. Gunnar Liestøl's *Kon-Tiki Interactive:* The Introductory Overview. This interactive overview surrounds an image of the globe with seven circular images, representing Thor Heyerdahl and six of his expeditions. These images serve as icons, previews, and conceptual overviews. Clicking on any one of them halts sound and animation and opens an interview for the subject it represents. (Used by permission of Gyldendal Publishers.)

Gunnar Liestøl's *Kon-Tiki Interactive,* which I shall discuss in more detail below, surrounds an image of the globe by seven circular images, each of which animates in turn, Thor Heyerdahl and six of his expeditions (Figure 18). Not all the overviews that wish to avoid hierarchy or linearity need to have this kind of circular format. Unlike the Liestøl and *Victorian Web* overviews, those for the hypertext section of *The Cyberspace Web* and all materials in the Postcolonial Literature Web do not emphasize centrality. Instead, taking the

A. S. Byatt Overview as an example, we find the topic of the document above three rows of five square icons each (see Figure 17). This arrangement, which also avoids the linearity of a table of contents, has the advantage of permitting one to employ some of the same icons both in overviews and at the foot of each screen.

Chronologies represent another form of sitemap or overview, which one can easily create using two-column tables in HTML. They offer a means of clearly organizing materials or even entire courses that have a strong chronological orientation. Any timeline with links in fact serves as an overview for the materials it joins. Although timelines provide a means of organization particularly convenient to authors, remember that they may simplify complex relationships and do little to compel the interest of a reader unacquainted with the subject.

Images of natural objects, like the photograph of a cell or maps, provide a kind of naturally occurring concept map that authors can easily apply. Attaching links to labels in technical diagrams similarly provides an obvious way of enriching conventional information technology. These kinds of overviews, incidentally, exemplify a perfect use for World Wide Web image maps. Perhaps my favorite is a map of Italy showing major Italian websites: click on the tiny square representing a particular city, and a link takes you to its website.

If hypermedia is characterized by connectivity, to realize its potential one must employ devices that emphasize that quality. Lists, tables of contents, and indices, though still of significant use, do not work in this manner, but one may wish to use them in addition to other kinds of graphic organizing devices, as does the elegant *Kon-Tiki Interactive* CD-ROM, which parallels its circular overview with an interactive outline. Mousing down on its individual elements, say, that for the Kon-Tiki itself, produces a list of eleven items (Figures 19–20).

When converting text documents originally created for book technology for presentation on hypermedia, one may occasionally use the document itself as its own overview. Any document in a hypermedia system with more than a few links in essence serves as a sitemap since, once opened, it provides the immediate center and reference point for the reader's next act of exploration. The author of educational materials, particularly those involving literary texts or those that place primary emphasis on the details of a text, may therefore wish to take advantage of this quality of hypermedia. Section 7 of *In Memoriam* (see Figure 9) exemplifies a brief text document that functions as its own overview or sitemap in a Storyspace web, and each section of the heavily annotated HTML version of "Hudson's Statue" by Carlyle func-

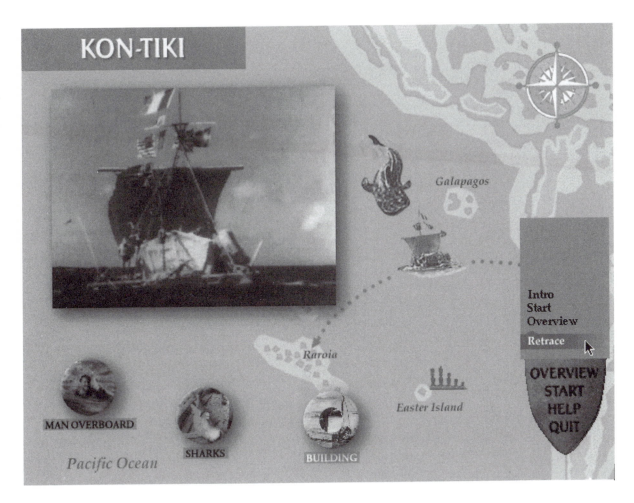

Figure 19. The Help Function in the Kon-Tiki CD-ROM. Moving the mouse near the ray stone shield halts all sound and video while simultaneously revealing a menu of choices. (Used by permission of Gyldendal Publishers.)

tions in the same manner. One must take care not to overdo this kind of heavily linked text document on the Web, which has few orienting devices, since linked text alone does not always provide very clear indications of where its links take the reader. In a scholarly or critical presentation of a text, such as the Pepys's *Diary* Weblog or "Hudson's Statue," in which the links clearly take one to annotations and commentary, these heavily linked lexias can function in this way, in large part because the nature of the document indicates the kind of links that will attach to it. In contrast, some heavily linked opening screens of personal sites on the Web, though occasionally amusing, often appear completely chaotic.

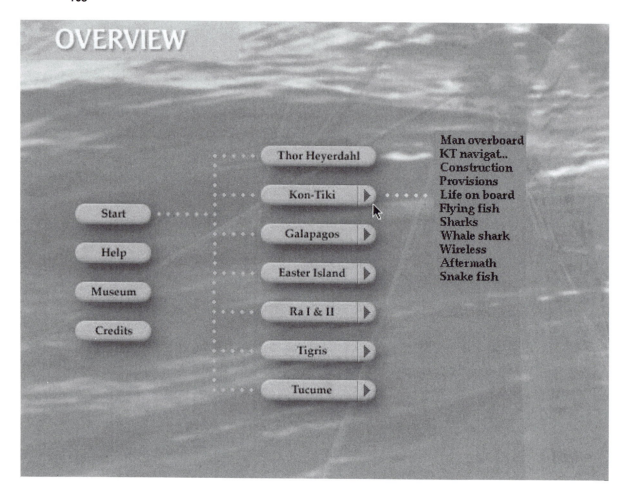

Figure 20. Kon-Tiki Overview. Selecting any item in this interactive overview produces a sublist of items. (Used by permission of Gyldendal Publishers.)

Whatever kind of overviews or local sitemaps one chooses, one should accommodate—and encourage—different styles of hypertext reading by providing as many as is convenient for each subject, and one should also expect that individual lexias, particularly in information hypertexts, will link to multiple overviews. Thus, an essay comparing women's issues in Graham Swift's *Waterland* and A. S. Byatt's *Possession* would link to the literary relations document for each work but also to those for themes, gender matters, and techniques as well.

Closely related to overviews and directories are those documents that serve as gateways between courses or bodies of materials in separate disciplines.

Such gateway lexias are particularly useful on the World Wide Web when a link brings the reader from the present website to another. The most common form of such gateways appears in the separate documents containing lists of links to other websites. Another example of such a transitional lexia is *The Victorian Web*'s brief introduction to the University Scholars Program (USP), National University of Singapore, which funded several postdoctoral and senior research fellows whose work appears on the site. As a means of identifying their work, icons representing the USP appear throughout *The Victorian Web*. Instead of linking these icons directly to the USP site, which would confuse readers—who might wonder, "Why am I reading about Singapore?"—I have linked them to a lexia that describes the USP and its role in supporting the site; links within that lexia then bring anyone who wants to know more about the USP to its homepage. Even within a small section of a single website, such as that formed by materials on a single author, concept, or event, such introductory transitional lexias prove useful. For example, when creating links from an icon or subject heading for one author's relation to other writers, one can either link to a local sitemap listing all relevant essays or one can link first to a general introduction; this latter approach works particularly well as a means of introducing complex relations not evident from a sitemap or of indicating a special concentration of materials in one area.

Gleamware. In addition to describing some effective software solutions to meet the needs of the hypertext author, permit me to propose something like a wish list. Computer users often refer to promised projects as so much vaporware, meaning that a product or research project that someone has presented as already existing is in fact little closer to reality than a plan or a promise. Let's go even farther back—from promise to desire. When I was much younger, I remember hearing the expression that mentioned a time when someone "was just a gleam in" their eyes of their parents. Let us consider gleamware or wishware.

Such an example of gleamware would be semiautomatically generated sitemaps and crossroads documents in HTML that would permit reader-authors on the Web to produce such intermediary documents by combining complex searches with elegant templates. At the moment of writing, no World Wide Web browser has the one-to-many linking that I believe so crucial to creating a fully multiple hypertext. Therefore, to translate materials originally created in systems that have such linking or to emulate them, authors find themselves forced to expend an enormous amount of time and effort manually creating—and maintaining—link menus. The implications of such difficulty will become clear when I report that the multiple links that required twenty to thirty minutes to create for an Intermedia or Storyspace overview—and less than half that time for Microcosm, using its more sophisticated generalized link-options—can take several days when translating such materials for the Web: one must go through the subset of documents that will link to the overview and manually create separate ones for literary relations, themes, biographical materials, and so on. Even if one already has an earlier version of one's materials in another software environment to remind one of possible links, it still takes hours of repetitive work—with the result that authors inevitably tend to avoid as much of it as possible and thereby produce a relatively flattened hypertext.

So here's my first two gleamware proposals, the first of which may already exist as a proprietary research tool in some large corporations in the computer industry. Imagine combining Macintosh OS 10.4's Spotlight feature or a commercially available search tool, such as On Location, with some C-programming, and a set of templates that would permit one to generate with minimum expenditure of time and effort a suboverview entitled, say, "Political Themes in Dickens" simply by calling up a menu and typing "Dickens," "themes," and "politics." An even better version—one that I have spoken about longingly since the last few years of the Intermedia project—involves automatically generated graphic representations of literary and other complex relations. In this example of gleamware, one would simply choose from a menu a literary relations (or similar) option, and one's authoring (reading?) system would combine a graphics engine, search tools, resulting indices, glossaries, chronologies, and synonym lists to produce automatically the kind of concept maps Paul Kahn created in the early *Dickens Web:* using synonym lists and chronologies, this Relations Map Generator—let's give it a properly stuffy name—places the chronologically earlier authors or texts toward one end of a chosen axis; earlier ones could appear, for example, at the left, at the top, or, in a three-dimensional representation, farther away. Authors or texts that I considered more important—either for reasons of some cultural standard (Shakespeare), relation to the author in question (Hallam to Tennyson, the Brownings to each other), or quantity of available commentary could be made to appear larger or in brighter colors. You get the idea. Let's take the gleam one step further: if one could produce such documents quickly enough—something that probably assumed preexistent indices—such documents could exist only dynamically, created each time one followed a link from an overview, and hence always current, always up-to-date.

Author-Created Orientation Devices: Marking the Edges. In the absence of such tools, what kinds of techniques can one use to assist readers? One device especially important to those creating materials in HTML involves using visual indications of a lexia's identity, location, and relation to others. These signals can take the form of header icons, color schemes, background textures, linked icons that appear at the foot of lexias, or all in combination. Such devices play a crucial role on the Web, where readers may arrive at any document via a search engine, entering at what could be the middle of a planned sequence or set of documents. Without some such device even readers who find that a particular arrival lexia meets their needs and taste become frustrated because they cannot conveniently determine whether it forms part of a larger structure.

One of the most commonly used such devices is the header icon, which immediately informs the reader that a lexia belongs to a particular web or subweb. For example, in *The Victorian Web* a blue-and-white header element appears immediately following the lines providing title and author. At the left of this icon, which is a third of an inch high and 7 inches wide, appears a black-and-white image of Queen Victoria followed by the words "The Victorian Web" and a white line extending the remaining length of the header. Using an editor, such as Dreamweaver or BBEdit, which permits easily making global changes—that is, changing all occurrences of a word or phrase in an entire set of documents rather than having to do them one at a time— makes inserting such elements extremely easy to do. Whereas *The Victorian Web* employs a single header icon, some of the other websites I maintain, such as that on recent postcolonial literature, uses a different one for each major division or subweb. This second web has separate sections for anglophone literature of Great Britain, the Indian subcontinent, Africa, and Australia and New Zealand, and therefore employs different headers as well as other devices for each section. Similarly, Adam Kenney's *Museum*, a Web version of anthology-fiction like *The Decameron*, organizes itself around a series of individual narrators and uses an image at the top center of each lexia to identify different narrative arcs.

Other devices include color schemes as well as background textures, and combined with footer icons they make an effective means of simultaneously orienting the reader while indicating the permeable borders of both the lexia and the larger units to which it belongs. For example, an essay from *The Victorian Web* that compares the railway swindlers in Trollope's *The Way We Live Now* and Carlyle's "Hudson's Statue" has five footer icons, one for the main

Victorian overview followed by one each for Trollope, his novel, Carlyle, and his text (Figure 21). The first three icons denote increasing specificity, thus indicating that the document contributes to the web as a whole, to those materials concerning Trollope, and to those about this particular novel. In contrast, the five icons, taken together, indicate that the lexia in question simultaneously participates in two subwebs or directories. These icons thus serve to orient readers by clearly stating how the lexia in which they find themselves relates to one or more large categories—in this example, five separate ones.

Furthermore, because links attach to each of these icons, clicking on them brings readers to a sitemap for these larger categories. Attaching links to the icons, in other words, makes them devices of navigation as well as orientation. In one sense these devices mark the edges of one or more groups or categories to which the lexia belongs or with which it associates, but the most important function involves not so much delimiting an edge or border of a document as indicating its relation to, or membership in, one or more subwebs. The effect of this congeries of devices, therefore, is to orient readers who find themselves in a particular lexia by clearly indicating its relation to others, its (intellectual) place within a web.

This combination of headers, color schemes, and linked footer icons works particularly well in large or particularly complex collections of interlinked lexias, such as those created by participants in courses or departments. The *Cyberspace and Critical Theory* web, for example, contains not only course materials and links to many websites external to Brown but also to a collection of elaborate individual student projects, some consisting of more than one hundred lexias and graphic elements. In this situation such identification and bordering schemes prove particularly useful because they inform readers that they have arrived at a discrete document set. Following a link from the print section of the information technology overview brings one to Amanda Griscom's *Trends of Anarchy and Hierarchy: Comparing the Cultural Repercussions of Print and Digital Media,* her World Wide Web translation of a substantial honors thesis comparing the seventeenth-century pamphlet wars in England and the periodical press that succeeded them with the situation on the Internet today. In contrast to the black background and white and yellow text of the Infotech overview, *Trends of Anarchy* confronts the reader with a pale yellow background, black text, and a light pastel header announcing the title and author of the entire piece. Like many student contributions to the cyberspace, this one contains a link to the web overview only on its contents page; the other lexias contain only footer icons to the contents, next page, and works cited. Since the reader can enter portions of this subweb from various

In contrast, Carlyle focuses on the English public's blind admiration for Hudson, the powerful railway king. He criticizes them for allowing a corrupt financier to be "[mounted] on the highest place you can discover in the most crowded thoroughfare." Carlyle argues the people's desire to erect a statue to Hudson is itself lamentable.

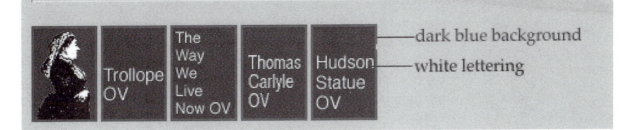

1. First three icons indicate increasing specificity: As one moves from left to right, one moves down directory structure.

2. In contrast, the five icons indicate that the present document simultaneously participates in two subwebs or directories.

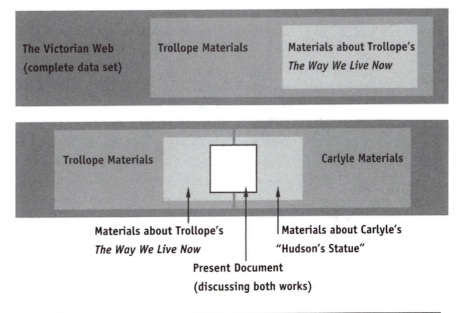

Figure 21. Footer Icons in World Wide Web Documents. This example from *The Victorian Web* shows how linked icons at the bottom of a lexia can indicate its simultaneous participation in several subwebs or document sets, thereby orienting the reader.

overview headings that indicate discussions of McLuhan, scribal culture, media in the seventeenth century, and so on, the reader needs to know the separateness—as well as the entire scope—of this subweb.

In concluding this section, I have to emphasize that on the World Wide Web the borders and limits of these hypertext documents, their edges, as it were, clearly have to be understood only as fictions, as agreed-on convention, since both links and search engines easily cross these proposed margins. The header graphic, for example, announces both the existence of a web (or major section of one) and implicitly proclaims its boundaries—only documents containing that header belong to this web—but on the Internet these claims are at best provisional, at worst an obvious fiction, since links to or from the lexias on any site make any assertion of boundaries into a gesture or wish or hope, particularly when, as in any large and complex web (site), the documents do not possess the kinship endowed by author function, for they have been produced by more than one author or entity. In sets of lexias created by a single author one can posit limits—that is, pretend they exist—more easily than one can for webs that both draw extensively on quoted passages and images created by others and also link to other sites as well. Nonetheless, we need such classifications in order to read. But the crossing of such textual (non)borders is one of the characteristics of hypertextuality, one completely analogous to the way links both permit one to employ a folder structure and yet not be confined by it.

This Text Is Hot. Readers of hypermedia need some indication of where they can find links and then, after they have found them, where those links lead; finally, after they have arrived at a new lexia, they need some justification why they have been led there. All these issues raise the question of the degree to which specific systems, authors, or both working together require an active, even aggressive reader. In examining a range of solutions to the first problem—how to indicate the presence of hot (or linked) text—I shall follow my usual procedure and begin by surveying the means thus far employed to do so and then suggest ways authors can write with and against their systems. As always, a major theme will be to suggest how World Wide Web authors can benefit from lessons provided by other forms of hypermedia.

In examining some ways existing hypertext systems signal the presence of linked text or images, I shall begin with least successful examples and proceed to better ones. Let us begin with Intermedia, though this time because its solution, though clear and unambiguous, proved too visually intrusive. All versions of this system employed a link marker in the form of a small horizontal rectangle within which appears an arrow. This icon appeared automatically above and to the left of any section of linked text and could not be moved; in graphics documents, however, the authors could place it wherever they wished. Like many elements in Intermedia, the link marker worked synergistically with other system features. Clicking once on it, for example, highlighted the icons representing destination lexias in the Web View, and by going to the main menu, readers could learn both anchor extent (the extent of linked text) and anchor description (the label the author attached to it). Unfortunately, placing the icon above linked text proved too intrusive, for it distorted the document's leading—the spacing between lines—a particularly annoying effect when one placed links in a print text, say, a poem by Tennyson.

World Wide Web browsers offer a slightly better solution to the problem of how to inform readers about the location of links. As is well known, browsers conventionally indicate the presence of links with color and underscored text: Mosaic and Netscape established the convention that unlinked text appears in black against a light gray background, blue underlined text indicates the presence of a link, and magenta underlined text indicates a link that one has previously followed. I am of course describing the default version—that, in other words, which one receives if neither reader nor author customizes these elements; authors can choose entirely different color schemes or choose to make regular and linked text the same color, thus employing only underlining to indicate link presence.

Although the conventional HTML approach seems less visually intrusive than Intermedia's annoying link marker, its manner of signaling link presence, like that of the earlier system, produces a visual hodgepodge in alphanumeric text. The simple fact is that the links and the colors are always present, and for certain purposes their presence becomes too annoying. Although annoying in written text, such permanently displayed link markings sometimes work well with graphic elements, since a colored outline doesn't always intrude on an icon the way a combination of color and underlining do on text. Many intensely graphic sites, however, employ HTML options that permit authors to turn off such color variation, providing no visual cues to the location of links at all, because they know that in current browsers, when a mouse moves over a link, it turns into the image of a hand. As those who use the Web have become more sophisticated, designers increasingly assume that when users encounter a screen without any obvious links, they will move their cursors over images and other screen elements that they have learned are likely to serve as link anchors. In contrast, earlier websites often used linked icons and, unsure whether readers would know what to do, they added linked text, usually in very small type, immediately below them. An increasingly sophisticated audience, in other words, no longer finds the absence of obvious clues disorienting.

In contrast to basic HTML and Intermedia, Microcosm, like several other hypertext systems, does not permanently display indications of links. As we have already observed when examining Microcosm's rich assortment of link types, this system invites active, even aggressive readers who interrogate the text they encounter (see "Forms of Linking," chapter 1). This approach, which removes any possibility of visually marring the appearance of literary texts, appears in the way it provides information about the presence and extent of linked material. In keeping with Microcosm's encouragement of the active reader, users who come upon a word or phrase that they believe likely to serve as an a link anchor have several choices: they can double click on it, perform the equivalent action by choosing "Follow Link" from a menu, ask about link extent from a menu, or create their own links with the "Compute Links" function. For many applications, particularly educational and informational ones, the Microcosm approach strikes me as wonderfully appropriate to the medium. My only suggestion would be to follow Hypercard and Storyspace and make a simple key combination show both the presence and extent of author-created links. Storyspace's use of frames that surround hot text when readers hold down Option and Apple (or Command) keys simultaneously proves a particularly valuable feature and one that I would like to see both Microcosm and HTML viewers emulate. Changing the cursor from an arrow to a hand to indicate that readers have encountered hot areas in a document has many advantages, yet it still requires in many cases that readers grope blindly around the screen, and in large text documents omitting any signs of existing links creates confusion; transient indications of links that readers could activate by simple operations, such as a key combination, would be useful as an additional feature.

Airlocks, Preview Functions, and the Rhetoric of Departure. All these system-based devices that we have just observed constitute the first, and simplest, part of any rhetoric of departure, for they inform readers that they can leave a text stream for somewhere or something else. Not surprisingly, most readers do not feel comfortable jumping off into limbo. Although much hypertext fiction and poetry plays with surprise and disorientation as desired aesthetic effects, other kinds of hypertext writing require some way of giving readers an idea of what links will do.

Such preview functions—what Mark Bernstein has called an "airlock"—serve to both inform and reassure readers, and when systems do not provide adequate information, authors must find their own ways to obtain it. The just-discussed HTML convention of changing the color of anchors indicating links that have already been followed exemplifies one kind of valuable system support. When one mouses over hot text (an anchor), most web browsers also show the destination URL, though unfortunately not the title as well, in a panel at the bottom of the viewer window—though this feature does not seem to work with documents using frames.[3]

The point is that readers need a general idea of what to expect before they launch themselves into e-space. Help them, therefore, by making text serve as its own preview: phrase statements or pose questions that provide obvious occasions for following links. For example, when an essay on Graham Swift or Salman Rushdie adds links to phrases like "World War I" or "self-reflexive narrators," readers who follow them should encounter material on these subjects.

In addition, whenever possible provide specific information about a link destination by directly drawing attention to it, such as one does by creating text- or icon-based footer links. Another precise use of text to specify a link destination takes the form of specific directions. For example, in *The Victorian Web* to which student-authors contributed differing interpretations of the same topic, say, labor unrest in *North and South* or gender issues in *Great Expectations,* functioning as an editor, I have added notifications of that fact. Thus, at the close of essays quoting and summarizing different contemporary opinions about strikes and labor unrest, I have added "Follow for another contemporary view," a device that should be used sparingly, and lists of related materials, which are particularly useful when indicating bibliographical information and documents on the same subject.

Such careful linking becomes especially important in writing hypertext for the World Wide Web, since current browsers lack one-to-many-linking (see Figure 3), and it does not seem likely after more than a decade that they will ever incorporate it. This apparently minor lack has devastating consequences for authors, who have to create manually the link menus that other systems generate automatically. Without one-to-many links, readers and writers lose the crucial preview function they provide. I find that the effect of being reminded of branching possibilities produces a different way of thinking about text and reading than does encountering a series of one-to-one links sprinkled through a text.

The Rhetoric of Arrival. Many non-Web hypertext systems use various means to highlight the reader's point of arrival, thus permitting links into portions of longer lexias. Intermedia, for example, surrounded the destination anchor with a marquee or moving dotted line that traversed a rectangular path around the intended point of arrival; a single mouse click turned off the marquee. Storyspace, in contrast, employs a rectangular block of reverse video around arrival anchors. Unfortunately, thus far, although World Wide Web authors can use the <A NAME> anchor feature to bring the reader to a particular portion of a document, no browser shows the exact extent of the arrival anchor. Instead, HTML just opens the arrival lexia at the line in which the anchor appears, something extremely useful for bibliographical citations and other lists.

The difficulty in the World Wide Web is exacerbated by the fact one often links to documents over which one has no control, and hence cannot insert an anchor. One way of accommodating those who link from outside involves using the identifying color schemes and headers described earlier. If one can obtain permission from the document's author or owner, one could place an anchor there. Similarly, if one can obtain permission to do so, one could copy and incorporate the arrival lexia within one's own web. Although such an approach, which I have used in *The Victorian Web*, occasionally proves useful, it strikes me as basically inefficient and contrary to the spirit of the World Wide Web's dispersed textuality.

Converting Print Texts to Hypertext. Before considering the best ways to hypertextualize printed matter, we might wish to ask why one would want to bother. After all, a somewhat sympathetic devil's advocate might begin, it's one thing to expend time and effort developing new modes of reading and writing, but why modify the book, which is in so many ways a perfectly good text-delivery machine? For many nonliterary uses, the answer seems obvious, since linked digital text permits an adaptability, speed of dissemination, and economy of scale simply not possible with print. Maintenance manuals for large, complex machines, like airplanes, parts catalogues, and many other uses of the codex form of text presentation seem better served in electronic form. For these reasons in some scholarly or scientific fields, such as high energy physics, the digital word has increasingly replaced the physical, and the most important form of publication takes place online. This movement away from printed text has certainly happened in the workplace, where people who formerly used printed schedules, parts catalogues, tax and real estate information, and delivery forms now read on screen; even courier and package delivery services now have customers sign electronic pads, as do the check-out girls at many supermarkets. Those of us who work with books every day and who enjoy doing so may not have noticed that for a growing number of people, printed matter, not just books, plays an ever smaller role in their work day. The implications of this change seem obvious: in coming years, the printed book will eventually become, for many people, an increasingly exotic object in much the same way that beautiful illuminated manuscripts eventually appeared exotic to many raised on reading books. Perhaps, since as Bolter and Grushin argue, no information medium ever dies out completely, the codex book will be reserved for classic novels or recreated with on-demand publication; in an increasingly digital world, whatever happens, future readers will not experience books as we do.

But why in literature and in humanities education, our devil's advocate might continue, would we want to take works originally conceived for print and translate them into hypertext? Particularly given the comparably primitive state of on screen typography, why take a high-resolution object like a book and transform it into blurrier words on flickering screens? Now that the World Wide Web promises to make all of us self-publishers, these questions become particularly important. I would answer: we translate print into digital text and then hypertextualize it for several obvious reasons: for accessibility, for convenience, and for intellectual, experiential, or aesthetic enrichment impractical or impossible with print.

When I began to work with hypermedia a decade ago, the combination of a desire to create materials best suited for reading in an electronic environment and the need to avoid possible copyright infringement led my team to create all materials from scratch, but soon enough teaching needs drove us to include hypertext translations of print works. These needs will be familiar to anyone teaching today: works around which I had planned portions of a course suddenly went out of print. Placing otherwise unavailable documents within a hypertext environment allowed us to create an economical, convenient electronic version of a reserve reading room, one that never closes and one in which all materials always remain available to all readers who need them.

Now that the World Wide Web can link together lexias whose source code resides on different continents—the texts of many classic Victorian novels, for example, reside on a server in Japan—such accessibility provides an even stronger impetus to hypertextualize otherwise unavailable materials. This ease of accessibility from a great distance means that more readers can use one's text—and, in return, that one can hope to find texts translated by others for one's own use.

Books and articles on the Web take two very different forms, the first of which is the Web version that closely follows the print paradigm and uses the Internet solely as a means of making texts available—hardly an unworthy goal. In contrast, the second approach attempts with varying degrees of success to translate works created for print into hypertext, thereby exploring the possible modes of scholarly publication in a digital world. By far the most common approach thus far is the Web version that preserves some formatting from the original print version but ignores many characteristic advantages of the new medium. The many PDF versions of scholarly and scientific articles downloadable from the Web exemplify this approach, as does Project Gutenberg, which tries to provide as many digitized versions of printed texts as is possible. Project Gutenberg embraces both primary and secondary texts, but its mission doesn't allow distinguishing between them. More scholarly textbases, like the Women Writers Project, necessarily have as their chief goal the preservation of as much information as is practicable about the physical form of often rare and usually inaccessible printed books; text encoding, rather than hypertextualization, understandably has the highest priority. Other projects, like Mitsuharu Matsuoka's *Victorian Literary Studies Archive* at Nagoya University take texts by 150 British and American authors of the nineteenth century—it includes, for example, two dozen books by Dickens and another half dozen about him—and join them to Masahiro Komatsu's Hyper-Concordance; although the *Archive* does not create hypertext translations of these works, it takes advantage of their digitization to create on-demand corpus-wide searches.

As valuable as these print preservation projects are, they do not help us answer the question, What will happen, and what has already happened, to the scholarly or critical book on the Web? Phil Gyford's translation of Samuel Pepys's *Diary* into a Weblog, at which we looked in chapter 3, exemplifies a new scholarly genre that took form outside the academy. Sites like *Slashdot* and many smaller ones on technical subjects, such as software for Weblogs or digital photography, are understandable, given the nature of early adopters: *Slashdot's* motto is "News for Nerds. Stuff that Matters." One does not expect scholarship in the humanities, with its long-established hostility to collaborative publication, to come online in such a radically collaborative form, and, yes, it turns out it that did not take place with any support from academic institutions, most of the members of which, I'm sure, do not know it exists. In contrast, new media studies (not surprisingly) has embraced the blog as a scholarly genre, attaching them to interviews and book reviews published online.

These Weblogs produce a kind of scholarship and criticism, perfectly valid, which centers on short forms—the essay and review. What about the scholarly book or monograph? In an attempt to answer this question, which has been nagging at me for a decade, I decided to carry out an experiment, testing my proposition that hypermedia provides literary scholars with the equivalent of a scientific laboratory. Therefore, as some of my own books on Victorian art, literature, and religion have gone out of print, I retrieved copyright from my publishers, translated them into HTML, and placed them on *The Victorian Web;* since the appearance of *Hypertext 2.0* more than a dozen scholars have contributed one or more of their works to this enterprise. As I did so, I began to take advantage of characteristics of hypertext that justify translating a book into a web, the most basic of which is the synergy that derives from linking materials together. Reasons of economy and scale had prevented including illustrations of the many paintings mentioned in the original print version of my edition of the letters of John Ruskin and the Victorian artist, W. Holman Hunt, but once I had created web versions of this correspondence, my book on that artist, and articles from *Art Bulletin* and other journals, I found that these texts all worked together better than they had alone. A footnote providing the source of a letter could now, for example, lead to the text of the letter itself. Even more important, if an illustration was available anywhere in the text, it became available everywhere. Texts needing illustrations particularly benefit from electronic presentation, since a digital image, which is a matter of codes rather than physical marks on physical surfaces, multiplies while taking up no additional space. Using an image fifty times within one text or set of texts requires no more storage space or other resources than does using it once. Digitization thus permits the reuse of the image at several fixed places in a text; hypertextualization permits the image to be called up from numerous points as the reader finds its presence of use. In World Wide Web viewers, which temporarily store images downloaded from a network in a cache, reusing the same image takes much less time than it did obtaining it in the first place.

A final reason for translating a book into an electronic environment involves adding capacities not possible in print. Hypermedia translations of print texts in mathematics, sciences, music, history, and the arts have already appeared, employing sound, animation, video, and simulation environments. Let us look at some instances of these when proposing some general rules for employing sound and motion within alphanumeric text.

Assuming that you have a text that demands hypertextualization, how should you go about carrying it out? Since I have thus far converted several dozen books into various hypermedia systems, I believe the best way to answer that question would be to summarize my experience and use it to draw some general guidelines. Furthermore, since some of these electronified books exist in two or more different hypertext environments, we can observe the degree to which minor differences in hardware and software influence hypertextualization.

First, one has to obtain an accurate digital version of the text to be converted. For my earlier books and articles, written before I began to work with computing, I used OminPage Professional, software for scanning text and then interpreting the resultant image into alphabetic characters—an often time-consuming process. Since I had written the first version of this book in Microsoft Word, working with it proved to be fairly easy. Converting the text for Intermedia required only saving it in a particular format (RTF) and then creating links within Intermedia. Working in Storyspace proved even easier because this system imports Word documents, automatically translating footnotes into linked lexias; the one chapter translated in HTML used the Storyspace export function to create a basic linked working text to which I then added header and footer icons. The DynaText version required adding SGML tags and manually adding coding for links.

In adapting the printed text for all four kinds of hypertext systems, I found I had to make decisions about the appropriate length of lexias. In each case, I took chapters already divided into sections and created additional subdivisions. Whereas print technology emphasizes the capacity of language to form a linear stream of text that moves unrelentingly forward, hypermedia encourages branching and creating multiple routes to the same point. Hypertextualizing a document therefore involves producing a text composed of individual segments joined to others in multiple ways and by multiple routes. Hypermedia encourages conceiving documents in terms of separate brief reading units. Whereas organizing one's data and interpretation for presentation in a print medium necessarily leads to a linear arrangement, hypermedia, which permits linear linking, nonetheless encourages parallel, rather than linear, arguments. Such structures necessarily require a more active reader. Since a major source of all these characteristics of hypermedia derives from these linked reading units, one has to create hypermedia with that fact in mind. Therefore, when creating webs, conceive the text units as brief passages in order to take maximum advantage of the linking capacities of hypermedia.

Whatever its ultimate effect on scholarly and creative writing, hypermedia today frequently contains so-called legacy text—texts, like the 1992 version of *Hypertext*, originally created for delivery to the reader in the form of a printed book. Such materials combine the two technologies of writing by attaching linked documents, which may contain images, to a fixed steam of text. Anyone preparing such materials confronts the problem of how best to preserve the integrity of the older text, which may be a literary, philosophical, or other work whose overall structure plays an important role in its effect. The basic question that someone presenting text created for print technology in hypermedia must answer is, Can one divide the original into reading units shorter than those in which it appeared in a book, or does such presentation violate its integrity? Some literary works, such as sonnet sequences or Pascal's *Pensées*, seem easily adapted to hypermedia since they originally have the form of brief sections, but other works do not seem adaptable without doing violence to the original. Therefore, when adapting documents created for book technology, do not violate the original organization, though one should take advantage of the presence of discrete subsections and other elements that tend to benefit from hypertextualization. However, when the text naturally divides into sections, these provide the basis of text blocks. The hypermedia version must contain linkages between previous and following sections to retain a sense of the original organization.

Converting Footnotes and Endnotes. The treatment of notes in the four hypertext versions of the first version of this book provides an object lesson about the complexities of working in a new kind of writing environment. It reminds us, in particular, how specific writing strategies depend on a combination of equipment and often apparently trivial features of individual systems, some of which militate against what seem to be intrinsic qualities of hypertext. For example, as we have already observed in "Reconfiguring Text," hypertextualization tends to destroy the rigid opposition between main and subsidiary text, thereby potentially either removing notes as a form of text or else demanding that we create multiple forms of them. Certainly, in hypertextualizing some of my own works, longer footnotes and endnotes containing substantive discussions became lexias in their own right.

Briefer notes that contain bibliographical citations embody a more complex problem that has several different solutions, each of which creates a different kind of hypertext. If one wishes to produce a hypertext version of a printed original that remains as close as possible to it, then simply converting endnotes into a single list of them makes sense; using the <A Name> tag, each link will lead to the appropriate item. Another approach, which produces an axially structured hypertext, involves placing each bibliographical note in its own lexia and linking to it. Using simple HTML, authors can make this lexia open in a number of ways. Following standard HTML procedure, the note replaces the text in which the link to the note appears; one can create a return link to the main text, or rely on knowledgeable readers to use the browser's "Back" button. In addition, one can leave the body of the document on screen by using the "target = '_blank'" option in the link (A HREF), which makes the note open in a new window; authors who take this route, opening annotations in a separate window, often include instructions to close the window to return. Unfortunately, both of these convenient approaches produce an unattractive document in which a sentence or two appears at the very top of a large, otherwise empty window. A third approach uses HTML frames to bring up the text of the note in a column next to the main text, and yet a fourth uses HTML tables to place the text in the margin, thereby recreating the effect of some eighteenth-century books; tables are, however, very difficult to use when one of the columns has large blank areas. Perhaps the most elegant solution employs Java scripts to create small pop-up windows for each note. The problem here involves the nature of one's audience: Java scripts notoriously do not work in all browsers or even in all versions of the same one. If you create your materials for the widest possible audience, which includes many users with old versions of Netscape and Internet Explorer, you will have to forgo using this elegant, if time-consuming, solution, but if you direct your materials at a single educational or commercial institution that has announced standards for supported hardware and software, you can use anything that works there, though you may lose people working at home.

Whatever option you choose (and all have advantages and disadvantages), try not to use superscript numbers to indicate the presence of links. Not only does it prove difficult to mouse down on the small target provided by a single tiny character, but, more important, numbered notes only make sense when readers consult them in a list, and placing notes in separate lexias destroys this list. I find that one can almost always use a relevant phrase as an anchor for a note to a link. When the note contains bibliographical information, one can link to a relevant phrase, such as "(source)" or "(bibliographical materials)," placed at the end of the sentence.

Wherever possible, the best and most obvious solution to the problem of representing annotations in Web documents involves converting all bibliographical notes to the current Modern Language Association (MLA) in-text citation form, whether one links all such citations to a list of references or just includes the relevant bibliographical items in each lexia; I prefer the latter approach.

In addition to dividing a print text into sections, adapting notes and bibliography, and adding header and footer icons, creating a hypermedia version requires adding features and materials that would be impracticable or impossible to have in a printed version. Thus, the Storyspace, Microcosm, and World Wide Web versions of both my Pre-Raphaelite materials and *Hypertext* contain a great many links that serve as cross-references and that provide additional paths through the text, and they have additional images, too. These webs also have elements not found in books, such as multiple overviews that permit traversing them in ways difficult, or impossible, in a print version. All the hypermedia translations of *Hypertext,* for example, contain overviews for both critical theory and hypertext, and various versions add ones for information technology, scribal culture, and individual theorists.

Perhaps the most obvious difference—in addition to links—between the hypertext and print versions lies in its size: the way links produce an open-ended, changing, multiply authored Velcro-text appears nowhere more clearly than in the fact that so much new material appears in *Hypertext in Hypertext* than in the print version. As one might expect from what I've already written, once I created Intermedia and Storyspace versions for my course on hypertext and literary theory, my students read them as wreaders—as active, even aggressive readers who can and did add links, comments, and their own subwebs to the larger web into which the print has version has transformed itself. Within a few years the classroom version contained five hundred of their interventions, criticizing, expanding, and commenting on the text, often in ways that take it in very different directions than I had intended. In addition to some fifty of these new lexias, *Hypertext in Hypertext* contains entries on individual theories and theorists from *The Johns Hopkins Guide to Literary Theory and Criticism,* edited by Michael Groden and Martin Kreiswirth, as well as materials by Gregory Ulmer and Jacques Derrida. To these materials, we added, with permission, some of Malcolm Bradbury's parodies of critical theory and all the reviews the book had received by the time we went into production. These new lexias, which constitute a subweb of their own, serve to insert other voices, not always in agreement with mine, into the expanded text. Throughout, the principle of selection was be a cardinal rule of hypertext adaptation—use materials only when they serve a purpose and not just because you have them. Hypertext writing, in other words, should be driven by needs and not by technology.

Rules for Dynamic Data in Hypermedia. The preceding pages have focused on writing hypertext with essentially static forms of data—words, images, diagrams, and their combinations. Kinetic or dynamic information, which includes animation and sound (animal cries, human speech, or music), raises additional problems because it imposes a linear experience on the reader. This strong element of linearity itself presents no fundamental difficulty, or even novelty, since as Nelson pointed out long ago, individual blocks of text, particularly those with few links, offer a linear reading path.

The difference between dynamic and static data lies in the fact that it is so importantly time-bound. Speech or visual movement potentially immerses readers in a linear process or progression over which they have relatively less control than they do when encountering a static document, such as a passage of writing. One can stop and start one's reading at any point—when the phone rings, a child cries, or a thought strikes one. Turning one's attention away from time-bound, linear media, in contrast, throws one out of one's position within a linear stream, and this place cannot be recovered, as it can with writing, simply by turning attention back to the text, since one's point, or location, or place within the text has moved on. Therefore, when one follows a link from a text discussing, say, mitosis to digitized animation of a cell dividing or from a work of criticism to a scene from Shakespeare, one cues the beginning of a process. Such dynamic data place the reader in a relatively passive role and turns hypermedia into a broadcast, rather than an interactive, medium.

Many websites and CD-ROMs employ Quicktime movies accompanied by sound to present materials often more efficiently and more enjoyably encountered as text. Using the talking heads approach in which someone filmed from the chest or neck upward looks out of the screen and talks to us can occasionally prove an effective strategy, particularly when establishing a person's appearance and character seems important, but it expends two kinds of valuable resources: storage capacity and, because it occupies time, the reader's patience, or at least forbearance. Too many creators of HTML and CD-ROM materials seem grossly unaware of the fact that one of the key advantages of written language lies precisely in abstractness and indirection that permit communicating important information with great economy.

Systems designers and hypermedia authors have to empower readers in at least two ways. First, they must permit readers to stop the process and exit the environment easily. Second, they must indicate that particular links lead to dynamic data. One may use labels or icons for this purpose, and one may also connect the link-indicator to an additional document, such as a menu or command box, that gives users precisely the kind of process-information they can activate. Such documents in the form of a control panel, which permits the reader to manipulate the process to the extent of replaying all or part of a sequence, make the reader more active. If one employs talking heads or voiceovers, one must, of course, allow readers to stop them in midsequence. (Voyager's *Freak Show* CD-ROM makes a witty play on these features by having its Master of Ceremonies or Ringmaster respond with different expressions of annoyance each time we cut him off in midsentence.)

The one digital form that does not create problems for hypermedia is the fundamentally controllable multimedia document created by Apple's Quicktime VR (Virtual Reality) or rival software like Ipix and Live Picture. These kinds of software, whose creations are easily inserted into HTML, produce two different kinds of manipulatable three-dimensional images, the so-called spherical and cylinder panoramas. In the first, one finds oneself placed in a three-dimensional space within which one can rotate 360 degrees by using the mouse to stop, start, and control the speed of rotation or, using a zoom function, move closer or farther away. *The World Book Encyclopedia's* two CD-ROMs, for example, include dozens of Ipix scenes in the format they call "bubble views," including St. Mark's Square, Venice; Stonehenge; the Maya ruins at Palenque; the Coliseum in Rome; and the Zojoi Temple in Japan. Or if one wants a Web example, go to NASA's Mars Pathfinder site for a panorama of the Sagan Memorial Station, or *A Wrinkle in Time: a collaborative synchronized effort by QTVR producers around the globe* to create a hundred panoramic views at the same time on December 21, 1997 (see bibliography).

The other kind of Quicktime VR (the cylinder panorama) takes the form of a virtual object that users can turn 360 degrees, examining it as they wish, as well as zooming in and out. This kind of image, possible only with computing, has great value when representing three-dimensional objects online. *The Victorian Web*, for example, contains a Quicktime VR image of an unattributed bronze statue of a young woman that I believe was created by Alfred Drury (1856–1944). While carrying out research that might lead to an attribution, I visited various English photo-archives and collections of sculpture without finding any particularly convincing evidence. Several years later, after I had created a rotatable Quicktime VR image of the statue, I came upon a photograph of Drury's 1896 bronze, *Griselda*, in a 1980 catalogue from Christopher Wood Gallery in London. Observing the way in which the sculpture depicted the pleats and folds around the shoulder of the *Griselda*, I opened the Quicktime VR image of the unattributed statue, using my mouse to rotate it until I positioned the statue as seen from the same vantage point as the photograph of *Griselda*. At that point, it became obvious that both sculptures had been executed by the same person. Obviously, anyone interested in Victorian and Edwardian sculpture would prefer to have both works in the same physical space and be able to move them about until one could look at them from the same vantage point; anyone who has ever studied sculpture of this period immediately realizes how unlikely would be the opportunity to make such a comparison. Instead, the researcher usually has to use published photographs, each of which of necessity is taken from one position, and even if one visits major museums, one often finds that sculptures are displayed in such a way (often in a corner or against a wall) that one cannot obtain the view one wants. With Quicktime VR, one can. Anyone interested in sculpture clearly cares about the materiality, the sheer mass and surface of the object in question, and so one really wants access to the original objects with the ability to move and touch them and gaze at them for a long time under different kinds of light conditions. If one cannot have the actual object, then Quicktime VR is the next best thing; it is certainly superior to my 35 mm slides, which make *The Ghent Altarpiece* and a tiny woodengraving appear to be on the same scale.

Hypertext as Collage Writing

Most current examples of hypertext take the form of texts originally produced by the hypertext author in and for another medium, generally that of print. In contrast, this section on collage writing derives from a hypertext, though it incorporates materials ultimately derived from printed books, too. On Tuesday, June 7, 1994, at 17:01:54 Eastern Standard Time, Pierre Joris, a faculty member at the State University of New York, posted some materials about collage on a electronic discussion group called Technoculture. (I have discussed the first year of Technoculture's existence in "Electronic Conferences and Samiszdat Textuality: The Example of Technoculture," in the 1993 MIT volume, *The Digital Word,* which I edited with Paul Delany.) Joris wished to share with readers of this e-conference a gathering of texts on the subject he had delivered as a combination of an academic paper and performance art while in graduate school. His materials seemed to cry out for a hypertext presentation, and so after moving them from my mailbox to a file on the Brown University IBM mainframe, I transferred them—in the jargon, "downloaded them"—in a single document via a phone line to a Macintosh whirring away in my study at home. Next, I opened them in Microsoft Word, and, passage by passage, quickly copied the individual elements of "Collage between Writing and Painting," pasting each into a separate writing space or lexia in a new Storyspace web and then linked them together. Along the way, I created the following opening screen (or analogue to a book's title page):

COLLAGE BETWEEN WRITING AND PAINTING

Pierre Joris

George P. Landow

being an assemblage starring

Kurt Schwitters & Tristan Tzara
with special guest appearances by
Georges Braque &
Pablo Picasso

and also featuring
dedicated to . . .

This opening screen, which also serves as a combination overview, information map, contents page, and index, contains links from the obvious places—such as, for example, all the proper names it lists. Clicking on "Collage" takes one either to one possible terminal point of the web or to a definition of the term from *Le Petit Robert*. Since this dictionary definition, which mentions Picasso and Braque, serves as a another ready-made overview or crossroads document, I linked various words in it to permit readers to traverse Joris's materials in multiple ways. "COLLAGE," for example, leads to a dozen and a half mentions of the term, and the names of the artists link to illustrations of their work. Because I created this web largely as an experiment and not for publication, I did not have to worry at the moment about copyright issues and therefore scanned monochrome images of Braque's *Le Courrier* and Picasso's *Still Life with Chair Caning* and linked them to the names of the artists. At the same time I added H. W. Janson's discussions of collage, linking them as well. Finally, I created a list of thirty authors whose statements Joris included in "Collage between Writing and Painting," linking this list to the phrase "also featuring" on the title screen.

At this point, some of the similarities between hypertext and collage will have become obvious. Having first appropriated Joris's materials by placing them in a web and then adding materials to it that they seemed to demand, I found that, like all hypertexts, it had become open-ended, a kind of Velcrotext to which various kinds of materials began attaching themselves. First, I included a discussion of Derrida and appropriation, after which I added definitions of hypertext and a list of qualities that it shares with collage. Next, I added several dozen screenshots, or pictures of how the screen appears while reading, of various hypertext webs; these came from a since-published web that served as an introduction to the hypertext anthology, *Writing at the Edge*. Then, I added a dozen photographs, each involving issues of representation, illusion, simulation, or subject and ground. Finally, I added a new title page for *Hypertext and Collage: being in part, an appropriation of "collage between writing and painting."*

After using this web to deliver my contribution to the August 1995 Digital Dialectic conference at the Art Center College of Design, I discovered I would have to transform it into a more or less traditional essay if it were to be part of the planned volume. These pages thus represent a translation of the *Hypertext and Collage Web*. When I write "translation," I cannot help thinking of the Italian maxim *"traddutore = traditore"* or "translator = traitor." Converting the essay from one information technology to another, I continually encountered the kind of reduction that one encounters translating—or representing—something in three (or more) dimensions within a two-dimensional medium. An examination of the differences between the two versions will take us a way into understanding the reasons for describing hypertext as collage writing.

The online version of the *Oxford English Dictionary* defines collage, which it traces to the French words for pasting and gluing, as an "abstract form of art in which photographs, pieces of paper, newspaper cuttings, string, etc., are placed in juxtaposition and glued to the pictorial surface; such a work of art." The *Britannica Online* more amply describes it as the

artistic technique of applying manufactured, printed, or "found" materials, such as bits of newspaper, fabric, wallpaper, etc., to a panel or canvas, frequently in combination with painting. In the 19th century, *papiers collés* were created from papers cut out and put together to form decorative compositions. In about 1912–13 Pablo Picasso and Georges Braque extended this technique, combining fragments of paper, wood, linoleum, and newspapers with oil paint on canvas to form subtle and interesting abstract or semiabstract compositions. The development of the collage by Picasso and Braque contributed largely to the transition from Analytical to Synthetic Cubism.

This reference work, which adds that the term was first used to refer to dada and surrealist works, lists Max Ernst, Kurt Schwitters, Henri Matisse, Joseph Cornell, and Robert Rauschenberg as artists who have employed the medium.

In *The History of World Art,* H. W. Janson, who explains the importance of collage by locating it within the history of cubism, begins by describing Picasso's *Still Life* of 1911–12: "Beneath the still life emerges a piece of imitation chair caning, which has been pasted onto the canvas, and the picture is 'framed' by a piece of rope. This intrusion of alien materials has a most remarkable effect: the abstract still life appears to rest on a real surface (the chair caning) as on a tray, and the substantiality of this tray is further emphasized by the rope." According to Janson, Picasso and Braque turned from brush and paint to "contents of the wastepaper basket" because collage permitted them to explore representation and signification by contrasting what we in the digital age would call the real and virtual. They did so because they discovered that the items that make up a collage, "'outsiders' in the world of art," work in two manners, or produce two contrary effects. First, "they have been shaped and combined, then drawn or painted upon to give them a representational meaning, but they do not lose their original identity as scraps of material, 'outsiders' in the world of art. Thus their function is both to represent (to be part of an image) and to present (to be themselves)" (522–23).

Hypertext writing shares many key characteristics with these works of Picasso, Braque, and other cubists, particularly their qualities of juxtaposition and appropriation. Some of these qualities appear when one compares the hypertext and print versions of my discussion. First of all, despite my division of this essay into several sections and the use of figures that a reader might inspect in different orders, this essay really only allows one efficient way of proceeding through it. In contrast, the original hypertext version permits different readers to traverse it according to their needs and interests. Thus, someone well versed in twentieth-century art history might wish to glance only briefly at the materials on collage before concentrating first on the materials about hypertext. Someone else more acquainted with hypertext could concentrate on the materials about collage. Others might wish to begin with one portion of the discussion, and then, using available links, return repeatedly to the same examples, which often gather meaning according to the contexts in which they appear.

Another difference between the two forms of "my" discussion of this subject involves the length of quoted material and the way the surrounding texts relate it to the argument as a whole. Take, for example, the passage I quoted above from Janson's *History of World Art*. In the Storyspace version the passage is several times longer than in the print one, and it appears without any introduction. The object here is to let the quoted, appropriated author speak for himself, or, rather, to permit his text to speak for itself without being summarized, translated, distorted by an intermediary voice. To write in this manner—that is to say, to copy, to appropriate—seems suited to an electronic environment, an environment in which text can be reproduced, reconfigured, and moved with very little expenditure of effort. In this environment, furthermore, such a manner of proceeding also seems more honest: the text of the Other may butt up against that by someone else; it may even crash against it. But it does seem to retain more of its own voice. In print, on the other hand, one feels constrained to summarize large portions of another's text, if only to demonstrate one's command (understanding) of it and to avoid giving the appearance that one has infringed copyright.

These two differences suggest some of the ways in which even a rudimentary form of hypertext reveals the qualities of collage. By permitting one to make connections between texts and text and images so easily, the electronic link encourages one thus to think in terms of connections. To state the obvious: one cannot make connections without having things to connect. Those linkable items must not only have some qualities that make the writer want to connect them, they must also exist in separation, apart, divided. As Terence Harpold has pointed out in "Threnody," most writers on hypertext concentrate on the link, but all links simultaneously both bridge and maintain separation (174). This double effect of linking appears in the way it inevitably produces juxtaposition, concatenation, and assemblage. If part of the pleasure of linking arises in the act of joining two different things, then this aesthetic of juxtaposition inevitably tends toward catachresis and difference for their own end, for the effect of surprise, and sometimes surprised pleasure, they produce.

On this level, then, all hypertext webs, no matter how simple, how limited, inevitably take the form of textual collage, for they inevitably work by juxtaposing different texts and often appropriating them as well. Such effects appear frequently in hypertext fiction. Joshua Rappaport's "The Hero's Face" (one of the webs included in *Writing at the Edge*) uses links, for example, to replace what in earlier literary writing would have been an element internal to the text; that is, the link establishes a symbolic as well as a literal relationship between two elements in a document. In "The Hero's Face," after making one's way through a series of lexias about the members of a rock band, their experiences on tour, and their musical rivalry—all of which might seem little more than matters of contemporary banality—the reader follows a link from a discussion of the narrator's seizing the lead during one performance and finds herself or himself in what at first appears to be a different literary world, that of the Finnish epic, the *Kalelava*.

Following Rappaport's link has several effects. First, readers find themselves in a different, more heroic age of gods and myth, and then, as they realize that the gods are engaged in a musical contest that parallels the rock group's, they also see that the contemporary action resonates with the ancient one, thereby acquiring greater significance as it appears epic and archetypal. This single link in *Hero's Face,* in other words, functions as a new form of both allusion and recontextualization. Juxtaposing two apparently unconnected and unconnectable texts produces the pleasure of recognition.

Such combinations of literary homage to a predecessor text and claims to rival it have been a part of literature in the West at least since the ancient Greeks. But the physical separation between texts characteristic of earlier, nonelectronic information technologies required that their forms of linking—allusion and contextualization—employ indicators within the text, such as verbal echoing or the elaborate use of parallel structural patterns (such as invocations or catalogues). Hypertext, which permits authors to use traditional methods, also allows them to create these effects simply by connecting texts with links. David Goldberg's web essay, "New Perverse Logic: The Interface of Technology and Eroticism in J. G. Ballard's *Crash* and William Gibson's *Neuromancer*" (1996), uses HTML frames to accomplish a similar form of juxtaposition without links. Clicking on various topics from the opening screen opens a two-column document, in one half of which appear discussions of Baudrillard and Gibson, virtual textuality and Ballard, and two passages from the novelist.

Hypertext here appears as textual collage—*textual* referring to alphanumeric information—but more sophisticated forms of this medium also produce visual collage as well. Any hypertext system (or, for that matter, any computer program or environment) that displays multiple windows produces such collage effects. Multiwindow systems, such as Microcosm, Storyspace, Intermedia, Sepia, and the like, have the capacity to save the size and position of individual windows. This capacity leads to the discovery of what seems a universal rule at this early stage of e-writing: authors will employ any feature or capacity that can be varied and controlled to convey meaning. All elements in a hypertext system that can be manipulated are potentially signifying elements. Controlled variation inevitably becomes semiosis. Hypertext authors like Stuart Moulthrop have thus far written poems in the interstices of their writing environments, creating sonnets in link menus, and sentences in the arrangements of titles of lexias in the Storyspace view.

Inevitably, therefore, authors make use of screen layout, tiled windows, and other factors to . . . write. For example, in an informational hypertext, such as *The In Memoriam Web,* tiling of documents constructs a kinetic collage whose juxtaposition and assembling of different elements permits easy reference to large amounts of information without becoming intrusive (see

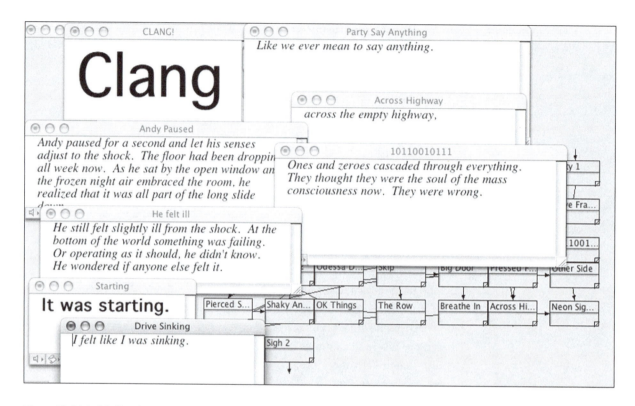

Within the figure, the following text appears in the overlapping windows:

CLANG!

Clang

Party Say Anything

Like we ever mean to say anything.

Across Highway

across the empty highway,

Andy Paused

Andy paused for a second and let his senses adjust to the shock. The floor had been dropping all week now. As he sat by the open window and the frozen night air embraced the room, he realized that it was all part of the long slide down...

10110010111

Ones and zeroes cascaded through everything. They thought they were the soul of the mass consciousness now. They were wrong.

He felt ill

He still felt slightly ill from the shock. At the bottom of the world something was failing. Or operating as it should, he didn't know. He wondered if anyone else felt it.

Starting

It was starting.

Drive Sinking

I felt like I was sinking.

Figure 22. Digital Collage in Hypertext Narrative: Nathan Marsh's *Breath of Sighs and Falling Forever*. Marsh has arranged the texts that make up his web so that some lexias show in their entirety, others only in part. In making their way through this fiction, readers encounter multiple narrative lines. The web continually changes the juxtaposition of texts as the narrator navigates in the text. In the course of reading, one is repeatedly returned to the lexia "Clang!" which opens with the sound of an explosion, but the meaning of the word changes according to the lexia that one has read immediately before encountering it.

Figure 11). In addition to employing the set placement of the windows, readers can also move windows to compare two, three, or more poems that refer back and forth among themselves in this protohypertextual poem.

Turning now to another work of hypertext fiction, one sees that in Nathan Marsh's *Breath of Sighs and Falling Forever* lexias place themselves around the surface of the computer monitor, making the screen layout support the narrative as one crosses and recrosses the tale at several points (Figure 22). In *The In Memoriam Web* the collage-effect of tiling, separate windows, and juxtaposed text arises in an attempt to use hypertext technology to shed light on qualities of a work created for the world of print (see Figure 9.) Here this story arises out of the medium itself. In making their way through this fiction readers encounter multiple narrative lines and corollary narrative worlds both joined and separated by ambiguous events or phenomena. At certain points readers cannot tell, for example, if one of the characters has experienced an earthquake tremor, a drug reaction, or a powerful illumination. Has the floor actually fallen, or are we supposed to take a character's experience as figurative? Certainly, one of the first lexias readers encounter could suggest any and all of these possibilities: "Andy paused for a second and let his senses adjust to the shock. The floor had been dropping all week now. As he sat by the open window and the frozen night air embraced the room, he realized that it was all part of the long slide down." Clicking on this brief lexia leads one to "Clang!," which opens with a loud sound and displays its single word in 80-point type. As one reads one's way through "Breath of Sighs" one repeatedly returns to "Clang!" but finds that it changes its meaning according to the lexia that one has read immediately before encountering it.

Marsh has arranged each of the texts that make up his web so some lexias show in their entirety, others only in part. As one reads through this web, one encounters a continually changing collage of juxtaposed texts. Two points about hypertext writing appear in Marsh's web. First, we realize that such collage writing produces a new kind of reading in which we must take into account not only the main text but also those that surround it. Second, this emphasis on the increasing importance of the spatial arrangement of individual lexias leads to the recognition that writing has become visual as well as alphanumeric; or since visual layout has always had a major impact on the way we read printed texts, perhaps it would be more accurate to say that in hypertext (where the author controls more of the layout) hypertext writing requires visual as well as alphanumeric writing. Marsh's web exemplifies a form of hypertext fiction that draws on the collage qualities of a multi-window system to generate much of its effect.

Despite interesting, even compelling, similarities, hypertext collage obviously differs crucially from that created by Picasso and Braque. Hypertext and hypermedia always exist as virtual, rather than physical, texts. Until digital computing, all writing consisted of making physical marks on physical surfaces. Digital words and images, in contrast, take the form of semiotic codes, and that fundamental fact about them leads to the characteristic, defining qualities of digital infotech: virtuality, fluidity, adaptability, openness (or existing without borders), processability, infinite duplicability, capacity for being moved about rapidly, and finally, networkability. Digital text is virtual because we always encounter a virtual image, the simulacrum, of something stored in memory rather than encounter any so-called text itself or physical instantiation of it. Digital text is fluid because, taking the form of codes, it can always be reconfigured, reformatted, rewritten. Digital text is hence infinitely adaptable to different needs and uses, and since it consists of codes that other codes can search, rearrange, and otherwise manipulate, digital text is also always open, unbordered, unfinished and unfinishable, capable of infinite extension. Furthermore, since it takes the form of digital coding, it can be easily replicated without in any way disturbing the original code or otherwise affecting it. Such replicability in turn permits it to be moved rapidly across great spaces and in so doing creates both other versions of old communication, such as the bulletin board, and entirely new forms of communication. Finally—at least for now—all these other qualities of digital textuality enable different texts (or lexias) to join together by means of electronic linking. Digitality, in other words, permits hypertextuality.

The connection of the fundamental virtuality of hypertext to the issue of collage becomes immediately clear as soon as one recalls the history of collage and the reasons for its importance to Picasso, Braque, Schwitters, and other painters. As Janson explains, collage arose within the context of cubism and had powerful effects because it offered a new approach to picture space. Facet cubism, its first form, still retained "a certain kind of depth," and hence continued Renaissance perspectival picture space. "In collage Cubism, on the contrary, the picture space lies in front of the plane of the 'tray'; space is not created by illusionistic devices, such as modeling and foreshortening, but by the actual overlapping of layers of pasted materials" (522–23). The effect of collage cubism comes from the way it denies much of the recent history of Western painting, particularly that concerned with creating the effect of three-dimensional space on a two-dimensional surface. It does so by inserting some physically existing object, such as Picasso's chair-caning and newspaper cuttings, onto and into a painted surface. Although that act of inclusion certainly redefines the function and effect of the three-dimensional object, the object nonetheless resists becoming a purely semiotic code and abrasively insists on its own physicality.

The collage of collage cubism therefore depends for its effect on a kind of juxtaposition not possible (or relevant) in the digital world—that between physical and semiotic. Both hypertext and painterly collage make use of appropriation and juxtaposition, but for better or worse one cannot directly invoke the physical within the digital information regime, for everything is mediated, represented, coded.

The final lexia in this grouping, however, moves this more traditional form of virtuality to that found in the world of digital information technology, for it both repeats sections of all the images one may have seen (in whatever order), blending them with multiply repeated portions of a photograph of a Donegal, Ireland, sunset, and it also insists on the absence of any solid, physical ground: not only do different-sized versions of the same image appear to overlay one another but in the upper center a square panel has moved aside, thus revealing what the eye reads as colored background or empty space. In this photographic collage or montage, appropriation and juxtaposition rule, but, since all the elements and images consist of virtual images, this lexia, like the entire web to which it contributes, does not permit us to distinguish (in the manner of cubist collage) between virtual and real, illusion and reality.

This last-mentioned lexia bears the title "Sunset Montage," drawing on the secondary meaning of *montage* as photographic assemblage, pastiche, or, as the *OED* puts it, "the act or process of producing a composite picture by combining several different pictures or pictorial elements so that they blend with or into one another; a picture so produced." I titled this lexia "Sunset Montage" to distinguish the effect of photographic juxtaposition and assemblage from the painterly one, for in photography, as in computing, the contrast of physical surface and overlaying image does not appear. Upon hearing my assertion that hypertext should be thought of as collage writing, Lars Hubrich, a student in my hypertext and literary theory course, remarked that he thought *montage* might be a better term than *collage*. He had in mind something like the first *OED* definition of montage as the "selection and arrangement of separate cinematographic shots as a consecutive whole; the blending (by superimposition) of separate shots to form a single picture; the sequence or picture resulting from such a process." Hubrich is correct in that whereas collage emphasizes the stage effect of a multiwindowed hypertext system on a computer screen at any particular moment, montage, at least in its original cinematic meaning, places important emphasis on sequence, and in hypertext one has to take into account the fact that one reads—one constructs—one's reading of a hypertext in time. Even though one can backtrack, take different routes through a web, and come upon the same lexia multiple times and in different orders, one nonetheless always experiences a hypertext as a changeable montage.

Hypertext writing, of course, does not coincide fully with either montage or collage. I draw upon them chiefly not to extend their history to digital realms, and, similarly, I am not much concerned to allay potential fears of this new form of writing by deriving it from earlier avant-garde work, though in another time and place either goal might provide the axis for a potentially interesting essay. Here I am more interested in helping us understand this new kind of hypertext writing as a mode that both emphasizes and bridges gaps and that thereby inevitably becomes an art of assemblage in which appropriation and catachresis rule. This is a new writing that brings with it implications for our conceptions of text as well as reader and author. It is a text in which a new kind of connection has become possible.

Is This Hypertext Any Good? Or, How Do We Evaluate Quality in Hypermedia?

What is quality in hypertext? How, in other words, do we judge a hypertext collection of documents (or web) to be successful or unsuccessful, to be good or bad *as hypertext*? How can we judge if a particular hypertext achieves elegance or never rises above mediocrity? Those questions lead to another: What *in particular* is good about hypertext? What qualities does hypertext have in addition to those possessed by nonhypertextual forms of writing, which at their best can boast clarity, energy, rhythm, force, complexity, and nuance? What qualities, in other words, derive from a form of writing that is defined to a large extent by electronic linking? What good things, what desirable qualities, come with linking, since the link is the defining characteristic of hypertext? As I have argued earlier, the defining qualities of the medium include multilinearity, consequent potential multivocality, conceptual richness, and—especially where informational hypertext is concerned—reader-centeredness or control by the reader. Obviously, works in a hypertext environment that fulfill some or all of these potential qualities exemplify quality in hypermedia. Are there other perhaps less obvious sources of quality?

One question we must raise while trying to identify sources of quality in hypermedia is, To what extent do literary and informational hypermedia differ? In the following pages, I shall propose several possible ways to answer these questions, each of which itself involves a central issue concerning this information technology.

Individual Lexias Should Have an Adequate Number of Links. Since the link is the characteristic feature that defines hypertextuality, one naturally assumes that lexias containing a larger number of valuable links are better than those that have fewer. Of course, the emphasis here must be on "valuable." In the early days of the Web, one would often come upon personal homepages in which virtually every word other than the articles *the, a,* and *an* had links, many of which led to external sites only generally connected to the discussion at hand. Obviously, overlinking, like choosing poor link destinations, is bad linking. As Peter Brusilovsky and Riccardo Rizzo have pointed out in "Map-Based Horizontal Navigation in Educational Hypertext," the opposite problem—a lack of linking precisely in those places one would expect it to appear—characterizes much recent World Wide Web hypertext. Part of the problem here may come directly from the World Wide Web's use of unsuitable terminology derived from print technology, such as *homepage,* which locks neophyte users into an inappropriate paradigm. Brusilovsky and Rizzo correctly note that much hypertext today takes the form of passages of unlinked text surrounded by navigation links. Encountering these kinds of lexia, one receives the impression that the authors, who have dropped digitized versions of printed pages into an electronic environment, don't seem to grasp the defining qualities of hypermedia and use HTML chiefly as a text formatting system. They are still working, in other words, with and within the paradigm of the printed page and book.

The Victorian Web (victorianweb.org), an academic site I manage that receives as many as fifteen million hits a month, contains four basic kinds of documents: (1) overviews (sitemaps), (2) link lists, (3) simple two-column tables used primarily for chronologies as well as art works and text describing them, and, (4) lexias containing primarily text, though some may also include thumbnail images linked to larger plates. Most text documents on this site contain two to four navigation links in the form of linked icons that appear at the bottom of each lexia plus multiple text links that weave the lexia into a miniature hypertext network. Although I find myself unable to formulate any rule as to proper number of text links, I have observed two things: (1) lexias approximately one to two screens in length tend to have at least three text links, and (2) as new documents arrive, older lexias receive additional links.

The comparative lack of text links observed in much web-based hypermedia also appears in much hyperfiction, as many authors seem uninterested in using more than single links, which create an essentially linear flow. Caitlin Fisher's *Waves of Girls,* a web narrative that won the 2001 Electronic Literature Organization (ELO) prize for electronic fiction, exemplifies the comparatively rare literary hypertext that includes both framing navigational links and others in the body of the text. Thus, in the following brief example, the phrases "I was so sad," "our principal," "grade 5 boys . . . ," "making out really meant . . ." all lead to—that is, produce—new text (Figure 23). In addition to the navigation links that appear at the left of the screen, the main text also contains frequent opportunities to follow links, which lead to other narrative arcs.

Following the Link Should Provide a Satisfying Experience. Linking in informational hypermedia obviously has to work in a clear, coherent manner, but what produces this requisite coherence?[4] In other words, what should appear

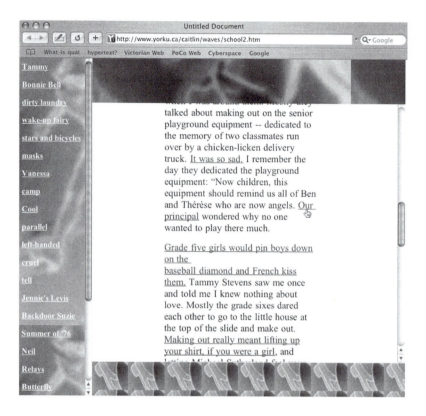

Figure 23. A Link-intensive Hyperfiction: Caitlin Fisher's *Waves of Girls*. In addition to the navigation links that appear at the left of the screen, the main text also contains frequent opportunities to follow links.

at the end of a link to satisfy the intellectual and aesthetic needs of the reader?[5] Let's take as an example what happens when one comes upon linked text in the midst of the following sentence in a lexia from *The Victorian Web* about the prose fantasies of William Morris: "Like John Ruskin, Morris creates prose fantasies permeated by his beliefs about political economics." What should one find at the end of the link attached to the name *John Ruskin?* For the reader of the present lexia, which discusses fantastic literature by Morris, the most useful link would produce a discussion of fantastic fiction by Ruskin, and in fact *The Victorian Web* has such a relevant document, "John Ruskin and the Literary Fairy Tale," one section of which explains the relations of his early fantasy to his later political writings. One might even hope that such a link would bring one to a comparison of the distinctive qualities of each author's writings in this mode, which this existing document does not. All these desired link-destinations, one notes, are implied by the wording of the sentence in which the linked text appears.

What happens, however, when such discussions are unavailable? What usually happens both in the websites I've examined and those I manage is that the link of the compared author—here Ruskin—goes to very basic or general information about that figure. Notice that such a link to general information, which may provide a kind of basic identification of the figure for neophytes in the field, is not necessarily a bad link. In fact, for certain users, particularly those new to a certain field or subject, such a link destination might prove very useful. Still, most users of documents about quite specific topics require information that directly illuminates the main subject at hand (in this case, Ruskin's fairy tale). The fact is, though, that such specific link destinations are far more rare than the more general, glossary-type ones.

Obviously, one would prefer to give readers a choice of information, in this case providing both general and very specific information, in part because such a choice offers a richer, more user-centered embodiment of hypertextuality. Unfortunately, the World Wide Web, which at present allows only links from a word or phrase to a single destination, does not offer one of the most useful kinds of linking—the one-to-many or branching link that offers the reader a choice of destinations at the point of departure. One solution is to link the anchor—here "John Ruskin"—to another document, which has to be manually created, that offers multiple choices. Depending on the subject of the lexia in which this name appears, the link list or area sitemap at the end of such a link can take the form of lists of links to biographical information about "John Ruskin," those leading to his influence on various authors, and so on. Another approach to handling links to several destinations, not always possible to implement, requires adding phrases that might provide multiple anchors in the departure sentence. Thus, one could link general information to the figure's name (John Ruskin) and specific information only to phrases, such as "permeated by his beliefs," that lead the reader to expect a very specific discussion at the destination lexia.

The Pleasures of Following Links in Hyperfiction and Poetry. Since much hyperfiction and poetry aims to produce reader disorientation, however transient, the informational hypertext features of reader empowerment, multiple approaches, and clarity might not appear particularly important to it. Instead, the qualities of surprise and delight characterize such success, for with hyper-

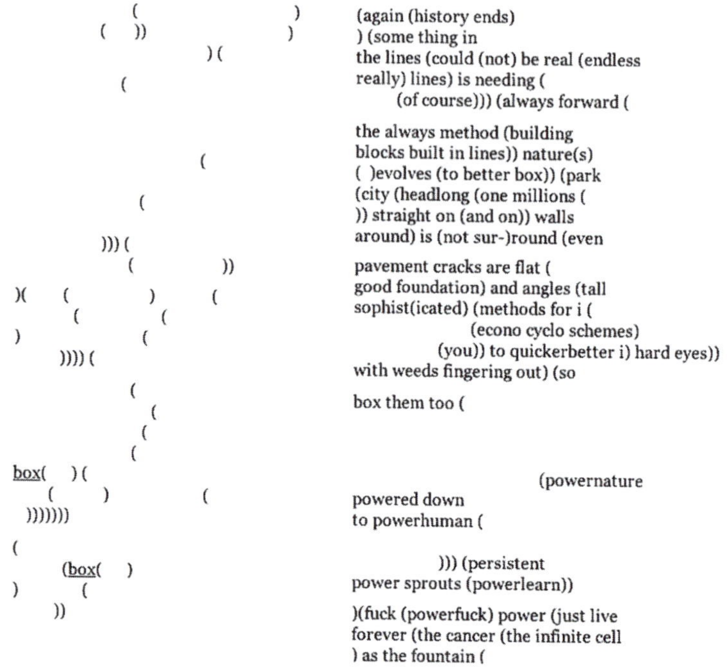

```
        (                    )        (again (history ends)
    (   ))                   )        ) (some thing in
                    ) (               the lines (could (not) be real (endless
        (                             really) lines) is needing (
                                         (of course))) (always forward (

                                      the always method (building
                (                     blocks built in lines)) nature(s)
                                      ( )evolves (to better box)) (park
        (                             (city (headlong (one millions (
                                      )) straight on (and on)) walls
      ))) (                           around) is (not sur-)round (even
        (          ))
  )(   (      )      (                 pavement cracks are flat (
        (          (                  good foundation) and angles (tall
  )          (                        sophist(icated) (methods for i (
    )))) (                                  (econo cyclo schemes)
                                             (you)) to quickerbetter i) hard eyes))
        (                             with weeds fingering out) (so
          (
          (                           box them too (
          (
  box(    ) (                                                  (powernature
      (    )          (               powered down
    ))))))))                          to powerhuman (
  (
      (box(    )                              ))) (persistent
  )      (                            power sprouts (powerlearn))
      ))                              )(fuck (powerfuck) power (just live
                                      forever (the cancer (the infinite cell
                                      ) as the fountain (
```

Figure 24. Two Stages of Reading and Ian M. Lyons's *(box(ing))*. At the left appears the screen one encounters early in one's reading; at right, the screen after the reader has brought forth words and phrases by clicking his or her mouse.

fiction and poetry the question must be, not does following the link chiefly satisfy an intellectual need but does following the link produce surprise and delight? Instances of such pleasing results of following links appear in Stephanie Strickland's *Vniverse* (see Figure 11) and Ian M. Lyons's *(box(ing))*, both of which produce text ex nihilo. When one moves one's mouse over a predetermined area (near a parenthesis in *(box(ing))* and within the night sky in *Vniverse*) and then clicks, text appears.[6] Thus, when the reader opens *(box(ing))*, little appears on the screen other than multiple gray parentheses scattered across a white background (Figure 24). Lyons explains, "The placing of the parentheses" was intended to "convey nested levels of associative meaning . . . arranged hierarchically; that is, if I opened one parenthetic set and then opened a second, this second set I always made to close before the first. For example: $(_1 \ldots (_2 \ldots)_2 \ldots)_1$." Lyons explains that "the piece's parenthetically obsessive syntax closely resembles that used in the entirely outmoded programming language, LisP (more recently reincarnated under the name Scheme)." Clicking on the screen within some parentheses and outside others incrementally produces text. Lyons's poem, which he implemented in HTML, Storyspace, and Visual basic, was, he tells us, originally written to be read on paper with the intention of questioning "hierarchical modes of organization" found in post-Chomskian linguistics and implicitly confounded by hypertext, since, as Nelson has pointed out, the shortcomings of classification systems, all of which require hierarchies, explain the need of hypermedia in the first place.[7] The pleasures of reading *(box(ing))*, I propose, come from the discoveries of text the reader produces and of the meanings of that quite difficult text.

Stephanie Strickland's *Vniverse* (see Figure 11), a much more complex project than *(box(ing))*, represents a comparatively rare example of literary hypermedia that aims at producing both delighted surprise and the virtues associated with information hypermedia—reader empowerment and multivocality, or multiple approaches to a single general subject.[8] Upon opening *Vniverse*, one encounters a night sky—a black screen speckled with stars—in which the central portion rotates. A small circle appears at top right and a slightly smaller one appears diagonally opposite at lower left. Moving one's mouse across the sky halts the rotation and reveals various constellations. Meanwhile instructions scroll across the bottom of the screen: "Scan the stars . . . click once or click twice . . . click the darkness." Clicking on darkness brings forth a constellation, a particular star with its assigned number, and text that appears when one keeps one's mouse over the point at which one clicked. Typing a number in the top right-hand circle produces the star with that number and its surrounding constellation. Like many hypermedia projects that employ Flash and similar software, *Vniverse* boasts animated text. Unlike many such projects, it also emphasizes a high degree of reader control.

Coherence. Rich linking, plus a substantial degree of reader control, thus appear to characterize success in both informational and literary hypermedia. Another necessary quality, I propose, is some sort of crucial coherence.

Since hypertext fiction and poetry often employ disorientation effects for aesthetic purposes, coherent and relevant linking might not seem to be necessary, but I suspect it's simply that coherence not take as obvious forms as it does in information hypermedia. For example, our experience of reading pioneering hyperfiction, such as Michael Joyce's *afternoon*, proves definitively that much of what we have assumed about the relations of coherence to textuality, fixed sequence, and the act of reading as sense-making is simply false. Reading *afternoon* and other fictional narratives shows, in other words, that we can make sense of—that is, perceive as coherent—a group of lexias even when we encounter them in varying order. This inherent human ability to construct meanings out of the kind of discrete blocks of text found in an assemblage of linked lexias does not imply either that text can (or should) be entirely random, or that coherence, relevance, and multiplicity do not contribute to the pleasures of hypertext reading. Movement in *afternoon* from a lexia containing, say, the conversation of two men to one containing that of one of their wives may at first appear abrupt (and hence random or without any relevance), but continued reading establishes the essential coherence of the link between the two lexias: the movement between the one containing the men speaking and the second containing the women can be repeated, thus establishing a pattern like cinematic cross-cutting. Similarly, the next lexia one encounters can reveal that the words of one pair of speakers serve as the context, the backstory, for the others.

Coherence as Perceived Analogy. In linking, this necessary coherence can also take the form of perceived analogy—that is, the link, the jump across the textual gap, to some extent reifies the implied connection (implied link) found in allusions, similes, and metaphors. For an example, let us look at another early Storyspace narrative, Joshua Rappaport's *Hero's Face*, which shows how linking can serve as a new form of textual allusion. In *Hero's Face*, which relates the struggles for musical supremacy in a rock band, one particular link transports one from adolescent rock'n'roll to an entirely different, and very unexpected, world of ancient epic. Most of the story consists of lexias about the people in the band and the relationships among them. In one crucial lexia the narrator describes the first time he "climbed serious lead"—seized control of the music in midperformance—and realized that the experience resembles the feelings he has had while mountain climbing: "There comes a moment when all of a sudden you look behind you and you're out eight or ten feet from your last piece, which adds up to a twenty-foot fall onto the dubious support of some quickly-wedged chunk of metal in a crack—you look behind you, and it's just straight down, eighty or a hundred feet, and your belayer barely visible there at the bottom waiting for you to peel off—every muscle pumped up to bursting, as you realize that it is the mere strength of fingers and arms and your innate sense of balance keeping you up in the air." After readers encounter this comparison of musical improvisation to mountaineering, they come upon a link that functions as a second analogy, for following this link brings one to the world of the Finnish epic, the *Kalelava:*

The old Vainamoinen sang:
the lakes rippled, the earth shook
the copper mountains trembled
the sturdy boulders rumbled
the cliffs flew in two
the rocks cracked upon the shores.
He sang young Joukahainen—
saplings on his collar-bow
a willow shrub on his hames
goat willows on his trace-tip
sang his gold-trimmed sleigh
sang it to treetrunks in pools
sang his whip knotted with beads
to reeds on a shore

Following Rappaport's link has several effects. First, readers find themselves in a different, more heroic age of gods and myth, and then, as they realize that the gods are engaged in a musical contest that parallels the rock group's, they also see that the contemporary action resonates with the ancient one, thereby acquiring greater significance since it now appears epic and archetypal. This single link in *Hero's Face,* in other words, functions as a new form of both allusion and recontextualization.

In hyperfiction, Michael Joyce invented this form of reified comparison or allusion when he had links transport readers from his story to passages from Plato's *Phaedo,* Vico's *New Science,* Basho's *The Narrow Road through the Provinces,* and poems by Robert Creeley and others. Perhaps the ultimate source here is Julio Cortázar's *Hopscotch* (to which Joyce alludes in the lexia entitled "Hop Scotch"). Frequently used, such juxtapositions-by-linking produce the kind of collage writing that appears to be very typical of hyperfiction and poetry.

Such combinations of literary homage to a predecessor text and claims to rival it have been a part of literature in the West at least since the ancient Greeks. But the physical separation between texts characteristic of earlier, nonelectronic information technologies required that their forms of linking—allusion and contextualization—employ indicators within the text, such as verbal echoing or the elaborate use of parallel structural patterns (such as invocations or catalogues). Hypertext, which permits authors to use traditional methods, also permits them to create these effects simply by connecting texts. When successful, such linking-as-allusion creates a pleasurable shock of recognition as the reader's understanding of the fictional world suddenly shifts.

Does Hypertext Have a Characteristic or Necessary Form of Metaphoric Organization? The creation of coherence in linking via implied analogy can characterize not just the relation between two lexias but also an entire hypertext. The kind of textuality created by linking encourages certain forms of metaphor and analogy that help organize the reader's experience in a pleasurable way. Some of the most successful hyperfictions, such as Shelley Jackson's *Patchwork Girl,* employ powerful organizing motifs, in this case scars and stitching together that function as commentaries on gender, identity, and hypertextuality. Stitches and scars, which have obvious relevance in a tale involving Dr. Frankenstein and one of his monsters, become metaphorical and create unity and coherence for the entire assemblage of lexias. At an early crux in the narrative ("Sight"), Jackson creates a branching point at which the reader must choose between two lexias, both of which emphasize the analogous relationships among writing, reading a hypertext, and sewing up a monster ("written," "sewn"). Jackson's witty plays on these topics all have a role in a hyperfiction that exposes the way we create and experience texts, hypertexts, gender, and identity.

One can also create unifying metaphors or analogies that do not refer to hypertext, the medium itself. David Yun's *Subway Story* is a work of hyperfiction that employs metaphors that inform the narrative in "nonreflexive" modes. *Subway Story* employs the organizational metaphor of the map for the New York subway system: it includes both a map of that system and a lexia for each of its stations. Yun has created a lexia for every stop on the subway, and he has used the paths of the individual trains as link paths that create narrative arcs. As Stefanie Panke pointed out to me when I asked her why she thought it an example of good hypertext fiction, "*Subway Story* is an extraordinary hypertext because of the application of a spatial metaphor that allows a navigation that is somehow 'linked' to the story itself. It is a beautiful example for a metaphor that works because it is a part of (and not apart from) the storytelling."

Gaps. As should be obvious by now, good hypertext—quality in hypertext— depends not only on appropriate and effective links but also on appropriate and effective breaks or gaps between and among lexias. Terence Harpold long ago pointed out that Derridean gaps, the presence of which requires linking in the first place, have just as much importance in hypertext as do links themselves. N. Katherine Hayles has more recently explained that "analogy as a figure draws its force from the boundaries it leapfrogs across. Without boundaries, the links created by analogy would cease to have revolutionary impact" (93), and the same is true of the hypertext link. Without good—by which I mean effective and appropriate—separations one cannot have good links. Like the epic hero who requires an adequate antagonist to demonstrate his superiority, linking requires a suitable gap that must be bridged. We have all read hypertexts in which following a link produces a text that seems to follow what came before in such obvious sequence, the reader wonders why the author simply didn't join the two. We've all encountered relatively poor or ineffectual gaps by which I mean those breaks in an apparently linear text that appear arbitrary: the gap, the division between two texts, appears unnecessary when the link does nothing more than put back together two passages that belong together when no other paths are possible.

Hyperfiction and poetry can have two very different kinds of gaps, the first being those bridged or surmounted by links, the second being those that remain, well, gaps because nothing in the software environment joins the two texts or lexias. Whereas the first kind of gap, that joined by links, seems obvious because we encounter it every time we follow a link, the other is not. As an example of the second I am thinking of entire sections or narrative arcs in works like *Patchwork Girl* that remain separate and separated in the reader's experience and yet may be joined by allusion or thematic parallels. Thus, in *Patchwork Girl* gatherings of lexias about the stitched-together nature of the female Frankenstein monster reside in a different folder or directory than those comprising Shelley Jackson's collage of lexias composed of various texts from Jacques Derrida, L. Frank Baum, and Mary Shelley. These discrete sections join in variations on the themes of text, stitched-together-ness, coherence, origins, and identity. As this example of gaps unjoined by links makes clear, not all connections in effective hypertext require electronic connections—like nonhypertextual prose and poetry, hypertext also makes use of allusions, metaphors, and implicit parallels. The real question turns out to be, then, How does one decide when to make the potential connection, relation, or parallel explicit by means of an electronic link and when to leave connections, relations, or parallels implicit?

Individual Lexias Should Satisfy Readers and Yet Prompt Them to Want to Follow Additional Links. Hypertext is, after all, still text, still writing, and we find it difficult to distinguish many of the qualities of other good writing from writing with links. In other words, excellence in hypertext does not depend solely on the link. To an important extent, the text that surrounds the link matters, too, because the quality of writing and images within an individual lexia relate to one key hypertextual quality—its ability to make the reader simultaneously satisfied enough with the contents of a particular lexia to want to follow a link from that lexia to another. The problem that any writer faces—whether the writer of hyperfiction or of stories intended for print— can be defined simply as how to keep the reader reading. Making readers want to continue reading seems much easier in print text for a variety of reasons: knowing the genre signals, readers know what to expect; looking at their place in a physical text, they know how much more they have to read; without choices demanded by linking, readers have essentially one choice— to continue reading or to put down the story, novel, or poem.

Particularly in these early days of the history of these new technologies and associated media, readers have a more difficult time deciding whether to keep reading. The text they read must persuade them to go on by the essential, traditional, conventional means—that is, by intriguing, tantalizing, satisfying, and above all entertaining them. In a hypertext lexia the reader must encounter text that is simultaneously, perhaps paradoxically, both satisfying and just unsatisfying enough: in other words, the current lexia readers encounter has to have enough interest, like any text, to convince them to keep reading, and yet at the same time it must also leave enough questions unanswered that the reader feels driven to follow links in order to continue reading. In the terms of Roland Barthes, the lexia must include sufficient plot enigmas or hermeneutic codes to drive the reader forward. This demand explains why the opening lexia of Michael Joyce's classic *afternoon,* perhaps the first and still one of the most interesting hyperfictions, takes the form of such ornate metaphorical prose. Here, for example, is the second paragraph in *afternoon's* opening lexia ("begin"):

octopi and palms of ice—rivers and continents beset by fear, and we walk out to the car, the snow moaning beneath our boots and the oaks exploding in a series along the fenceline on the horizon, the shrapnel settling like relics, the echoing thundering off far ice. This was the essence of wood, these fragments say. And this darkness is air. By five the sun sets and the afternoon melt freezes again across the blacktop into crystal

The rich, sensual metaphoric style of this lexia promises readers a lush reading experience and therefore makes them want to keep reading, but this section is also self-contained enough to cohere as a separate lexia. As anyone who has read *afternoon* knows, not all its lexias have this richness—some

Figure 25. A Hyperfiction Sitemap: Jackie Craven's *In the Changing Room*.

are quite bare and brief—but Joyce does employ this style elsewhere, for example, in "Staghorn and starthistle."

The Reader Can Easily Locate and Move to a Sitemap, Introduction, or Other Starting Point. Can the reader easily return to documents or images encountered in a previous session? Such a requirement obviously pertains more to informational or discursive hypertext than to hyperfiction or poetry, though some fictions, such as Jackie Craven's *In the Changing Room* (Figure 25), employ a sitemap, in this case, consisting of the names of each of eight characters. Whereas Craven's sitemap takes the form of a typical HTML set of labeled links, Deena Larsen's *Stained Word Window* (1999; Figure 26) uses an active (or "hot") sitemap at screen left (on a black background) to bring up text at the right. Simply mousing over a word, such as "faces," "in," "understanding," or "windows," produces brief patches of free verse that contain links, and one can always return to the beginning or opening lexia because Larsen provides a linked footer icon that brings one back to it. Texts that invite a more active,

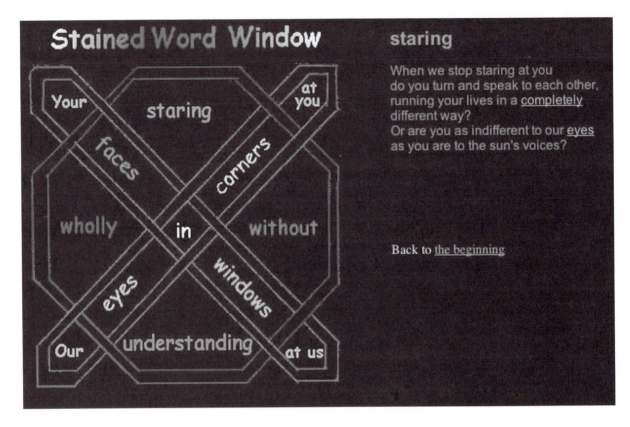

Figure 26. An Active Sitemap. Deena Larsen's *Stained Word Window* uses an active or "hot" sitemap: simply mousing over different portions of it makes poems appear at screen right.

even aggressive reader need, like informational hypertext, such devices, since the reader's orientation, rather than disorientation, plays a major role.

The Document Should Exemplify True Hypertextuality by Providing Multiple Lines of Organization. In a hypertext, whether fiction, poetry, or informational, one generally does not expect individual lexias to follow one another in linear fashion. True, linear sequences do have their use: Vannevar Bush–style trails require linear sequences, and authors of fiction use them to create a main (or default) axis for a narrative from which one can easily depart. Perhaps surprisingly, much hypertext narrative thus far takes the form of narrative loops or paths in which most of the lexias follow one another in a linear fashion, thus creating a series of self-contained stories. Of course, an electronic document may work quite well and yet not work hypertextually in any complex or interesting way. One can, for example, have hypertexts in which linking only serves to join an index to individual sections. To be clear, let's remind ourselves that hypertextuality—or excellence in hypertext, whatever we decide that might be—obviously is important in judging a hypertext as hypertext, but it need not necessarily play an important role in other forms of digital arts and literature. Here I'm concerned only with the problem of quality in hypertext.

Steve Cook's "Inf(l)ections" and Jeff Pack's "Growing up Digerate" exemplify successful, richly linked discursive hypertexts. Cook's stands as experiment in new forms of academic writing whereas Pack's experiment autobiography provides three different kinds of organization that the reader can follow: (1) a linear path arranged chronologically, (2) a topic-driven reading facilitated by a sitemap in the form of an alphabetical list, and (3) a multilinear narrative provided by links scattered throughout the text of individual lexias. Jackie Craven's *In the Changing Room* similarly allows both linear narrative, permitting the reader to follow the story of a single character, or move among the eight characters, each one in effect being defined as a storyline, a narrative arc. As the introduction says, "Click on an underlined word, and the stories will merge and take new form. *Your path will not be straight.* Here in the Changing Room, all things are linked—and everyone is a reflection . . . of a reflection . . . of a reflection."

The Hyperdocument Should Fully Engage the Hypertextual Capacities of the Particular Software Environment Employed. In asking if an individual hypermedia project pushes the limits of the software it employs, one enters a minefield. In the first place, such a question implicitly assumes that the new, the experimental, has major value in itself, and even if one accepts this hypothesis, it might have validity only in the early stages of a genre or media form. Of course, at the present moment, all writing in hypertext is experimental since the medium is taking form as we read and write. Electronic linking, one of the defining features of this technology, can reconfigure notions of author, text, reader, writer, intellectual property, and other matters of immediate concern to those who design hypertext systems or author documents with them. Because hypertext fiction—writing at and over the edge—sets out to probe the limits of the medium itself, it acts as a laboratory to test our paradigms and our fundamental assumptions. A sample of hypertexts shows the ways they illuminate issues ranging from reader disorientation and authorial property to the nature of hypertext genres and the rules of electronic writing.

Within this project of writing-as-discovery, all elements in a hypertext system that can be manipulated can function as signifying elements. To provide an example of the creative use of system features, let us turn to a few very early examples from *Writing at the Edge* (1994), all of which were created in Eastgate Systems' Storyspace, a stand-alone hypertext environment available for both Windows and Macintosh platforms.

In addition to containing traditional elements such as fonts, graphics, sound, and color, Storyspace also supports the creative utilization of "screen real estate"—the tiling of windows and the order in which they appear and arrange themselves. Nathan Marsh's lexias in *Breath of Sighs* place themselves around the screen, making the screen layout support the narrative as one crosses and recrosses the tale at several points (see Figure 22). Marsh's work, which dates from 1993, provided an early demonstration that writing had become visual as well as alphanumeric. It also reveals that a single software feature, such as the ability to control window size and location, leads directly to a particular mode of writing—here writing as collage and montage in which the multiple-window format permits readers to move back and forth among overlapping lexias. This feature also encourages active readers, since they can easily move about among lexias, thus creating a kind of spatial hypertext.

Several other hypertexts from *Writing at the Edge* show the imaginative deployment of another system feature of the software—the Storyspace view, a dynamic graphic presentation of the arrangement of document organization. Storyspace, a hypertext environment that also functions as a conceptual organizer, permits authors to nest individual spaces (lexias) inside others, or to rearrange the hypertext's organization by moving lexias without breaking links. Some works, like Shelley Jackson's *Patchwork Girl* (see Figure 28), take advantage of this graphic organizational feature to structure hyperfiction by means of separate folders or directories. Others, like Ho Lin's *Nicely Done*, arrange all lexias on a single level and indicate discrete narrative lines. This hypertext novel, which links a murder story and the events of a professional football championship game, suggests its organization by arranging its lexias, all of which appear on the top level, in four parallel lines. Timothy Taylor's *LBJ*—Lazarus + Barabbas + Judas—takes graphic indications of narrative and conceptual organization farther than Ho Lin's *Nicely Done,* arranging its lexias in the form of three crosses, the central one of which has a circle (halo?) over it. Here, rather than indicating the narrative structure, Taylor implies graphically something about the subject and theme of his fiction. Like similar projects that Michael Joyce reproduces in *Of Two Minds* (38), Adam Wenger's *Adam's Bookstore,* which I discuss in the following chapter, uses a circular deployment of the graphic elements representing lexias in Storyspace view to indicate that his document can be entered—and left—at any point (see Figure 32). One of the most bravura examples of arranging lexia-icons in the Storyspace view appears in Marc A. Zbysznski's playful use of hundreds of them to create an image of a human face beneath a recycling symbol. Even the naming of lexias can provide opportunities for unexpected signification. Andrew Durden's playful arrangement of lexias in *Satyricon Randomly Generated* forms a grammatical sentence. Reading the titles of the upper-level folders reveals the following playful comment: "I / think this / lexia / is a good / start place." Stuart Moulthrop famously carried this playful use of system features much farther, creating sonnets within a menu of links!

As the previous examples suggest, hypertext environments have, if not precisely McLuhan's message in the medium, at least certain tendencies that derive from specific features of the software. The capacity to control size and location of multiple windows encourages collage-like writing that employs these features, just as the presence of one-to-many linking and menus of links that have a preview function encourage certain forms of branching. Both features and the limitations or constraints of these features encourage certain ways of writing, just as the fourteen-line sonnet encourages certain kinds of poetry.

Turning from Storyspace to HTML and the World Wide Web, by far the most widespread form of hypermedia today, one wonders if it, like other hypermedia environments, encourages certain modes of writing. HTML, which is basically an extremely simple text-formatting language that works on the Internet, has two defining features—first, the ability to insert links between lexias and, second, the ability to insert other media into individual lexias, originally just images but soon after sound, video, and animation created by Java scripts or Flash. The rapid spread of access to broadband connections to the Internet has transformed the World Wide Web from a simple system for linking text-representations into a multimedia platform. The implications of this change for anyone trying to determine the message in the medium are obvious: whereas earlier proprietary systems, such as Intermedia, Microcosm, HyperCard, Storyspace, Guide, and so on, had built-in, clearly defined characteristics, some of which provided clear limitations, the World Wide Web does not. Anyone working with basic HTML encounters certain obvious features, which may act as imitations. These include the absence of one-to-many linking, preview features, and preview functions, as well as the inability to place and control the size of windows. Anyone using Flash or Java in HTML documents, however, does not necessarily confront any of these limitations, though they may confront others, such as incompatibility with particular versions of browsers. Such freedom, such absence of limitations, brings with it the relative absence of those restraints that often both limit and inspire creativity.

Conclusion. All forms of writing at their best can boast clarity, energy, rhythm, force, complexity, and nuance. Hypertext and hypermedia, forms of writing largely defined by electronic linking, are media that possess the potential qualities of multilinearity, consequent potential multivocality, conceptual richness, and—especially where informational hypertext is concerned—some degree of reader-centeredness or control. Obviously, hypertexts that build on the chief characteristics of the medium succeed. In addition, as we have seen, examples of hyperfiction and hyperpoetry reveal other sources of quality: individual links and entire webs that appear coherent, appropriate gaps among lexia, effective navigation and reader orientation, pervasive metaphoricity, and the exploration—and testing—of the limits of the medium.

Reconfiguring Narrative

Approaches to Hypertext

Fiction—Some

Opening Remarks

Every digital narrative, we must remind ourselves, does not necessarily take the form of hypertext. A case in point appears in Christy Sheffield Sanford's visually elegant World Wide Web fiction, *Safara in the Beginning* (1996), which the author describes as "a web-novel" written in the spirit of classical tragedy about "a young African princess taken as a slave from Senegal to Martinique." The opening screen, the first of twenty-one successive lexias, explains that to the left of the "main textual body, Bible quotations and natural history descriptions echo Old Testament Christianity and Animistic traditions at their point of contact: mythopoetization of nature." Essentially, Sanford works with the powers of digital information technology to add colors, images, and some motion to narrative, but the HTML links function solely to provide sequence. Among her twenty-one lexias, she includes what she describes as "five filmic scenes using the close-up, time-lapse and other cinematic effects. These techniques enable the conflict-crisis-resolution model to have conciseness and scope." These cinematic effects appear, not as full-motion video but as a kind of a film script, though Sanford's romantic tragedy does use occasional animations above and to the left of the main text spaces. I cite this elegant project not to criticize its lack of hypertextuality but to remind us that the digital word and image, even on the World Wide Web, does not inevitably produce hypertextual narrative.

The examples of hypertext fiction at which we shall look in the following pages and have already examined in the earlier discussion of writing hypermedia suggest that even in this early stage hypertext has taken many forms, few of which grant readers the kind of power one expects in informational hypertext. As Michael Joyce, our first major author of hypertext fiction, has explained, the desire to create multiple stories out of a relatively small amount of alphanumeric text provided a major force driving in writing *afternoon*:

I wanted, quite simply, to write a novel that would change in successive readings and to make those changing versions according to the connections that I had for some time naturally discovered in the process of writing and that I wanted my readers to share. In my eyes, paragraphs on many different pages could just as well go with paragraphs on many other pages, although with different effects and for different purposes. All that kept me from doing so was the fact that, in print at least, one paragraph inevitably follows another. It seemed to me that if I, as author, could use a computer to move paragraphs about, it wouldn't take much to let readers do so according to some scheme I had predetermined. (*Of Two Minds*, 31)

From one point of view, then, such an approach merely intensifies the agenda of high modernism, using linking to grant the author even more power.

Other authors take a self-consciously postmodern approach, using the multiplicity offered by branching links to create a combinatorial fiction that in some ways seems the electronic fulfillment of the French group, Oulipo. For example, in its forty-nine fictional lexias Tom McHarg's *The Late-Nite Maneuvers of the Ultramundane,* one of the hypertexts included in *Writing at the Edge,* attempt to "veer toward a narrative . . . not entirely dependent upon linearity, causality, and probable characterization" ("On Ultramundane"). McHarg creates seven lexias for each day of the week, each a variation or transformation of the other. Choosing the first Monday, for example, the reader encounters the following narrative (which represents the first half of the lexia):

Dwight awoke at 3:15 a.m. to find his girlfriend, Johnette, attempting to conceal a bomb under his pillow. "You've waken me up," he said. "And I've discovered your treachery."

"The only treachery is yours," said Johnette.

"I'm only sleeping," said Dwight. "You're the one planting bombs."

"Perhaps you deserve it," said Johnette.

"But I love you," said Dwight.

"Then why do you accuse me of treachery?" said Johnette.

"It's obvious," said Dwight. "You planned to murder me as I lie here dreaming of our sex."

"You weren't dreaming," said Johnette. "I was watching your eyes."

"Perhaps not, but at least I wasn't trying to murder you," said Dwight.

"I only meant to scare you into loving me more," said Johnette.

"With a bomb ?" said Dwight.

"You need to love me a lot more," said Johnette.

And so it goes. Each variation introduces a different weapon, a different betrayal, as *Ultramundane* explores how "a fictional text must be stretched, skewered, and sliced if it is to exploit the freedoms and accept the responsibilities offered by hypertext technology and its new writing spaces." Thus, on the first Tuesday a friend stands at the foot of their bed with a gun, on the first Wednesday the hero's hair has all fallen out, and on Sunday Dwight returns to find their home on fire. Like *afternoon, The Late-Nite Maneuvers of the Ultramundane* combines into different narratives, many about sex and violence, producing effects according to the route one follows through it. Otherwise, this web, whose tone and content suggest the influence of Coover's print fiction, contrasts entirely with Joyce's work. My point here is not that one should prefer either the crystalline richness and emotional intensity of *afternoon* to McHarg's PoMo playful, removed sense of literature as its own laboratory, or that *Ultramundane* is in some way more hypertextual. Rather, I wish to emphasize that, like print fiction, that produced in the form of linked lexias can take many forms.

In some the author compounds her or his power; in others, such as that exemplified by Carolyn Guyer's *Quibbling,* the author willingly shares it with readers. Similarly, in some fictional webs, such as *afternoon,* readers construct or discover essentially one main narrative; in others, like *Semio-Surf, Freak Show,* or *Ultramundane,* one comes upon either a cluster of entirely separate stories, or else one finds narrative segments out of which one weaves one or more of them. A third opposition appears between those stories that, however allusive, consist largely of fictional lexias, and those like *Patchwork Girl* and *Semio-Surf,* that weave together theory and nonfiction materials with the narrative. A fourth such opposition separates fictional hypertext entirely derived from the author's "own" writing and those, like *Patchwork Girl* and Stuart Moulthrop's *Forking Paths,* that their authors wrote to varying degrees in the interstices of other works.

Hypertext narrative clearly takes a wide range of forms best understood in terms of a number of axes, including those formed by degrees or ratios of (1) reader choice, intervention, and empowerment, (2) inclusion of extralinguistic texts (images, motion, sound), (3) complexity of network structure, and (4) degrees of multiplicity and variation in literary elements, such as plot, characterization, setting, and so forth. Following the lead of Deleuze and Guattari, I prefer to think of the organizing structures in terms of ranges, spectra, or axes along which one can array the phenomena I discuss rather than in terms of diametric oppositions, such as male/female or alphanumeric versus multimedia text. I avoid such polarities because, particularly in the case of hypertext fiction and poetry, they hinder analysis by exaggerating difference, overrating uniformity, and suppressing our abilities to perceive complex mixtures of qualities or tendencies. Another reason for emphasizing a spectrum of possibilities when discussing hypertext lies in the fact that neither end of any particular spectrum is necessarily superior to the other. For example, hyperfiction that demands intervention by readers, or otherwise empowers them, will not on those grounds alone turn out to surpass hyperfiction that employs links to solidify—indeed amplify—the power of the author.

Hypertext and the Aristotelian

Conception of Plot

Hypertext, which challenges narrative and all literary form based on linearity, calls into question ideas of plot and story current since Aristotle. Looking at the *Poetics* in the context of a discussion of hypertext suggests one of two things: either one simply cannot write hypertext fiction (and the *Poetics* shows why that could be the case) or else Aristotelian definitions and descriptions of plot do not apply to stories read and written within a hypertext environment. At the beginning of this study, I proposed that hypertext permits a particularly effective means of testing literary and cultural theory. Here is a case in point. Although hypertext fiction is quite new, the examples of it that I have seen already call into question some of Aristotle's most basic points about plot and story. In the seventh chapter of the *Poetics,* Aristotle offers a definition of plot in which fixed sequence plays a central role:

Now a whole is that which has beginning, middle, and end. A beginning is that which is not itself necessarily after anything else, and which has naturally something else after it; an end is that which is naturally after something itself either as its necessary or usual consequent, and with nothing else after it; and a middle, that which is by nature after one thing and also has another after it. (1462)

Furthermore, Aristotle concludes, "a well-constructed Plot, therefore, cannot either begin or end at any point one likes; beginning and end in it must be of the forms just described. Again: to be beautiful, a living creature, and every whole made up of parts, must not only present a certain order in its arrangement of parts, but also be of a certain definite magnitude" (1462). Hypertext therefore calls into question (1) fixed sequence, (2) definite beginning and ending, (3) a story's "certain definite magnitude," and (4) the conception of unity or wholeness associated with all these other concepts. In hypertext fiction, therefore, one can expect individual forms, such as plot, characterization, and setting, to change, as will genres or literary kinds produced by congeries of these techniques.

When I first discussed hyperfiction in the earlier versions of this book, the novelty, the radical newness, of the subject appeared in the fact that almost all the sources cited were unpublished, in the process of being published, or published in nontraditional electronic forms: these sources include unpublished notes on the subject of hypertext and fiction by a leading American novelist, chapters in forthcoming books, and prerelease versions of hypertext fictions. Now, a few years later, substantial numbers of examples of both hypertext fiction and critical discussions of the subject have appeared. Following the strategy used in previous chapters, I shall therefore take almost all of my examples from widely available work, using whenever possible material published on the World Wide Web.

Quasi-Hypertextuality in

Print Texts

One approach to predicting the way hypertext might affect literary form has pointed to *Tristram Shandy, In Memoriam, Ulysses,* and *Finnegans Wake* and to recent French, American, and Latin American fiction, particularly that by Michel Butor, Marc Saporta, Robert Coover, and Jorge Luis Borges (Bolter, *Writing Space,* 132–39). Such texts might not require hypertext to be fully understood, but they reveal new principles of organization or new ways of being read to readers who have experienced hypertext. Hypertext, the argument goes, makes certain elements in these works stand out for the first time. The example of these very different texts suggests that those poems and novels that most resist one or more of the characteristics of literature associated with print form, particularly linear narrative, will be likely to have something in common with new fiction in a new medium.

This approach therefore uses hypertext as a lens, or new agent of perception, to reveal something previously unnoticed or unnoticeable, and it then extrapolates the results of this inquiry to predict future developments. Because such an approach suggests that this new information technology has roots in prestigious canonical texts, it obviously has the political advantage of making it seem less threatening to students of literature and literary theory. At the same time, placing hypertext fiction within a legitimating narrative of descent from "great works," which offers material for new critical readings of print texts, makes those canonical texts appear especially forward-looking, since they can be seen to provide the gateway to a different and unexpected literary future. I find all these genealogical analyses attractive and even convincing, but I realize that if hypertext has the kind and degree of power that previous chapters have indicated, it does threaten literature and its institutions as we know them. One *should* feel threatened by hypertext, just as writers of romances and epics should have felt threatened by the novel and Venetian writers of Latin tragedy should have felt threatened by the *Divine Comedy* and its Italian text. Descendants, after all, offer continuity with the past but only at the cost of replacing it.

One interesting approach to discussing hypertextual narrative involves deducing its qualities from the defining characteristics of hypertext—its non- or multilinearity, its multivocality, and its inevitable blending of media and modes, particularly its tendency to marry the visual and the verbal. Most who have speculated on the relation between hypertextuality and fiction concentrate, however, on the effects it will have on linear narrative. In order to comprehend the combined promise and peril with which hypertextuality confronts narrative, we should first recall that narratology generally urges that narration is intrinsically linear and also that such linearity plays a central role in all thought.[1] As Barbara Herrnstein Smith argues, "there are very few instances in which we can sustain the notion of a set and sequence of events altogether prior to and independent of the discourse through which they are narrated" ("Narrative Versions," 225).

Hayden White states only a particularly emphatic version of a common assumption when he asserts that "to raise the question of the nature of narrative is to invite reflection on the very nature of culture and, possibly, even on the nature of humanity itself . . . Far from being one code among many that a culture may utilize for endowing experience with meaning, narrative is a metacode, a human universal on the basis of which transcultural messages about the nature of a shared reality can be transmitted" (1–2). What kind of a culture would have or could have hypertextual narration, which so emphasizes non- or multilinearity, and what happens to a culture that chooses such narration, when, as Jean-François Lyotard claims, in agreement with many other writers on the subject, "narration is the quintessential form of customary knowledge" (*Postmodern Condition,* 18)? Lyotard's own definition of postmodernism as "incredulity toward metanarratives" (xxiv) suggests one answer: any author and any culture that chooses hypertextual fiction will either already have rejected the solace and reassurance of linear narrative or will soon find their attachment to it loosening. Lyotard claims that "lamenting the 'loss of meaning' in postmodernity boils down to mourning the fact that knowledge is no longer principally narrative" (26), and for this loss of faith in narrative he offers several possible technological and political explanations, the most important of which is that science, which "has always been in conflict with narratives," uses other means "to legitimate the rules of its own game" (xiii).[2]

Even without raising such broader or more fundamental issues about the relation of narrative to culture, one realizes that hypertext opens major questions about story and plot by apparently doing away with linear organization. Conventional definitions and descriptions of plot suggest some of them. Aristotle long ago pointed out that successful plots require a "probable or necessary sequence of events" (*Poetics,* 1465). This observation occurs in the midst of his discussion of *peripeteia* (or in Bywater's translation, peripety), and in the immediately preceding discussion of episodic plots, which Aristotle considers "the worst," he explains that he calls "a Plot episodic when there is neither probability nor necessity in the sequence of its episodes" (1464).

Answering Aristotle: Hypertext and the Nonlinear Plot

One answer to Aristotle lies in the fact that removing a single "probable or necessary sequence of events" does not do away with all linearity. Linearity, however, now becomes a quality of the individual reader's experience within a single lexia and his or her experience following a path, even if that path curves back on itself or heads in strange directions. Robert Coover claims that with hypertext "the linearity of the reading experience" does not disappear entirely, "but narrative bytes no longer follow one another in an ineluctable page-turning chain. Hypertextual story space is now multidimensional and theoretically infinite, with an equally infinite set of possible network linkages, either programmed, fixed or variable, or random, or both" ("Endings").

Coover, inspired by the notion of the active hypertext reader, envisions some of the ways the reader can contribute to the story. At the most basic level of the hypertext encounter, "the reader may now choose the route in the labyrinth she or he wishes to take, following some particular character, for example, or an image, an action, and so on." Coover adds that readers can become reader-authors not only by choosing their paths through the text but also by reading more actively, by which he means they "may even interfere with the story, introduce new elements, new narrative strategies, open new paths, interact with characters, even with the author. Or authors." Here, of course, Coover, who has used Intermedia and Storyspace in his hyperfiction classes, refers to the kind of hypertexts possible only with systems that permit readers to add text and links, and very few works of this sort have been attempted. Although some authors and audiences might find themselves chilled by such destabilizing, potentially chaotic-seeming narrative worlds,

Coover, a freer spirit, mentions "the allure of the blank spaces of these fabulous networks, these green-limned gardens of multiply forking paths, to narrative artists" who have the opportunity to "replace logic with character or metaphor, say, scholarship with collage and verbal wit, and turn the story loose in a space where whatever is possible is necessary."

Coover offers a vision of possibilities, and now that many instances of hyperfiction have seen publication, one can make some first guesses which of its suggestions seem most likely to be realized and which least. Although true that readers in certain systems—those with search tools or something like Microcosm's compute links functions—can follow "some particular character, for example, or an image, an action, and so on," the hypertext software used for writing most hyperfiction thus far does not make such reading easy or, in some cases, even possible. Similarly, Intermedia permitted reader-authors to enter the text freely, and students at Brown created works, such as *Hotel,* one version of which has since appeared in the web and recreated as a MOO (see Meyer, Blair, Hader, "*WAXweb*"). *Hotel* permitted any reader to add a new room to the fictional structure as well as change or even delete the work of others. Speaking with Coover in the Spring of 2004, I learned that he considered the experiment a failure because too many visitors vandalized the work of others. Just as the Internet later showed, cyber-utopian techno-anarchy did not fulfill the hopes of those who first envisaged it. I consider *Hotel* not so much a failure as an experiment using digital literature to test a theory. It was worth the effort.

The way readers follow links presents an even more fundamental issue in hyperfiction. In contrast to informational hypertext, which must employ rhetorics of orientation, navigation, and departure to orient the reader, successful fictional hypertext and poetry does not always do so with the result that its readers cannot make particularly informed or empowered choices. Webs created in systems like Storyspace that permit one-to-many links, link menus, and path names all provide authors with the power to empower the reader; that is, authors can write in such a way to provide the reader with informed choices. Taro Ikai's *Electronic Zen,* which we shall later examine in more detail (chapter 7), exemplifies a web that chooses to do so. Using Storyspace's link menus (rather than its richer ability to name links and create paths), Ikai named, as we have already observed, the first destinations "water" and "chef." After reading even a few of his brief lexias, readers realize that he has created two paths, one characterized by Zen meditations and the other by details of the speaker's mundane existence. With that information, readers can now choose which path they wish to follow.

Similarly, in *Quibbling*, which employs the Storyspace view, Carolyn Guyer permits readers at any point to leave individual lexias and pursue the characters and narratives suggested by the names of individual folders and folders within folders. In *Victory Garden* Stuart Moulthrop uses path names, overviews, and other devices to encourage readers to make wise choices. Many hypertext fictions published thus far show, in contrast, that authors either prefer authorial power, reader disorientation, or both. Rather than concluding too quickly that Coover's vision has motes or blurs, we should recognize two things: first, that there will be—indeed, there already are—as many kinds of hyperfiction as occur in print; there is probably no one ideal; and, second, that fictional hypertext has different purposes, modes, and effects than its informational and educational forms.

Print Anticipations of Multilinear

Narratives in E-Space

Doing away with a fixed linear text therefore neither necessarily does away with all linearity nor removes formal coherence, though it may appear in new and unexpected forms. Bolter points out that

in this shifting electronic space, writers will need a new concept of structure. In place of a closed and unitary structure, they must learn to conceive of their text as a structure of possible structures. The writer must practice a kind of second-order writing, creating coherent lines for the reader to discover without closing off the possibilities prematurely or arbitrarily. This writing of the second order will be the special contribution of the electronic medium to the history of literature. (*Writing Space*, 144)

In "Poem Descending a Staircase," William Dickey, a poet who works with hypertext, similarly suggests that authors can pattern their hypertexts by creating links that offer several sets of distinct reading paths: "The poem may be designed in a pattern of nested squares, as a group of chained circles, as a braid of different visual and graphic themes, as a double helix. The poem may present a single main sequence from which word or image associations lead into subsequences and then return" (147). Hypertext systems that employ single directional as opposed to bidirectional linking make this kind of organization easier, of course, but fuller and freer forms of the medium also make such quasi-musical organization possible and even inevitable. The main requirement, as Paul Ricoeur suggests, becomes "this 'followability' of a story," and followability provides a principle that permits many options, many permutations (1: 67).

Another possible form of hypertextual literary organization involves parataxis, which is produced by repetition rather than sequence. Smith explains that in literary works that employ logical or temporal organization, "the dislocation or omission of any element will tend to make the sequence as a whole incomprehensible, or will radically change its effect. In paratactic structure, however (where the principle of generation does not cause any one element to 'follow' from another), thematic units can be added, omitted, or exchanged without destroying the coherence or effect of the poem's thematic structure." According to Smith, "'variations on a theme' is one of the two most obvious forms that paratactic structure may take. The other one is the 'list.'" The main problem with which parataxis, like hypertext, confronts narrative is that any "generating principle that produces a paratactic structure cannot in itself determine a concluding point" (*Poetic Closure,* 99–100).

Since some narratologists claim that morality ultimately depends on the unity and coherence of a fixed linear text, one wonders if hypertext can convey morality in any significant form or if it is condemned to an essential triviality. White believes the unity of successful narrative to be a matter of ideology: "Narrativity, certainly in factual storytelling and probably in fictional storytelling as well, is intimately related to, if not a function of, the impulse to moralize reality, that is, to identify it with the social system that is the source of any morality that we can imagine" (14). Writing as a historian and historiographer, White argues that such ideological pressure appears with particular clarity in the "value attached to narrativity in the representation of real events," since that value discloses a desire to endow "real events" with a necessarily imaginary "coherence, integrity, fullness, and closure" possible only in fiction. The very "notion that sequences of real events possess the formal attributes of stories we tell about imaginary events," insists White, "could only have its origin in wishes, daydreams, reveries" (20, 23). Does this signify or suggest that contemporary culture, at least its avant-garde technological phalanx, rejects such wishes, daydreams, and reveries? White's connection of plot and morality suggests several lines of inquiry. One could inquire if it is good or bad that linear narratives inevitably embody some morality or ideology, but first one should determine if rejecting linearity necessarily involves rejecting morality. After all, anyone taking seriously the fictional possibilities of hypertext wants to know if it will produce yet another form of postmodernist fiction that critics like John Gardner, Gerald Graff, and Charles Newman will attack as morally corrupt and corrupting (McHale, 219). If one wanted liberation from ideology, were such a goal possible, nonideological storytelling might be fine. But before concluding that hypertext produces either ideology-free miracles or ideology-free horrors, one should look at the available evidence. In particular, one should examine prehypertext attempts to create nonlinear or multilinear literary forms and evaluate the results.

A glance at previous experiments in avoiding the linearity of the printed text suggests that in the past authors have rejected linearity because it falsified their experience of things. Tennyson, for example, as we have already observed, created his poetry of fragments in an attempt to write with greater honesty and with greater truth about his own experience. Moreover, as several critics have pointed out, novelists at least since Laurence Sterne have sought to escape the potential confinements and falsifications of linear narrative.

One does not have to look back at the past for examples. In his review of *Dictionary of the Khazars,* a work by the Yugoslavian Milorad Pavic that Robert Coover describes as a hypertext novel, he asserts that "there is a tension in narrative, as in life, between the sensation of time as a linear experience, one thing following sequentiality (causally or not) upon another, and time as a patterning of interrelated experiences reflected upon as though it had a geography and could be mapped" ("He Thinks," 15). Nonlinear form, whether pleasing to readers or even practically possible, derives from attempts to be more truthful rather than from any amorality. Many contemporary works of fiction explore this tension between linear and more spatial sensations of time that Coover describes. Graham Swift's *Waterland* (1983), for instance, questions all narrative based on sequence, and in this it agrees with other novels of its decade. Like Penelope Lively's *Moon Tiger* (1987), another novel in the form of the autobiography of a historian, *Waterland* relates the events of a single life to the major currents of contemporary history.

Using much the same method for autobiography as for history, Swift's protagonist would agree with Lively's Claudia Hampton, whose deep suspicion of chronology and sequence explicitly derive from her experience of simultaneity. Ricoeur suggests that "the major tendency of modern theory of narrative—in historiography and the philosophy of history as well as narratology—is to 'dechronologize narrative,'" and these two novelists exemplify a successful "struggle against the linear representation of time" (1: 30). Thinking over the possibility of writing a history of the world, Lively's heroine rejects sequence and linear history as inauthentic and false to her experience:

The question is, shall it or shall it not be linear history? I've always thought a kaleidoscopic view might be an interesting heresy. Shake the tube and see what comes out. Chronology irritates me. There is no chronology inside my head. I am composed of a myriad Claudias who spin and mix and part like sparks of sunlight on water. The pack of cards I carry around is forever shuffled and re-shuffled; there is no sequence, everything happens at once. (2)

Like Proust's Marcel, she finds that a simple sensation brings the past back flush upon the present, making a mockery of separation and sequence.

Returning to Cairo in her late sixties, Claudia finds it both changed and unchanged. "The place," she explains, "didn't look the same but it felt the same; sensations clutched and transformed me." Standing near a modern concrete and plate-glass building, she picks a "handful of eucalyptus leaves from a branch, crushed them in my hand, smelt, and tears came to my eyes. Sixty-seven-year-old Claudia . . . crying not in grief but in wonder that nothing is ever lost, that everything can be retrieved, that a lifetime is not linear but instant." Her lesson for narratology is that "inside the head, everything happens at once" (68). Like Claudia, Tom Crick takes historical, autobiographical narratives whose essence is sequence and spreads them out or weaves them in a nonsequential way.

The difference between quasi-hypertextual fictions and those in electronic form chiefly involves the greater freedom and power of the hypertext reader. Swift decides when Tom Crick's narrative branches and Lively decides when Claudia Hampton's does, but in Stuart Moulthrop's hypertext version of Borges's "Forking Paths" and in Leni Zumas's *Semio-Surf* or Carolyn Guyer's *Quibbling*, the reader makes this decision. Important prehypertextual narrative has, however, also required such reader decision. One of the most famous examples of an author's ceding power to the reader is found in "The Babysitter," in which Robert Coover, like an author of electronic hypertext, presents the reader with multiple possibilities, really multiple endings, with two effects. First, the reader, who takes over some of the writer's role and function, must choose which possibility, if any, to accept, and second, by encountering that need to decide, the reader realizes both that no true single narrative exists as the main or "right" one and that reading traditional narrative has brainwashed him or her into expecting and demanding a single right answer and a single correct story line. Coover's story not only makes a fundamentally moral point about the nature of fiction but also places more responsibility on the reader. One may say of Coover's text, in other words, what Bolter says of Joyce's interactive hypertext—that "there is no single story of which each reading is a version, because each reading determines the story as it goes. We could say that there is no story at all; there are only readings" (*Writing Space*, 124).

Narrative Beginnings and Endings

As we have already observed in chapter 3, the problems that hypertext branching create for narrativity appear with particular clarity in the matter of beginning and ending stories. If, as Edward Said claims, "a 'beginning' is designated in order to indicate, clarify, or define a *later* time, place, or action," how can hypertext fiction begin or be said to begin? Furthermore, if as Said also convincingly argues, "when we point to the beginning of a novel . . . we mean that from that beginning in principle follows *this* novel" (5), how can we determine what novel follows from the beginnings each reader chooses?

Thus far most of the hypertext fictions I have read or heard described, like many collections of educational materials, take an essentially cautious approach to the problems of beginnings by offering the reader a lexia labeled something like "start here" that combines functions of title page, introduction, and opening paragraph. They do so for technological, rhetorical, and other reasons. Most authors who wrote in HyperCard, Guide, and Storyspace did not use these environments on networks that could distribute one's texts to other reading sites. To disseminate their writings, authors had to copy them from their own machines to some sort of transfer media—at first floppy discs, later Zip or similar disks, CD-ROMs, or little memory cards that plug into a USB port—and then give that physical storage device to someone with another computer. This use of non-networked (or stand-alone) machines encourages writers to produce stories or poems that are both self-contained and small enough to fit on a single disk. In addition, since most of these early hypertextual environments do not give the reader the power to add links, authors in them necessarily tend to consider their works to be self-contained in a traditional manner. Another reason for using the "start here" approach appears in some writers' obvious reluctance to disorient readers upon their initial contact with a narrative, and some writers also believe that hypertextual fiction should necessarily change our experience of the middle but not the beginnings of narrative fiction.

In contrast, William Dickey, who has written hypertext poetry using Apple's HyperCard, finds it a good or useful quality of hypertext poetry that it "may begin with any one of its parts, stanzas, images, to which any other part of the poem may succeed. This system of organization requires that that part of the poem represented on any one card must be a sufficiently independent statement to be able to generate a sense of poetic meaning as it follows or is followed by any other statement the poem contains" (147). Dickey, who is writing about poetic rather than fictional structure, nonetheless offers organizational principles that apply to both.

Beginnings imply endings, and endings require some sort of formal and thematic closure. Ricoeur, using the image of "following" that is conventionally applied to narratives that writers about hypertext also use to describe activating links, explains that "to follow a story is to move forward in the midst of contingencies and peripeteia under the guidance of an expectation that finds its fulfillment in the 'conclusion' of the story." This conclusion "gives the story an 'end point,' which, in turn, furnishes the point of view from which the story can be perceived as forming a whole." In other words, to understand a story requires first comprehending "how and why the successive episodes led to this conclusion, which, far from being foreseeable, must finally be acceptable, as congruent with the episodes brought together by the story" (1: 66–67).

In her classic study of how poems produce satisfying endings, Smith provides evidence that might prompt students of hypertext to conclude that it either creates fundamental problems in narrative and other kinds of literary texts or else that it opens them to an entirely new form of textuality. She explains that since "a poem cannot continue indefinitely" (*Poetic Closure*, 33), it must employ devices that prepare the reader for ending rather than continuing. These devices produce in the reader "the sense of stable conclusiveness, finality, or 'clinch' . . . referred to here as closure . . . Whether spatially or temporally perceived, a structure appears 'closed' when it is experienced as integral: coherent, complete, and stable" (2)—qualities that produce a "sense of ultimate composure we apparently value in our experience of a work of art" and that we label "stability, resolution, or equilibrium" (34). Unlike texts in manuscript or print, those in hypertext apparently can continue indefinitely, perhaps infinitely, so one wonders if they can provide satisfying closure.[3] Or to direct this inquiry in ways suggested by Smith's analysis of closure, one should ask what techniques might provide something analogous to that desirable "sense of stable conclusiveness, finality, or 'clinch.'"

Taking another clue from fiction created for print publication, one perceives that many prehypertext narratives provide instances of multiple closure and also a combination of closure with new beginnings. Both Charles Dickens's novels written specifically for publication in periodicals at monthly intervals and those by other nineteenth-century novelists intended for first publication in the conventional triple-decker form make use of partial closure followed by continuation. Furthermore, Trollope's Palliser series, Lawrence Durrell's *Alexandria Quartet,* Faulkner's works, and countless trilogies and tetralogies in both fantastic and realistic modes suggest that writers of fiction have long encountered problems very similar to those faced by writers of hypertext fiction and have developed an array of formal and thematic solutions to them. In fact, the tendency of many twentieth-century works to leave readers with little sense of closure—either because they do not learn of the "final" outcome of a particular narrative or because they leave the story before any outcome occurs—shows us that as readers and writers we have long learned to live (and read) with more open-endedness than discussions of narrative form might lead us to expect.

Coover proclaims that endings will and must occur even in infinitely expandable, changeable, combinable docuverses:

There is still movement, but in hyperspace it is that of endless expansion. "A" is, or may be, an infinite multiplicity of starting points, "B" a parenthetical "B" somewhere beyond the beyond, or within the within, yet clearly mapped, clearly routed, just somewhat less definite than, oh, say, dying. Which for all the networking maneuvers and funhouse mirrors cannot be entirely ignored. Sooner or later, whatever the game, the whistle is blown. Even in hyperspace, there is disconnection. One last windowless trajectory. ("Endings")

Hypertext fictions always end because readings always end, but they can end in fatigue or in a sense of satisfying closure. Writing of the printed text, Barbara Herrnstein Smith reminds us that "the end of the play or novel will not appear as an arbitrary cut-off if it leaves us at a point where, with respect to the themes of the work, we feel that we know all there is or all there is to know" (*Poetic Closure*, 120). If individual lexias provide readers with experiences of formal and thematic closure, they can be expected to provide the satisfactions that Smith describes as requisite to the sense of an ending.

Michael Joyce's *afternoon*

Michael Joyce, a hypertext author, is suspicious of closure. In Joyce's *afternoon*, a hypertext fiction in 538 lexias, the section appropriately entitled "work in progress" advises readers: "Closure is, as in any fiction, a suspect quality, although here it is made manifest. When the story no longer progresses, or when it cycles, or when you tire of the paths, the experience of reading it ends." In other words, Joyce makes the responsibility for closure, for stopping, entirely the reader's. When the reader decides he or she has had enough, when he or she wishes to stop reading, why then the story is over. Joyce continues, however: "Even so, there are likely to be more opportunities than you think there are at first. A word which doesn't yield the first time you read a section may take you elsewhere if you choose it when you encounter the section again; and what sometimes seems a loop, like memory, heads off in another direction." Reading the highly allusive *afternoon*, which has so many points of departure within each lexia as well as continually changing points of linkage, one sees what Joyce means[4] (Figure 27).

The successive lexias one encounters seem to take form as chains of narrative, and despite the fact that one shifts setting and narrator, one's choices produce satisfying narrative sets. Moving from section to section, every so often one encounters puzzling changes of setting, narrator, subject, or chronology, but two things occur. After reading awhile one begins to construct

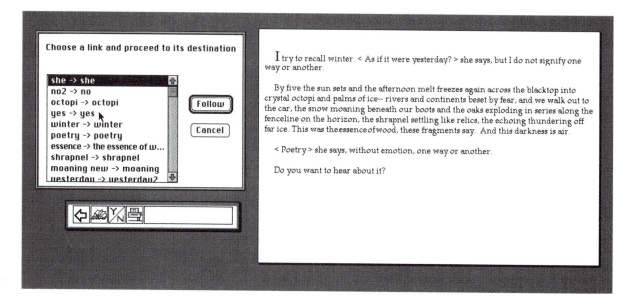

Figure 27. The Storyspace Page Reader: Michael Joyce's *afternoon*. The Page Reader, one of several ways of distributing a system's webs to readers without the authoring environment, has a moveable palette providing access to five functions: the arrow at left provides a backtracking function, and clicking on the book icon produces a menu of all links from the current lexia, such as appears at the left. In addition, readers can respond positively or negatively to questions in the text, such as we encounter here, by choosing the appropriate button, or they can search for links relating to a specific word by typing it in the space at right. Finally, readers can print individual lexias by using the icon at far right. (Courtesy of Eastgate Systems.)

narrative placements, so that one assigns particular sections to a provisionally suitable place—some lexias obviously have several alternate or rival forms of relation. Then, having assigned particular sections to particular sequences or reading paths, many, though not all, of which one can retrace at will, one reaches points at which one's initial cognitive dissonance or puzzlement disappears, and one seems satisfied. One has reached—or created—closure!

One might describe Joyce's hypertext fiction in the way Gérard Genette describes "what one calls Stendhal's *oeuvre*":

a fragmented, elliptical, repetitive, yet infinite, or at least indefinite, text, no part of which, however, may be separated from the whole. Whoever pulls a single thread must take the whole cloth, with its holes and lack of edges. To read Stendhal is to read the whole of Stendhal, but to read all of Stendhal is impossible, for the very good reason, among others, that the whole of Stendhal has not yet been published or deciphered, or discovered, or even written: I repeat, all the Stendhalian text, because the gaps, the interruptions are not mere absences, a pure non-text: they are a lack, active and perceptible as lack, as non-writing, as non-written text. (*Figures*, 165)

Genette, I suggest, describes the way a reader encounters the web of Joyce's hypertextual narrative. Even entering at a single point determined by the author, the reader chooses one path or another and calls up another lexia by a variety of means, and then repeats this process until she or he finds a hole or a gap. Perhaps at this point the reader turns back and takes another direction. One might just as well write something oneself or make present a remembered passage by another author in the manner that a book reader might begin a poem by Stevens, think of some parallel verses by Swinburne or a passage in a book by Helen Vendler or Harold Bloom, pull that volume off its shelf, find the passage, and then return to the poem by Stevens.

Whereas Genette's characterization of the Stendhalian oeuvre captures the reader's experience of the interconnectedness of *afternoon* and other hypertext fictions, his description of temporality in Proust conveys the experience of encountering the disjunctions and jumps of hypertextual narrative. Citing George Poulet's observation that in *À la Recherche du temps perdu* time does not appear as Bergsonian duration but as a "succession of isolated moments," he points out that similarly "characters (and groups) do not evolve: one fine day, they find that they have changed, as if time confined itself to bringing forth a plurality that they have contained in potentia from all eternity. Indeed, many of the characters assume the most contradictory roles *simultaneously*" (216). In other words, in *À la Recherche du temps perdu* readers find themselves taking leaps and jumping into a different time and a different character. In a hypertext narrative it is the author who provides multiple possibilities by means of which the readers themselves construct temporal succession and choose characterization—though, to be sure, readers will take leaps, as we do in life, on the basis of inadequate or even completely inaccurate information.

So many different contexts cross and interweave that one must work to place the characters encountered in them. Joyce's world, which also inevitably includes the *other* Joyce, has many moving centers of interest, including marriage and erotic relationships, sexual politics, psychotherapy, advertising, filmmaking and the history of cinema, computing, myth, and literature of all kinds. Reading habits one has learned from print play a role in organizing these materials. If one encounters a speaker's mention of his marriage and, in a later lexia, finds him at the scene of an automobile accident from which the bodies or injured people have already been removed, one might take the accident as an event in the recent present; the emotional charge it carries serves to organize other reported thoughts and events, inevitably turning some of them into flashbacks, others into exposition. Conversely, one could take that event as something in the past, a particularly significant moment, and then use it as a point of origin either that leads to other events or whose importance endows events it has not caused with a significance created by explanation or contrast or analogy. Our assistance in the storytelling or story-making is not entirely or even particularly random since Joyce provides many hooks that can catch at our thoughts, but we do become reader-authors and help tell the tale we read.

Nonetheless, as J. Hillis Miller points out, we cannot help ourselves: we must create meaning as we read. "A story is readable because it can be organized into a causal chain . . . A causal sequence is always an implicit narrative organized around the assumption that what comes later is caused by what comes before, 'post hoc, propter hoc.' If any series of random and disconnected events is presented to me, I tend to see it as a causal chain. Or rather, if Kant and Kleist are right, I must see it as a causal chain" (*Versions,* 127, 130). Miller, who silently exchanges a linear model of explanation for one more appropriate to hypertextual narrative, later adds: "We cannot avoid imposing some set of connections, like a phantasmal spiderweb, over events that just happen as they happen" (139).

Miller's idea of reading printed text, which seems to owe a great deal to gestalt psychology's theories of constructionist perception, well describes the reader-author demanded by Joyce's *afternoon* and other works of hypertext fiction. According to Miller, reading is always "a kind of writing or rewriting that is an act of prosopopoeia, like Pygmalion giving life to the statue" (186).[5] This construction of an evanescent entity or wholeness always occurs in reading, but in reading hypertext it takes the additional form of constructing, however provisionally, one's own text out of fragments, out of separate lexias. It is a case, in other words, of Lévi-Strauss's *bricolage,* for every hypertext reader is inevitably a *bricoleur.*

Such *bricolage,* I suggest, provides a new kind of unity, one appropriate to hypertextuality. As long as one grants that plot is a phenomenon created by the reader with materials the lexias offers, rather than a phenomenon belonging solely to the text, then one can accept that reading *afternoon* and other hypertext fictions produces an experience very similar to that provided by reading the unified plot described by narratologists from Aristotle to White and Ricoeur. White, for example, defines plot as "a structure of relationships by which the events contained in the account are endowed with a meaning by being identified as parts of an integrated whole" (9). Ricoeur similarly defines plot, "on the most *formal* level, as an integrating dynamism that draws a unified and complete story from a variety of incidents, in other words, that transforms this variety into a unified and complete story. This formal definition opens a field of rule-governed transformations worthy of being called plots so long as we can discern temporal wholes bringing about a synthesis of the heterogeneous between circumstances, goals, means, interactions, and intended and unintended results" (2: 8). According to Ricoeur, the metaphorical imagination produces narrative by a process of what he terms "predicative assimilation," which "'grasps together' and integrates into one whole and complete story multiple and scattered events, thereby schematizing the illegible signification attached to the narrative taken as a whole" (1: x). To this observation I would add, with Miller, that as readers we find ourselves forced to fabricate a whole or, as he puts it, integrate "into one whole and complete story multiple and scattered events, separate parts."

In his chapter on Heinrich von Kleist in *Versions of Pygmalion,* Miller provides us with an unexpectedly related model for this kind of extemporized construction of meaning-on-the-run. He quotes Kleist's claim that Mirabeau was "unsure of what he was about to say" (104) when he began his famous speech that ended "by creating the new French nation and a new parliamentary assembly" (105). The speaker posits a "syntactically incomplete fragment, says Kleist, without any idea . . . of where the sentence is going to end, [and] the thought is gradually 'fabricated'"; and Kleist claims that the speaker's feelings and general situation in some way produce his proposals. Disagreeing with him, Miller argues in the manner of Barthes that Mirabeau's revolutionary "thought is gradually fabricated not so much by the situation or by the speaker's feelings," as Kleist suggests, "but by his need to complete the grammar and syntax of the sentence he has blindly begun" (104). Structuralists and poststructuralists have long described thinking and writing in terms of this extemporized, in-process generation of meaning, the belief in which does so much to weaken traditional conceptions of self and author. Hypertext fiction forces us to extend this description of meaning-generation to the reader's construction of narrative. It forces us to recognize that the active reader fabricates text and meaning from "another's" text in the same way that each speaker constructs individual sentences and entire discourses from "another's" grammar, vocabulary, and syntax.

Vladimir Propp, following Veselovsky, long ago founded the "structuralist study of plot" and with it modern narratology applying notions of linguistic combination to the study of folk tales.[6] Miller, who draws upon this tradition, reminds us that fabricating folk tales, spoken discourses, and interpretative readings of print narratives follow an essentially similar process that entails the immediate, in-process construction of meaning and text. Miller's observations allow us to understand that one must apply the same notions to the activities of the reader of hypertext fiction. In brief, hypertext demands that one apply this structuralist understanding of speaker and writer to the reader as well, since in hypertext the reader is in this limited sense a reader-author. From this theory of the reader and from the experience of reading hypertext narratives, I draw the following, perhaps obvious but nonetheless important, conclusions. In a hypertext environment a lack of linearity does not destroy narrative. In fact, since readers always, but particularly in this environment, fabricate their own structures, sequences, and meanings, they have surprisingly little trouble reading a story or reading for a story.[7] Obviously, some parts of the reading experience seem very different from reading a printed novel or a short story, and reading hypertext fiction provides some of that experience of a new orality that both McLuhan and Ong have predicted. Although the reader of hypertext fiction shares some experiences, one supposes, with the audience of listeners who heard oral poetry, this active reader inevitably has more in common with the bard, who constructed meaning and narrative from fragments provided by someone else, by another author or by many other authors.

Like Coover, who emphasizes the inevitable connection of death and narrative, Joyce seems to intertwine the two. In part it is a matter, as Brian McHale points out, of avant-garde authors using highly charged subjects (sexuality, death) to retain readers' interest that might stray within puzzling and unfamiliar narrative modes. In part it is also a matter of endings: when the reader decides to stop reading *afternoon,* he or she ends, kills, the story, because when the active reader, the reader-author, stops reading, the story stops, it dies, it has reached an ending. As part of that cessation, that willingness to stop creating and interpreting the story, certain acts or events in the story become deaths because they make most sense that way; and by stopping reading the reader prevents other alternatives from coming into being.

Stitching Together Narrative, Sexuality, Self: Shelley Jackson's *Patchwork Girl*

Patchwork Girl, Shelley Jackson's brilliant hypertext parable of writing and identity, generates both its themes and techniques from the kinds of collage writing intrinsic to hypertext (Figure 28). Jackson, a published book illustrator as well as author, creates a digital collage out of her own words and images (and those of others, including Mary Shelley, Frank L. Baum,

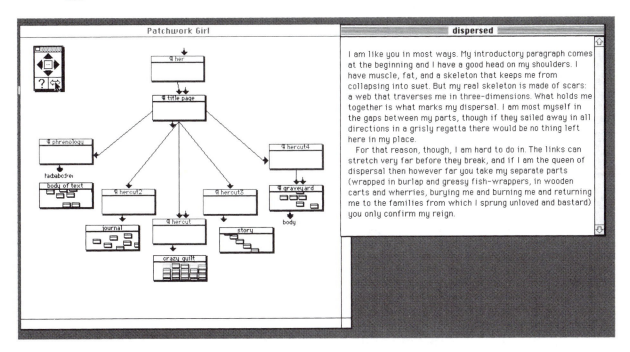

Figure 28. The Storyspace Reader: Shelley Jackson's *Patchwork Girl*. Readers can navigate the web (1) by simultaneously pressing option and ⌘ (or the command key in Windows) to discover linked text and then double clicking on it, (2) by mousing down on the double-headed arrow on the moveable palette (for default links), or (3) by exploring the Storyspace view, which consists of a folderlike arrangement of the web that authors can arrange as they wish. Unlike the Page Reader, which derives from the tiny size of original Macintosh screens, this form of Storyspace does not restrict documents to a cardlike format and permits scrollable text that readers can reconfigure. (Courtesy of Eastgate Systems.)

and Jacques Derrida) as she tells us about the female companion to Frankenstein's monster whose "birth takes place more than once. In the plea of a bygone monster; from a muddy hole by corpse-light; under the needle, and under the pen."

One form of collage in *Patchwork Girl* appears in the thirty-lexia section entitled "Crazy Quilt." As Karyn Raz explains in her section of "*Patchwork Girl* Comments," a student-created portion of the *Cyberspace, Hypertext, and Critical Theory Web*, "Each patch in Jackson's quilt is composed of various other patches, various other texts, from theoretical to fictional, from pop cultural to hearsay, sewn together to form either a sentence or paragraph" ("Patches"). The lexia entitled "seam'd" thus combines sentences from *Getting Started with Storyspace*, Frank L. Baum's *Patchwork Girl of Oz*, and Barbara

Maria Stafford's *Body Criticism: Imagining the Unseen in Enlightenment Art and Medicine:*

You may emphasize the presence of text links by using a special style, color or type-face. Or, if you prefer, you can leave needles sticking in the wounds—in the manner of tailors—with thread wrapped around them. Being seam'd with scars was both a fact of eighteenth century life and a metaphor for dissonant interferences ruining any finely adjusted composition. "The charm you need is a needle and thread," said the Shaggy Man.

As Raz points out, "Stitches, or links connect one patch to another, one text to another. Jackson seems particularly interested in examining the points of union between texts, such that 'being seam'd with scars' becomes a fact not only of eighteenth-century life but of hypertext writing, and indeed of any sort of creative process."

Fittingly, my discussion of Jackson's web has already taken on much of the appearance of collage itself. I first wrote a good portion of what follows for the *Electronic Book Review,* one of an increasing number of critical and scholarly World Wide Web periodical publications, but after students in my course on cyberspace and critical theory supplemented it with an HTML web in the form of some two dozen commentaries, I decided to find some way here to draw upon their work in a manner appropriate to print. In the *Cyberspace, Virtual Reality, and Critical Theory Web,* our lexias appear woven together, and I could *add to* my single lexia simply by using links. Here, following the conventions of print, I shall sum up and introduce these additional comments (which are now, of course, part of the "main" text, and hence no longer "additional"), citing some longer passages in the endnotes. Back to *Patchwork Girl* and hypertext collage.

Most of *Patchwork Girl*'s collage effects occur, not within individual patches or lexias but across them as we readers patch together a character and a narrative. Opening Jackson's web, we first encounter a black-and-white image of the stitched-together protagonist that she cuts and recombines into the images we come upon at various points throughout our reading. The first link takes us to her title page, a rich crossroads document, to which we return repeatedly, that offers six paths out: "a graveyard," "a journal," "a quilt," "a story," "broken accents," and a list of sources. The graveyard, for example, takes us first to a patchwork image created by cutting and rearranging the title screen, after which we receive some directions and then reach the head-stone, another overview or crossroads lexia that provides multiple paths; these paths take us to the lives of each of the beings, largely women, whose parts contributed to the Patchwork Girl.

According to Jason Williams, the graveyard section functions as a collage of "mini-narratives and fragmented character sketches" that serves as a "matrix for the meta-character and her story." Its removal from the narrative of the Patchwork Girl's own life avoids "interrupting the story's flow, and its compartmentalization encourages the application of its contents to the smaller narrative subsections. The graveyard itself focuses on the headstone and the list of the urns' contents, giving it a double-focus radial structure that unifies the parts without imposing a hierarchical order upon them. The introductory and concluding lexias temporally frame this structure and facilitate passage to the more linear sections of the text" ("Texture, Topology, Collage, and Biology in *Patchwork Girl*").

Jackson endows each tale, each life, encountered in the graveyard with a distinctive voice, thereby creating a narrative of Bakhtinian multivocality while simultaneously presenting a composite image of women's lives at the turn of the nineteenth century. The Everywoman Monster's left leg, we read,

belonged to Jane, a nanny who harbored under her durable grey dresses and sensible undergarments a remembrance of a less sensible time: a tattoo of a ship and the legend, Come Back To Me. Nanny knew some stories that astonished her charges, and though the ship on her thigh blurred and grew faint and blue with distance, until it seemed that the currents must have long ago finished their work, undoing its planks one by one with unfailing patience, she always took the children to the wharf when word came that a ship was docking, and many a sailor greeted her by name.

My leg is always twitching, jumping, joggling. It wants to go places. It has had enough of waiting.

Patchwork Girl makes us all into Frankenstein-readers stitching together narrative, gender, and identity, for as it reminds us: "You could say all bodies are written bodies, all lives pieces of writing." This digital collage-narrative assembles Shelley Jackson's (and Mary Shelley's and Victor Frankenstein's) female monster, forming a hypertext Everywoman who embodies assemblage, concatenation, juxtapositions, and blurred, recreated identities—one of many digital fulfillments of twentieth-century literary and pictorial collages. As the monster slyly informs us in a lexia one encounters early on,

I am buried here. You can resurrect me, but only piecemeal. If you want to see the whole, you will have to sew me together yourself. (In time you may find appended a pattern and instructions—for now, you will have to put it together any which way, as the scientist Frankenstein was forced to do.) Like him, you will make use of a machine of mysterious complexity to animate these parts.

In emphasizing the way we as her readers have to start out without a map or plan and then do a lot of the assembling ourselves, Jackson playful prepares us for the gaps and jumps we shall have to make.

In making us all readers in the mode of Dr. Frankenstein, she also strikes a Baudrillardian note. As David Goldberg argues, "Hypertext represents the fulfillment of the fantasy that Shelley proposes and Jackson revitalizes. A prevailing theme throughout the history of modern science and technology has been the simulation of life by artificial means. Frankenstein and his real-life predecessors . . . sought . . . to create new life, a copy without an original—Baudrillard's simulacrum," and reading *Patchwork Girl* offers "the opportunity to create a unique conformation of the text, of creating a copy without an original," something, one may add, characteristic of the collage form.[8]

Another source of such collage patchwork appears in the different link structures—what Jason Williams terms the link topologies—that characterize each section of the web. As Williams points out, that portion which tells the story of the Patchwork Girl herself relies on unified setting and chronological change. "Because this section emphasizes temporal dynamics, its link structure correspondingly parallels our normal linear perception of time, regularly progressing from past lexias forward. Mary Shelly's encounter outdoors with the monster and the more ambiguous bedroom scene behave similarly but take the peculiar cast of ancillary narratives, like apocryphal stories or appended myths—complete units that draw upon and support material from other units" ("Texture, Topology, Collage, and Biology in Patchwork Girl").

In contrast, that portion of the web containing the nonfictional components takes a more characteristically hypertext form of "paths that intersect at lexias containing similar subject matter. This arrangement permits a digressive textual interrogation in which the reader pursues attractive ideas down branching paths. This mode feels appropriate to nonfiction because it mirrors the normal scholarly process of following references between texts." Although the "Crazy Quilt" section follows a stricter, more limited sequence, "its clear grid layout [in the Storyspace view], the arrow keys, and the chunked arrangement of its content allow a grazing approach to reading."

According to Williams, each of these linking topologies patterns the sequence in which a reader experiences, constructs, or reconstructs the text: "Each corresponds to a temporal texture" created by the reader's perception of transitions between lexias "as smooth and determined, chaotic, or ornately interlocking." *Patchwork Girl* then combines these "linking textures and their composition into a meta-collage with a meta-texture":

Thematic, word-based links act as singular jumps between sections, but, ironically, a woven mass of them forms a canvas on which the author mounts scraps of structure. Links destabilize—or, more positively—stretch the text to flexibility, by pointing away from themselves, by suggesting the reader might read better elsewhere. But again ironically, this pulling apart lends the text its unity, because it permits meanings from separate subsections to bleed into one another through the cracks between them, permitting the text's colorings to mix throughout it.[9]

Finally, Williams concludes, he finds it less surprising that such qualities "appear so fundamental to hypertext and to *Patchwork Girl* than that earlier literary forms subdued them."

Having glanced at *Patchwork Girl*'s linking topologies and its collage-like features, let us next examine some of the ways they and its themes and techniques appear in its use of seams, sutures, links, and scars. As Tim McConville points out, cinema "theoreticians use the word suture to describe a film's ability to cover up cuts and fragments," thereby creating the appearance of "a fluid text that reads 'naturally.'" But because *Patchwork Girl*, "like all hypertext fiction, scoffs at the notion of a neat and tidy text," Jackson's patched-together protagonist defines herself by her scars. Thus, although, "like film, *Patchwork Girl* and all hypertext implement suture," unlike film, they do not do so "as a means of holding narrative together in one cohesive unit. Jackson uses sutures to tie various pieces together so that narrative may merely exist. After sutures have been set in place, the end result is a scar" ("Sutures and Scars").

And scars define the Patchwork Girl, *Patchwork Girl,* and, Jackson implies, all hypertext. In fact, according to Erica Seidel,

scars are analogous to hypertextual links. The monster's scars are intimate, integral, the essence of her identity. Similarly, the essence of hypertext is the linking, the private ways that the author chooses to arrange her piece, and the reader uses to meander through it. Just as the monster finds pleasure and identity in her scars,[10] good hypertext works are defined and distinguished by their unique linking structures. When Shelley and the monster become intimate, she first understands the significance of the monster's scars: "I see that your scars not only mark a cut, they also commemorate a joining." During this sexual encounter, Shelley genuinely identifies with the scars. "Her scars lay like living things between us, inscribing themselves in my skin. What divided her, divided me." Just as the stitchings of skin unite Shelley and the monster, hypertext links unite author and reader. ("The Hypertextuality of Scars")

Jeffrey Pack's mini-web discussing *Patchwork Girl* takes this analogy even farther, first looking at Webster's definitions of a scar as, "among other things, a 'mark left on the skin or other tissue after a wound, burn, ulcer, pustule, lesion, etc. has healed,' a 'marring or disfiguring mark on anything,' and 'the lasting mental or emotional effects of suffering or anguish.'" The first meaning, Pack explains, defines the scar "as a joining, that is, a visual signal that two pieces of skin that were not contiguous at one time now are. In this sense, a scar is the biological version of the seam, where Mother Nature (or, in the case of *Frankenstein* and *Patchwork Girl,* a human creator) sews flesh together in the same way a seamstress stitches together a quilt or the creator of a hypertext links texts together."

According to Pack, the next definition "presents the scar as a mark of disfigurement. Scars are ugly (in modern Western society, at least). They're jarring breaks in the otherwise even epidermis," and links similarly "disrupt, scar, an otherwise linear text." (Pack wrote his critique in HTML to be read with a World Wide Web viewer, and he thus added that the appearance of the link "is even similar; most graphical browsers will display a 'scar' beneath the links on this page, though a user can play the role of cosmetic surgeon and opt to conceal this disfigurement of the text if they so choose.")

The last definition

gives the scar a more abstract meaning; it is now a sign of trauma. In order for a scar to exist, the flesh must have been torn. The formation of a scar is a kludge: its appearance is the result of haphazard regeneration rather than orderly growth. The link is similarly a textual trauma; the transitions between sentences and paragraphs give way to (presumably) intuitive leaps between texts and ideas. The replacement (as opposed to the appending) of text caused by following an HTML link is disorienting to say the least; even the sudden appearance of another window (in an environment such as Storyspace) interferes with the reader's practiced down-and-to-the-right movement across a "page" of text.

Pack entitled his subweb, "Frankenfiction," and his examination of scars, links, and seams in *Patchwork Girl* emphasizes the way Jackson uses them to create a textual "monster."

Like Donna J. Haraway, Jackson rejoices in the cultural value of monsters. Traveling within Jackson's multisequential narrative, we first wander along many paths, finding ourselves in the graveyard, in Mary Shelley's journal, in scholarly texts, and in the life histories of the beings—largely women but also an occasional man and a cow—who provided the monster's parts. As we read, we increasingly come to realize an assemblage of points, one of the most insistent of which appears in the way we use our information technologies, our prosthetic memories, to conceive ourselves. Jackson's 175-year-old protagonist embodies the effects of the written, printed, and digital word. "I am like you in most ways," she tells us.

My introductory paragraph comes at the beginning and I have a good head on my shoulders. I have muscle, fat, and a skeleton that keeps me from collapsing into suet. But my real skeleton is made of scars: a web that traverses me in three-dimensions. What holds me together is what marks my dispersal. I am most myself in the gaps be-tween my parts, though if they sailed away in all directions in a grisly regatta there would be no thing left here in my place.

For that reason, though, I am hard to do in. The links can stretch very far before they break, and if I am the queen of dispersal then however far you take my separate parts (wrapped in burlap and greasy fish-wrappers, in wooden carts and wherries, burying and burning me and returning me to the families from which I sprung unloved and bastard) you only confirm my reign.

Hypertext, Jackson permits us to see, enables us to recognize the degree to which the qualities of collage—particularly those of appropriation, as-semblage, concatenation, and the blurring of limits, edges, and borders—characterize a good deal of the way we conceive of gender and identity.

Michael DiBianco points out, for example, that *Patchwork Girl* "addresses the issue of identity as it is inextricably linked to the author/subject relation-ship," particularly in relation to the narrator, who appears "as much a jumbled collection of disparate parts as her monster," something apparent in the way "Jackson continually incorporates different personas, different voices, at all levels of the text. There is a sense of unceasingly assuming new identities, trying them on briefly, then letting the hypertextual structure of the fiction erase them, only to be subsequently replaced by new identities" ("Commen-tary"). As the narrator puts it, "I hop from stone to stone and an electronic river washes out my scent in the intervals. I am a discontinuous trace, a dotted line." And: "I am a mixed metaphor. Metaphor, meaning something like 'bearing across,' is itself a fine metaphor for my condition. Every part of me is linked to other territories alien to it but equally mine."

Sooner or later all information technologies, we recall, have always con-vinced those who use them both that these technologies are natural and that they provide ways to describe the human mind and self. At the early stage of a digital information regime, *Patchwork Girl* permits us to use hypertext as powerful speculative tool that reveals new things about ourselves while at the same time retaining the sense of strangeness, of novelty.[11]

Quibbling: A Feminist
Rhizome Narrative

Quibbling, Carolyn Guyer explains, "is about how women and men are together, it tends slightly toward salacious, it is broadly feminist (so to speak), or, one could say it is the story of someone's life just before the beginning or a little after the end" ("Something"). I begin my discussion of *Quibbling* by directing attention to Guyer's emphasis on *are*—on a state of being rather than on narrative drive, for in fact the tale accumulates, eddies, and takes the form, as Guyer puts it, of a "lake with many coves." *Quibbling*'s dispersed set of narratives include those of four couples—Agnes and Will, Angela and Jacob, Hilda and Cy, and Heta and Priam—as well as a range of other characters, including several in a novel one of the characters is writing.

Quibbling sharply contrasts with Joyce's *afternoon,* which seems the electronic translation of high modernist fiction—difficult, hieratic, earnest, allusive, and enigmatic. In essence, the opposition comes down to attitudes toward sharing authorial power with readers, and Coover therefore well describes *Quibbling* as a "conventional, but unconventionally designed, romance by one of the most radical proponents of readerly interventions in hyperfictions" ("And Now," 11). In contrast to *afternoon,* which uses the resources of hypertext to assign even more authorial power relative to the reader, *Quibbling* tantalizes readers into wandering through its spaces in unexpected ways.

The contrast between the attitudes toward reader intervention taken by *Quibbling* and *afternoon* clearly appears in the versions of Storyspace they employ. Like *Patchwork Girl,* Guyer's hypertext fiction uses the Storyspace reader, which presents a single scrollable page, similar to that one encounters in the World Wide Web, along with the folder-like Storyspace view that permits readers to search in the innards of the text. Although Joyce originally used the Storyspace reader for some of the prepublication versions of *afternoon* (including one I illustrated in the first version), in the published version, he employed the simple Page Reader, which offers the reader far less power. In explaining her own choice, Guyer points out: "I want people to see the topographic structure itself, be able to go inside it and muck about directly. I want access left to the reader as much as possible." This choice means, as Coover correctly points out, that *Quibbling* "can be read by way of its multiple links, but it can also be read more 'geographically' simply by exploring these nested boxes as though they constituted a kind of topographical map" ("And Now," 11).

Similarly, Guyer's approach to linking reveals her to be far more willing to share power with the reader, as her changing attitude toward links suggests: "I've always felt dense linkage meant more options for the reader, and so greater likelihood of her taking the thing for her own. But this idea now seems wrong to me. Excessive linkage can actually be seen as something of an insult, and certainly more directive . . . In the end, I find I cannot bring myself to make the physical links that are inherent in the writing, that is, the 'obvious' ones (the motifs of glass, water, hands, color, walking, etc.)" (Journal).

Guyer's emphasis on an active reader, as opposed to simply a responsive, attentive one, relates directly to her conception of hypertext as a form of feminist writing. In fact, like *Patchwork Girl, Quibbling* makes us wonder whether hypertext fiction and, indeed, all hypertext is in some sort a feminist writing, the electronic embodiment of that *l'écriture feminine* for which Hélène Cixous called several decades ago. Certainly, like Ede and Lunsford, whose alignment of collaborative authorship with feminist theory we have already observed, Guyer believes that hypertext—an intrinsically collaborative form as she employs it—speaks to the needs and experience of women: "We know that being denied personal authority inclines us to prefer . . . decentered contexts, and we have learned, especially from our mothers, that the woven practice of women's intuitive attention and reasoned care is a fuller, more balanced process than simple rational linearity" (quoted by Greco from Joyce, *Of Two Minds*, 89).

According to Diane Greco, Guyer sees hypertext as the embodiment of "ostensibly female (or perhaps, feminine) characteristics of intuition, attentiveness, and care, all of which are transmitted from one woman to another via the universal experience of having a (certain kind of) mother. The opportunities for non-linear expression which hypertext affords coalesce, in this view, to form a writing that is 'female' in a very particular way: hypertext writing embraces an ethic of care that is essentially intuitive, complicated, detailed, but also 'fuller' and 'balanced'" (88). Reminding us that "some notable hypertexts by women, such as Kathryn Cramer's *In Small & Large Pieces* and Jane Yellowlees Douglas's *I Have Said Nothing,* feature violence, rupture, and breakage as organizing imagery," Greco remains doubtful of any claims that hypertext, or any other mode of writing, could be essentially female or feminine.[12]

Whether we agree with Guyer that hypertext fiction necessarily embodies some essential form of women's writing, we have to recognize that she has written *Quibbling* as a non-Aristotelian networked cluster of stories, moods, and narrative fragments that gather and rearrange themselves in ways that embody her beliefs about female writing. In her essay in *Leonardo,* she emphasizes both how her fiction web lacks conventional narrative and conventional aspirations to be literary: "It is hardly about anything itself, being more like the gossip, family discussions, letters, passing fancies and daydreams that we tell ourselves every day in order to make sense of things. These are not exactly like myths, or fairy tales, or literary fiction. They are instead the quotidian stream. In this sense, then, *Quibbling* is a work that tries not to be literary." Wending our ways through it, we encounter lexias that take the form of messages received via electronic mail, brief notes, and poetry as well as more usual narrative, description, and exposition. The links that join these lexias do not produce straight-ahead, or even eddying, narratives but instead generate an open montage-textuality, like that of *In Memoriam,* in which lexias echo one another, gathering meaning to themselves and sharing it with other, apparently unrelated patching of writing.

Guyer's basic approach appears in the way individual lexias follow one another and hence come to associate with each other. If, after reading the lexia entitled "walking w/Will," which relates an episode in the relationship of Agnes and Will, one double clicks on it to follow a link out, that action brings one to "following her," which relates how after their first date Priam secretly followed Hetta home to make sure she was safe; activating a link from this lexia brings one to an event or state in the relationship of Angela and Jacob. Reading along this link path, one perceives the somewhat analogous situations, thus finding similarity, though not identity, in the lives of different couples. In some cases, only by looking at the lexia's location in the web's structure (presented by the Storyspace view) can readers discern which couple they are reading. These stories take form, in other words, by gentle accretion as one lexia rubs up against another. In contrast to *afternoon,* in which our ignorance of a crucial event drives our reading, here no single core quickly comes to prominence as the necessary axis or center of all lexias. No single event endows the others with meaning. *Quibbling* seems far more a networked narrative in which the similar situations bleed back and forth across the boundaries of individual lexias, gradually massing meanings. "It is," as Guyer explains,

in that rhythmic sense of ebb and flow, of multi-directional change, of events that disappear before they are quite intelligible but somehow come to mean something, that *Quibbling* was made. In hindsight, I can see why water and its properties became one of the pervasive, propelling metaphors in the work. A lake with many coves is how I saw it. The coves being where we focus, where individuals exist, where things are at least partly comprehensible; the lake being none of that, but, naturally, more than the sum of the coves, or more than what connects them. As a metaphor, the lake and coves stand not just for the form of this hyperfiction, but hyperfictions generally, and yes, for life itself. ("Something")

Guyer has stated that when she encountered *Plateaus,* she recognized Deleuze and Guattari's ideas as something for which she'd long sought, and not surprisingly her conceptions of event and resolution in narrative are illuminated far more by their conceptions of nomadic thought than Aristotelian notions of plot. In *Quibbling,* she explains, "closure, resolution, achievement, the objects of our lives are inventions that operate somewhat like navigational devices, placemarkers if you will." Guyer's exposition of the ideas and attitudes that inform her fiction provides a valuable guide to her world of fluid narrative, a world of change and flux that has strong resemblances to that created by *In Memoriam,* a world in which "we go on like waves unsure of the shore, sometimes leaping backwards into the oncoming, but always moving in space-time, always finding someplace between the poles that we invent, shifting, transforming, making ourselves as we go" ("Buzz-Daze").

In discussing *Quibbling,* Guyer turns to Deleuze and Guattari's ideas of the smooth and the striated as a conceptual and fictional way of resolving problems created by the "nonexistent" sets of polarized abstractions in terms of which we lead our lives: "Female/Male, Night/Day, Death/Life, Earth/Sky, Intuitive/Rational, Individual/Communal . . . We make these things up! Deleuze and Guattari shift attention from polar oppositions to the constant transformations of one pole into the other. What's important to recognize is not the impossible duality of the poles, but what happens between them. You might say it's What We Learn, what we actually experience in space-time as we conceive ourselves, as we conceive space-time" ("Buzz-Daze").

Storyworlds and Other Forms of Hypertext Narratives

Many hypertexts, like *Quibbling* and *Ultramundane,* exemplify what Michael Innis, head of Inscape, Inc., termed a "storyworld." Storyworlds, which contain multiple narratives, demand active readers because they only disclose their stories in response to the reader's actions. Obviously derived from computer-based adventure gaming, these storyworlds, however, generally play down elements of danger or fighting monsters as a means of approaching some goal. *Uncle Buddy's Phantom Funhouse,* which John McDaid created in Hypercard, seems the first of this electronic genre whose more recent examples on CD-ROMs include Laurie Anderson and Hsien-Chien Huang's *Puppet Motel* (1996) and the Residents' *Freak Show* (1993) and *Bad Day at the Midway* (1995).

Like the extremely popular CD-ROM adventure game *Myst* (1993), these storyworlds reconfigure conflict and the role it plays in narrative and the reader's experience. All stories take the form of the conflict or the journey, and if one considers that distance serves as the antagonist in the journey narrative, then all stories turn out to involve various forms of conflict. The antagonist can be a personal opponent, a force, fate, or ignorance (or cognitive dissonance), which appear either as an internal state or as a relation between the self and environment. Whereas in both adventure narratives and adventure games the conflict requires some form of physical opposition, in the story-world that role is taken by mystery and enigma. The detective story becomes the paradigm for this electronic form—something already present in modernist and postmodern fictions such as *Absalom, Absalom!* and *Waterland*.

Like the detective, readers who find themselves within storyworlds must take an aggressive approach, even performing actions supposedly forbidden. In most narratives as in real life one learns it is considered bad form or even criminal behavior to interrogate, trespass, investigate behind the scenes. In hypertext storyworlds, one must do so or one encounters very little in the way of story or world. In both *Myst* and *Freak Show* one receives very little in the way of clues or instructions and must gradually piece together one's strategy, which involves recognizing the presence of clues and the attitude one must take toward the environment in which one comes upon them. When *Freak Show* begins, we find ourselves outside a side show tent, and using a computer mouse we move inside it and encounter a ringmaster who introduces us to what he terms "the world's most disturbing collection of human oddities"—Herman the Human Mole, Harry the Head, Jelly Jack, Wanda the Worm Woman, and so on. He pauses before the doorway or curtain that leads to each, providing a brief introduction in the manner of the carnival barker. Finding ourselves within this carefully rendered three-dimensional overview, we can pause before each of several possible choices, which also include a sampling of the Residents' music and a historical archive of freaks and deformities, listening to the barker's description. After each introduction, we can approach the relevant exhibit, thus prompting its display, or we can cut off the barker in midsentence, causing him to cry out, "Forget it!," "Okay! Okay," or a number of other expressions of irritation.

More important, this storyworld rewards readers who repeatedly disobey his irritated pronouncements that they cannot go behind the scenes. On the third attempt, readers discover that they can in fact enter a passage that takes them to the trailers inhabited by members of the sideshow. Entering each environment similarly rewards the active, intrusive, curious reader. Finding ourselves projected into the cursor (or reduced to it), we probe objects until they yield stories. Entering Herman the Human Mole's area, we find his wagon and can get a brief glimpse of him through the porthole-like window of his circus wagon. At this point, we can turn around, returning to the main tent, or nosily wander around until we find a way into his wagon. Probing this environment successfully, we eventually find Herman in hiding and he tells us his sad tale in the form of a set of primitive cartoons in what appears to be his personal style. The narratives of Herman and the other characters, each of which take very different forms, themselves have little of hypertextuality. They are simply the rewards of the reader's aggressive curiosity.

Readers or viewers of *Freak Show* find themselves in a situation quite different from that of the reluctant wedding guest whom Coleridge's Ancient Mariner forces to hear his tale. Here the reader-listener acts as the obsessive one, forcing the story out of a reluctant narrator, one who must be convinced by intrusive actions that the reader's obsessive curiosity matches his need to tell an explanatory narrative. Storyworlds, in other words, take the active, aggressive, intrusive critic as the paradigm of the ideal reader.

In what he terms *narrative archaeology,* Jeremy Hight moves the storyworld from a CD-ROM or the Internet to the physically existing city of Los Angeles. In his *34 North 118 West* project, "participants"—his word for what elsewhere would be readers or game players—"walk the streets of a city with a G.P.S. unit attached to a lap top computer" and at numerous "hot spots" can listen to recorded fictional narratives, thus experiencing a kind of augmented reality. Whereas virtual reality (VR) immerses the user within a world of represented data, augmented reality overlays information on top of the physical world in which one lives. Examples of augmented reality include images of an airplane's navigational devices projected on the windscreen so the pilot sees information superimposed upon the world through which he flies and wiring diagrams similarly projected upon the physically existing wires and cables so an airplane mechanic can more easily and accurately assemble or repair them.

The very idea of augmented reality prompts one to observe that stories always overlay and thus augment reality. Stories, written or heard, that we usually encounter differ from Hight's narrative archeology in one crucial way: we experience the narrative as removed from our physical world, and therefore as we enter the narrative world, we imaginatively and experientially leave our own to the extent to which we immerse ourselves in the story; when we return to our physically and emotionally existing world, we may bring the emotions, attitudes, and ideas of the story back with us and thus experience our everyday world in a somewhat different way. In written and oral narrative, whatever augmentation occurs happens after we experience the entire story. Hight wants to use his augmented reality to create something radically different by making the augmentation occur in the same place and time as the everyday physical world. He wants to create "an overlap experience in real time of experiencing two places at once."

He explains in his description of *34 North 118 West* that

voice actors read all written narratives to create an overlap in real-time experience of two places at once. The only visual is the map that tracks one's movement and shows hot spots and the distance readings on the G.P.S. unit . . . The key is the usage of sound. Walking the city with sounds from different points in time and metaphorical relationships with what is being seen allows the author to guide a fused experience of critical analysis and creative writing.

Like David Yun's Web-based *Subway Story*, *34 North 118 West* uses a map of a city as an overview that permits access to many narratives. The participant encounters the city plus a map superimposed upon it as an overview (or, in World Wide Web terms, a "sitemap"). In this narratively augmented physical world, "movement and reading," as Hight beautifully puts it, "brings a narrative of what was unseen and what has been lost in time, only for it to quiet again once passed."

One of Hight's most interesting points is that a "fictional narrative is an agitated space." Ever since Aristotle, students of narrative have understood that it involves disequilibrium and disturbance, for the antagonist, whether person, place, or thing that blocks the main character, in essence creates the story. With no obstacle there is no story. Instead of simply emphasizing the process—and hence the temporal, sequential aspect of narrative—Hight also conceives it in more spatial terms. A story, for him, is a storyworld; or perhaps one might say that narrative requires a world within which to take place. Furthermore, he points out that a "city is also an agitated space" that exists as "data and sub-text to be read in the context of ethnography, history, semiotics, architectural patterns and forms, physical form and rhythm, juxtaposition, city planning, land usage shifts and other ways of interpretation and analysis. The city patterns can be equated to the patterns within literature: repetition, sub-text shift, metaphor, cumulative resonances, emergence of layers, decay and growth." City spaces therefore provide an obvious way to reconfigure narrative, thus providing a means of experiencing the words of earlier inhabitants, including railroad workers and Latina women who worked nearby in the 1940s. Here are the fictional words, supposedly spoken in 1946, of a man who worked thirty-five years clearing the railroad tracks that ran through this space:

Those men, along the rails, tired. Death by train we called it . . . It was my job to assist . . . to help . . . kind words . . . or help clear the tracks after the impact . . . Such failures. My failures. Such small horrors. And it is not the most dramatic: an eye open tomato red with blood, a nose with ice covered nostril hairs that looked like a crab emerging from a shell, an ear lying by a man's feet like some dead wingless bird, a cheek punctured with teeth exposed, a wound open steaming in the snow. Those are so few, so specific, so clearly cut from men with faces I cannot help but still see. It is what never comes clear, not faces, not expressions, not the dignity of person, something that had a name. There is a sort of mutant slot machine, it comes to me at night: an odd collection, ever shifting, not bells and lemons but eyes, scars, blood, mouths, wounds, meat, an eye hanging alone gleaming wet and alien yet from some lost moment in 35 years, a nostril disconnected a failing island of memory from some dead man's face like an odd little lost cave. Those are the ones I truly failed.

Hight's narrative archeology reveals often obscured and forgotten layers of the past, thus augmenting present physical reality with lost voices. This kind of storyworld reveals an often opaque, meaningless physical cityspace to be a rich palimpsest of human meanings and experience. But, like *Myst* or *Freak Show,* his project requires the reader-listener-participant to explore a space to discover its stories; unlike these earlier storyworlds set in fictional spaces created by digital information technology, Hight demands that we traverse real, physically existing spaces augmented by this technology.

In the storyworld and noncombative adventure game, reader-viewers assume the positions of protagonist and their reward comes in the form of experience, not as a reward one might attain. Both these qualities involve or produce repetitive narrative structure as in the picaresque novel of old or much Japanese fiction, neither of which builds toward a single unique climactic movement. But this form of hypertext narrative does not so much do away with climaxes as emphasize multiple ones. We have already observed something akin to this tendency in Joyce's *afternoon* and Tennyson's *In Memoriam,* both of which achieve some sort of wholeness by formal terms. In Tennyson's case this derives from virtuoso formal closures, many of which pointedly do not coincide with intellectual or thematic ones, thereby making an individual lexia simultaneously self-sufficient and yet part of a larger whole because it demands closure elsewhere; the formal closure makes them end satisfactorily, the lack of intellectual closure joins each into a larger whole. Joyce's similar effects arise not so much from any formal closure—something harder to achieve in prose since one doesn't have what J. V. Cunningam used to called "the exclusions of a rhyme"—but from ornate, rich prose, each example of which in some sense satisfies. Thus in hypertext fiction, one needs a certain modicum of lexias that not only make sense when entered from multiple places but also satisfy, in some way seeming (partially) complete when they end or when one departs from them.

Computer Games, Hypertext, and Narrative

As these examples of gamelike storyworlds (or storyworld-like games) show, any discussion of narrative in digital forms leads sooner or later to the increasingly important topic of computer games. These take various forms, including story-worlds (*Myst*), simulations (*Sims*), first-person shooter (*Quake*), multiplayer (*Lineage*), and god-games. Although students of computer games compare them to print and hypertext narratives, I believe that the most useful point of comparison is instead to *hypertext as a medium* and not to hypertext narrative.[13] All computer or video games have five important similarities to hypertext. First, the player's actions—clicking a mouse or manipulating a similar device, such as a joystick—determines what the player encounters next. Second, like hypertext, games rely on branching structure and decision-points. Since the places in the video game where the player acts produce potentially different results, they appear structurally identical to hypertext's branching links. If one defines the production of different results by user's choice (whether alpha-numeric texts or actions), then hypertext becomes, as Aarseth claims, a subset of ergodic text. Third, games, like hypertext fictions (but unlike print narra-tive), are meant to be performed, and fourth, they are meant to be performed multiple times. Fifth (and this may only be a trivial point of convergence), the record of a game player's actions, like the experience of reading a hypertext, appears linear since both the players of games and readers of texts make their way through a series of choices in linear time; of course, the range of possible actions, of roads not taken, themselves constitute a branching or multilinear structure but one that is not immediately available to players and readers.

Janet Murray, who draws on Aristotle's *Poetics,* makes several compelling observations about the relation of gaming and narrative, the first of which is that both share two basic structures—those of the contest and puzzle. More-over, "stories and games are like one another in their insularity from the real world, the world of verifiable events and survival-related consequences" (3). Murray unfortunately also introduces the misleading term *cyberdrama* for computer games, which Michael Mateas expands on when trying to relate them to the *Poetics.* This approach has fundamental problems. Aristotle, we recall, distinguished between narrative in which an author relates a story (the *Iliad*) and drama in which actors show us a story (*Oedipus Rex*). The immer-sive video game, in which we take part as actors, is a third, fundamentally different mode. Saying that video games are like drama seems not much different from saying a cow is like a frog, except that, well, it's bigger, and it's a mammal, and it doesn't live in the water.

In contrast to the self-proclaimed Aristotelians, who argue that literary and cinematic studies of narrative have much to tell us about games, another group led by Espen Aarseth, Markku Eskelinen, and Raine Koskimaa argue that computer and other games require a new discipline—ludology. Aarseth, who introduced the concept of ergodic text, explains the fiercely contested battle "over the relevance of narratology for game aesthetics": "One side argues that computer games are media for telling stories while the opposing side claims that stories and games are different structures that are in effect doing opposite things" (45). "The traditional hermeneutic paradigms of text, narrative and semiotics are not well suited to the problems of a simulational hermeneutic" (54). Celia Pearce presents the ludologist's case in forceful, if measured, prose:

Because computer game theory is a relatively new discipline, much of which has emerged thus far has come from other disciplines absorbing game theory into their purview. It seems axiomatic that there must always be a phase where established media seek to "repurpose" their existing "assets" for use in the new medium. Most notably, film and literary theorists have begun to discuss game theory within their own idiosyncratic frameworks. These disciplines have much to add to the discourse on games, particularly when the discussion is centered on narrative. However, they are missing a fundamental understanding of what games are about . . . The result is a kind of theoretical imperialism. ("Towards a Game Theory of Game," in *First Person*, 143–44)[14]

In contrast, Eric Zimmerman argues, in "Narrative, Interactivity, Play, and Games," "as we observed with chess, games are in fact narrative systems. They aren't the only form that narrative can take, but every game can be considered a narrative system" (*First Person*, 160). In fact, most writers who compare games to narratives take chess as an example of a game that *cannot* be a narrative. Zimmerman, however, decides it is one, but I suspect that he confuses the experience of someone observing a game with that of the player.

Some of those who claim that stories and associated narratological theory provide the best way to understand computer games make the error of assuming that if a game includes any sort of a story, then narrative is a defining characteristic of games. There are, however, plenty of precedents for essentially non-narrative forms that include narrative. Victorian writers of nonfiction, such as Thomas Carlyle, John Ruskin, and Henry David Thoreau, all employ narratives, created characters, and dialogue within argumentative prose.[15] Markku Eskelinen, the most aggressively outspoken of the ludologists, therefore makes the crucial point in "Towards Computer Game Studies" that "a story, a backstory or a plot is not enough. A sequence of events enacted constitutes a drama, a sequence of events taking place a performance, a sequence of events recounted a narrative, and perhaps a sequence of events produced by manipulating equipment and following formal rules constitutes a game" (*First Person*, 37). Distinguishing between games and narratives, Eskelinen further explains that "in games, the dominant temporal relation is the one between user time and event time and not the narrative one between story time and discourse time" (37).

Even those theorists who insist that games are a form of narrative recognize that games and stories have major differences. "A story," Janet Murray explains, "has greater emphasis on plot; a game has greater emphasis on the actions of the player" ("From Game-Story to Cyberdrama," in *First Person*, 9). Furthermore, as Eskelinen points out, "information is distributed and regulated very differently in games than in narratives" (39). Pearce points to a third difference, namely, that "games tend to favor abstracted personas over 'developed' characters with clear personalities and motivations" (146). Despite the many disagreements between the two groups, they all accept two of Murray's major points—that agency is crucial to computer games and narrative has at least *something* to do with them.[16]

Aarseth, the founding father of studies of nonliterary digital textuality, makes what seems to me to be the crucial point that computer games characteristically involve simulation:

The computer game is the art of simulation. A subgenre of simulation, in other words. Strategy games are sometimes misleadingly called "simulation games," but all computer games include simulation. Indeed, it is the dynamic aspect of the game that creates a consistent gameworld. Simulation is the hermeneutic Other of narratives; the alternative mode of discourse, bottom up and emergent where stories are top-down and preplanned. In simulations, knowledge and experience is created by the player's actions and strategies, rather than recreated by a writer or moviemaker. ("Genre Trouble" in *First Person*, 52)

Aarseth's emphasis on simulation as a key element in games appears supported by an early nonludic hypermedia simulation project I saw demonstrated in 1988. To train trauma surgeons for conditions in military field hospitals, a group from the Uniformed Services University of the Health Sciences developed powerful scenarios, using a computer and video disc (Henderson). Video discs were then the latest thing, and although now generally obsolete, they provided valuable lessons directly applicable to the use of computer animation and video. In this training system, a surgeon would sit before a computer that was attached to a television and a video disc player and encounter the following scenario. A session begins when the surgeon, who has been assigned to a field hospital, arrives suitcase in hand, goes to the hospital tent to present himself to the officer in charge, expecting a warm greeting as a new member of a team. Inside the tent, as he greets his commanding officer, who is in surgical garb, he hears the sound of approaching helicopters. His superior, all business, orders him to drop his bags and get to work, since the short-handed unit needs him to begin immediately. In the scenario I saw, two medics bearing a wounded soldier on a stretcher appear, telling him that the patient who has no apparent wound and whom triage had therefore classified as not requiring immediate treatment has stopped breathing. They ask him what to do. At this point a clock appears at upper screen right, and its second hand begins to move. If the role-playing trainee makes a mistake and orders an x-ray, the corpsmen respond angrily that there's no time for that and the patient will die.[17] The clock keeps moving—the physician has a fixed limit, say, 120 seconds before his patient dies. If the trainee finds the right solution in time—the patient has a collapsed lung caused by a barely discernible wound—the patient lives. At this point the surgeon in charge, the trainee's new commanding officer, appears and, depending on his actions, praises him, welcoming him to the unit, responds rather more coldly, or, if the patient has died, bawls him out, ordering him to improve his skills.

Viewing this project demonstrated several points of value to anyone considering the relation of narrative, simulation, and games. First, the simulation did not have to achieve anything like a complete reality-effect to immerse the user (and onlookers) in the situation. Although the acting and production values were not of the highest quality, these lacks did not reduce the tremendous emotional effect of the simulation exercise. As part of the project, the researchers filmed physicians using the system and recorded their blood pressure before and during each session; their pressure shot up, they perspired, and in other ways they acted as if they were confronting an actual medical emergency. The fundamental connection of the scenario to the user's profession and self-image as a professional immediately produced a reality effect, a fact that reminds us the amount of authentic detail unconnected to the main enterprise—here making a correct diagnosis and saving the patient's life— plays only a minor role in the effectiveness of the simulation.

A second point: although this simulation has important narrative elements, they obviously play only a secondary role, setting the stage for the defining feature of the simulation, the surgeon's *choices*. Finally, this simulation takes the form of a game, although the player's professional investment in the outcome produces an earnestness only occasionally associated with game play.

In conclusion, although computer games have something to tell us of relevance to digital text and art, virtual reality, and educational simulations, they do not seem closely enough related to hypertext to tell us much about it. Video games have received their own field of study, and it is from this new discipline that we can expect insights about how they work and their social and political implications.

Digitizing the Movies: Interactive versus Multiplied Cinema

Although computers have affected cinema as dramatically as they have affected verbal text, at one crucial point—the relation of each medium to its audience—hypermedia and cinema appear fundamentally opposed. Since hypertext requires reader choice, it therefore fundamentally conceives of its audience, unlike that for cinema, as an audience of one. This statement of course presupposes that by "cinema" we want or expect it to remain a form intended for group audiences. Cinema, however, might divide into two forms, one remaining essentially identical to that now enjoyed in theaters, and another intended for single viewers. Given the financial success of both single-player computer games and DVD versions of films first shown in theaters, one very well could find a large audience for this second kind of virtual cinema.

Even if we compare traditional theatrical cinema to hypertext, it reveals important points of convergence. First of all, computer technology has so changed the ways we compose, edit, and even conceive of filmmaking that we can now accurately speak of digital (or virtual) cinema in the same way that we speak of digital writing and digital textuality. Computers have affected cinema in at least four ways, the first of which involves the near-universal use of digital technologies to edit footage produced by nondigital cinematic technology; this first form also includes the increasingly popular use of digital, rather than analogue, cameras to shoot film footage. The second effect of digital technology on cinema involves using computer-manipulated images. Such digitally created imagery ranges from manipulating individual frames to creating substantial sequences or even entire films with computer animation. Working on individual frames, for example, graphic artists employ software like Photoshop to remove visual evidence of the safety wires that permit actors to perform dangerous or otherwise impossible actions. These specially edited sections are then combined with nondigitally produced footage. Safety, convenience, and relative ease of manipulation often lead to employing computer animation for elaborate special effects, including flying over or zooming into cityscapes. These two contributions of computing to cinema have already had major economic effects on both Hollywood and independent filmmaking.

The two other ways computers have affected filmmaking bear more directly on its relation to digital and hypertext narrative, since they include two essentially new forms of digital cinema. The first, hypertext cinema—cinema closely analogous to hypertext narrative—theoretically permits the audience to choose narrative direction at key points in the story. Both the trauma surgery simulation and Janet Murray's MIT French language-teaching project, which exemplify this particular mode of virtual cinema, exemplify educational hypermedia that closely resembles computer-based adventure games. In Kristoffer Gansing's study of what he calls "the 'imaginary' genre of interactive film" (51), he claims that adventure games hold "a position as the *mainstream* of interactive cinema" (54), though he is more interested in the possibilities revealed by various forms of CD-ROM art films and net-based cinema. Gansing describes the "mini-genre . . . already made up of 'database narratives' which utilizes associative interaction with audio and video sequences—often collected from linear films. The user deconstructs and constructs his/her own version from a given set of materials that is called upon through experimental interfaces sometimes combined with elements of randomization" (55).

Like hypertext cinema, the second branch of digital filmmaking, which I term *randomized* or *multiple cinema,* also conceives of the film as essentially divided into a significant number of discrete sections. Unlike the hypermedia form, however, either the filmmaker or a computer program decides the order in which the audience views the segments. Since I have not seen Gansing's examples, I'll discuss instead Ian Flitman's *Hackney Girl* (2003) and that part of Diego Bonilla's *A Space of Time* that he calls *Stream of Consciousness* (2004). *Hackney Girl* (Figure 29), an example of net cinema, presents a varying number of randomized sections of film, presenting them in a different sequence each time viewed; I encountered between 139 and 143 sections, and the full viewing time also varied but remained somewhere around fifteen minutes. The viewer first encounters a large black screen on which appears a collage of as many as seven small windows, only one of which contains video. Some of these windows are monochrome, others color. Every version I saw fairly quickly made clear that London, departing airplanes, a young Englishman, his Turkish wife or girlfriend, and her cat would be the main subjects. The first version I encountered told a tale of the young woman's arrangements to leave London for Istanbul. Another began with scenes of London

Figure 29. Multiple Cinema Online: Ian Flitman's *Hackney Girl*. In this screenshot five thematically related images appear in the Web browser while another loads at the lower right. Only one image at a time activates as video, and as soon as one video segment ends, the number and arrangement of panels change. (Courtesy of Ian Flitman.)

life and then moved to many images of the couple's cat, a third began with the young woman walking down Chatsworth Road, London E5, going into a café, returning home, and discussing her cat's reaction to her departure while her young man, who remains offscreen, tells her about Freud's theory of the Oedipus complex, after which a series of scenes in a pub appeared. Later, in one window the young woman touches up her lipstick before Hagia Sophia in Instanbul while others show the building alone, people putting their shoes on apparently after visiting it, shoes awaiting their owners, and a juxtaposed scene in the couple's London flat. Next followed scenes in Istanbul, and it ended with a screen containing four images of the young woman plus a street scene in Turkey. After seeing one version, I already knew the main characters and settings in others. Nonetheless, I'm not sure that differing versions suggest the same chronological endpoint, since I thought in one that the young woman traveled alone and in another she was with her companion. According to Flitman's directions, viewers can change windows by hitting a key combination, but I seemed unable to activate specific windows of my choice. In this example the element of randomization proved far more important than any viewer intervention. I found *Hackney Girl* visually very interesting, even though I never encountered the usual kind of narrative in which a character overcomes an antagonist and reaches a goal, and despite the fact that certain events, such as packing, playing with the cat, and arriving in Instanbul occur repeatedly, they form more of a mosaic narrative than an orthodox one. Like *afternoon* and other enigmatic hypertexts, *Hackney Girl* demonstrates that readers or viewers can construct a coherent narrative from small chunks that they encounter in varying orders.

Diego Bonilla's *A Space of Time* (Figure 30) which comes on a CD-ROM, contains two forms of digital cinema constructed out of some of the same materials. *Stream of Consciousness,* like *Hackney Girl,* takes the form of randomized or what I would prefer to call multiplied cinema, and a viewing occupies between forty minutes and two hours. *Limbo,* the second part of *A Space of Time,* exemplifies a form of interactive cinema that has much in common with an adventure game centered on exploration rather than combat. Using Quicktime VR, Bonilla has created what he describes as "a virtual tour of a century-old building as a narrative device to tell the story." The viewer enters and explores a large, multistoried empty building, using the features of Quicktime VR to obtain 360-degree views of each of its parts. Moving one's mouse, the viewer finds hot spots that when clicked on move one forward, so one can climb stairs and explore each room. This portion of *Limbo,* therefore, provides an example of virtual reality hypertext. The user's mode of movement through its spaces feels very much like that used in *Myst* and *Riven,* the main difference between them being that Bonilla's work employs digital photography rather than computer graphics.

Limbo begins with three brief videos intended to set the scene, after which the viewer can explore the building. The most interesting part of *Limbo* appears in the way it branches to filmic segments taken from *Stream of Consciousness:* clicking on a photograph of a young woman on a wall produces a home movie of a young child, presumably the girl in the photograph. Entering a room on the first floor, one discovers half a dozen television monitors floating in midair; clicking on any one of them activates a video of an older woman, apparently a real estate agent, talking about the history of the building.

Figure 30. Integrating Virtual Reality and Video. In Diego Bonilla's *Limbo* section of *A Space in Time*, the viewer moves through a Quicktime VR environment, climbing stairs, moving through corridors, and entering various rooms, always able to rotate 360 degrees. Upon entering one of the large rooms in the empty building, the viewer encounters one or more floating figures, each of which functions as a link that activates segments of video taken from the project's second part, *Stream of Consciousness*. (Courtesy of Diego Bonilla.)

Similarly, upon entering certain rooms on the upper floors, the viewer sees one or more people floating in the air (Figure 30); clicking on them either launches videos or animated poetry supposedly written by Pandora (or Panda). Many of the videos present young people who have spent the night in the building at other times and places. In one group of these videos Panda and friends confront a local talk show host and attack the way advertising damages culture. In another, a young woman rants—Pandora's word—about the existence of websites offering beautiful Russian brides to ugly American (and other) men. Yet others depict the police interviewing a homeless man while sitting in a diner. Throughout, as the CD cover explains, Panda and her friends "attack the most wicked result of Capitalism. In addition to her criticism of advertising, Panda, her friends and colleagues express their contempt for the blind use of technology, the commercialization of love and sentiment, current fanatical faith in science, and ever-increasing speed of so-called 'progress.'" The dialogue, as one can gather from Bonilla's description, is very heavy handed, making these prosperous young people sometimes sound like seventy-five-year-olds. Some of the acting is also not very good, but the major problem with this nonetheless fascinating project is that the rants remain just that—passionate harangues—and that they seem to have no connection to the figures flying in the otherwise empty rooms. I found exploring the spaces of *Limbo* enjoyable and think the clickable floating fantastic figures a delightful conceit.

Bonilla's two experiments in digital cinema conceive their relation to their audience very differently. Whereas hypercinema like *Limbo* makes the single, isolated active reader its model audience, multiplied cinema like *Stream of Consciousness* is intended for a more conventional, essentially passive, group audience. Since a hypertextualized cinema treats passages of cinema like lexias in verbal hypertext, linking them with or without branching, it places major emphasis on audience choice, but it's difficult to imagine audiences composed of more than one or two people being able to make such choices. Murray's French language project, in which the student chooses key narrative arcs, obviously was intended for a single student in a language laboratory. The second form of digital cinema avoids this issue altogether, by emphasizing, not viewer choice and control, but narrative richness since it defines itself by creating multiple permutations of a limited set of bits of time-bound cinema.

One approach to such a single-person hypercinema appears in the *Hyper-Cafe* interactive video project created by Nitin Sawhney, David Balcom, and Ian Smith at the Georgia Institute of Technology (Figure 31). Their combination of digital video and hypertext, the authors explain, "places the user in a virtual cafe, composed primarily of digital video clips of actors involved in fictional conversations . . . You enter the Cafe, and the voices surround you. Pick a table, make a choice, follow the voices. You're over their shoulders looking in, listening to what they say—you have to choose, or the story will go on without you" (1). This experiment in multiple narrative was created in part to explore and extend the rhetoric of hypermedia. In particular, Sawhney, Balcom, and Smith successfully devise a means of permitting choice—and hence branching—in so fundamentally linear an information technology as video.

Some work in interactive video concentrates on creating branching within (or from) a narrative line, often by enabling the user to switch from the position of one character to another. In Greg Roach's *The Wrong Side of Town* (1996), for example, a woman on a business trip—the exposition is provided by a telephone call to her husband at home—goes to a diner for supper, encounters a beggar, orders her meal, and leaves. The viewer can experience

Figure 31. *HyperCafe*'s Overview Screen. This project employs an overview suited to its interactive materials, effectively mediating between the linear drive of video and the reader's desire for control. Although viewers cannot halt the videos, as they can in *Kon-Tiki Interactive*, they can choose among them. (Used by permission of the authors.)

the encounter with the beggar from the vantage point of either person, and once the protagonist enters the diner, one can experience her interaction with a waiter and waitress from each of their positions as well as from hers. The creators of *The Wrong Side of Town* use this opportunity in a clever, if heavy-handed, way to produce a Rashomon-like divergence of events coded according to class and gender positions.

Bad Day at the Midway, an interactive CD-ROM created by some members of the team that produced *Freak Show,* takes the ability to exchange positions even farther by employing it as a means of switching not just vantage points but entire lines of narrative. Although a storyworld rather than a full hyper-video, it exmplifies a work in which the areas one can explore, the facts one discovers, and the dangers that threaten one all depend on one's characters. Thus, although the little boy "Bobby" cannot gain access to certain dangerous heights, he also seems safe from the predations of a homicidal maniac who inhabits the midway. In contrast, each of the adults has the capacity to make more discoveries than does Bobby, but each also exists in greater danger as well.

"The time-based, scenario-oriented hypermedia" (2) of *HyperCafe* takes a different approach, for rather than finding oneself within a narrative where one discovers choices, one begins by encountering a field of competing narratives—essentially a video version of the situation one has in certain static hyperfictions, such as *Adam's Bookstore,* in which one begins by choosing a lexia as one's starting point (Figure 32). As Sawhney, Balcom, and Smith explain: "In *HyperCafe,* the video sequences play out continuously, and at no point can they be stopped by actions of the user. The user simply navigates through the flow of the video and links presented . . .The camera moves to reveal each table (3 in all), allowing the user 5–10 seconds to select any conversation. The video of the cafe overview scene plays continuously, forwards and then backwards, until the user selects a table" (2). Once viewers make a choice, they enter a particular scene, which then offers particular narrative lines.

The designers of the HyperVideo environment in which *HyperCafe* exists created three forms of linking—the temporal form we have already observed, what they term "spatial link opportunities," and a third kind, interpretive textual links. Viewers learn about the presence of spatial links by means of three "potential interface modes: flashing rectangular frames within the video, changes in the cursor, and/or possible playback of an audio-only preview of the destination video" (4). Finally, viewers learn about the choices they can make from text that scrolls across the screen, and these texts can take the form of random bits of dialogue or scripts for particular scenes. "Text intrudes on the video sequences, to offer commentary, to replace or even displace the videotext. Words spoken by the participants are subverted and rewritten by words on the screen, giving way to tensions between word and image" (4). Building upon the research of workers at MIT and the Universities of Amsterdam and Oslo, *HyperCafe* demands active, intrusive reader-viewers who build narratives by following links. The result, as the authors make clear, is that individual lexias participate in various story lines, or, as they put it, "narrative sequences may 'share' scenes" (5), and for this reason they specifically compare the kind of narrative found in *HyperCafe* to that of Joyce's *afternoon.*

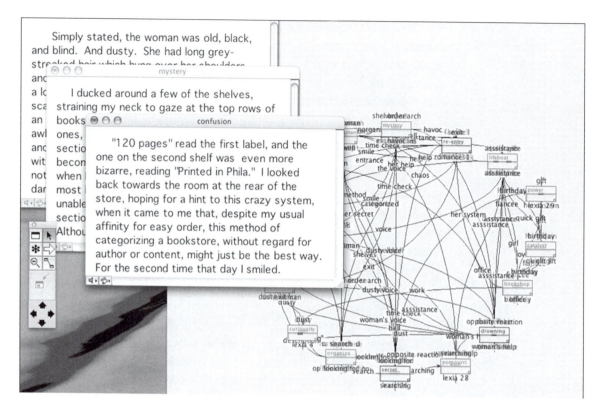

Figure 32. The Structure of Networked Hypertext. In creating a narrative that the reader can enter and leave at any point, Adam Wenger took advantage of the graphic capacities of Storyspace to arrange the individual lexias of *Adam's Bookstore* in the form of a circle or large polygon. (For the sake of legibility, I have increased the font size in three of the lexias in relation to the Storyspace view; in the original, one can easily read the titles of all the lexia icons.) (Used by permission of the author.)

Film theory does not have much in common with hypertext theory, but the one point at which writings on cinema converge with hypertext theory concerns the viewer's encounter with dramatic changes of direction after a cut. Ever since Sergei Eisenstein, film theorists have discussed the effect of changing scenes—the cinematic equivalent to arriving at a new hypertext lexia. At first glance, Eisenstein's writings on film would not seem to provide a very promising place to begin, since it is Eisenstein, after all, who complains that contemporary filmmakers have seemingly "forgotten . . . that role set it-self by every work of art, *the need for connected and sequential exposition of the theme, the material, the plot, the action,* the movement within the film se-quence and within the drama as a whole . . . The simple matter of telling *a connected story* has been lost in the works of some outstanding film masters"

(3). Taken out of context, Eisenstein sounds like twenty-first-century skeptical critics of hypernarrative, who claim readers and viewers cannot follow
abrupt changes of direction. Nonetheless, despite this apparent insistence on
linearity, Eisenstein's more fundamental emphasis that fragments form an
assemblage that makes sense to the audience (and not only in montage) shows
us someone who recognizes that segmentation, discontinuity, and gaps do
not destroy narrative but are crucial to it. First of all, he recognizes that the
universal human "tendency to bring together into a unity two or more independent objects or qualities is very strong" (5). This tendency appears in what
Eisenstein calls "the basic fact" that "the juxtaposition of two separate shots
by splicing them together resembles not so much a simple sum of one shot
plus another shot—as it does a *creation*" (7). In fact this human need for
order underlies "perception *through aggregation*" (16).

This pioneer of cinema shares a second point with writers on hypertext,
for he claims, though not very convincingly, that requiring spectators to
construct order and meaning out of fragments provided by the filmmaker
empowers them: "the spectator," Eisenstein claims, "is drawn into the creative act in which his individuality is not subordinated to the author's individuality, but is opened up throughout the process of fusion with the author's
intention, just as the individuality of a great actor is fused with the individuality of a great playwright in the creation of a classic scenic image" (33). Moreover, he not only asserts that this discontinuous form produces audience creativity; he also sees the audience in some sense as performing the text.
According to him, "it is precisely the *montage* principle, as distinguished
from that of *representation,* which obliges spectators themselves to *create*"
(35). Obviously, a theorist and practitioner who asserts that "modern esthet-
ics is built upon the disunion of elements" (95) anticipates some of the ideas
found in hypertext theory. Like hypertext theorists, he assumes that when the
audience encounters fragments and discontinuity, it will nonetheless manage to perceive (or construct) order, coherence, and continuity. Again, like hypertext theorists, he also claims that a constructivist art form shares some of
the author's creativity and power—though, to be sure, his claim appears little
more than that of theologians and New Critics, who asserted that enigma,
which exercises the mind, provides a means of forcefully conveying one's
points to the reader.

A more useful point at which film and hypertext theorists have something in common appears in what Christian Metz terms the "syntagma" of
film. As Clara Mancini has argued, Metz's classification and analysis of syntagma (including parallel, brace, alternative, descriptive, episodes, linear, and

so on) does prove useful because these categories directly involve how audiences of both media find coherence upon encountering the filmic or alphanumeric text—in other words, Metz and Lancini describe what I have elsewhere termed the *rhetoric of arrival*. Metz and other film theorists, like those concerned with the rhetoric of hypermedia, propose techniques that convince the audience that a newly encountered segment of a larger work is coherent—that is, that it has a comprehensible relation to something that the audience previously encountered. In this one sense, both forms of virtual cinema have much of importance to offer hypertext and hypermedia. Only single-person hypercinema, however, has close parallels with hypertext.

Is Hypertext Fiction Possible?

Is hypertext fiction—narrative composed and read within a hypertext environment that encourages branching story lines—possible? Some years back the *New York Times* printed a novelist's assertion that, contra what Bob Coover said, no one would *ever* write a hypertext novel. Admittedly, this statement appeared pretty silly on the face of it since at the time she wrote hundreds of examples of hypertext fiction had already been written and some famous ones had seen publication. Nonetheless, as one looks at the literature created in hypermedia that has appeared since then one wonders if perhaps that print-limited novelist might have unknowingly hit upon something important. One finds large numbers of digital poems in the form of animated text, hyperpoetry, and a combination of the two. Where amid all the digital literature that resides on the Web and within other hypertext environments offline is what Robert Coover has taught us to call hyperfiction?

No one doubts that digital literature, digital art, and fusions of the two flourish, and perhaps we are at the threshold of a new Lucasian age of literature. Georg Lukacs's *Theory of the Novel* (1923) proposed that each age has its own chief narrative form. Thus the classical ages had the epic, the medieval (or Christian) ages the chivalric romance, and the modern age the novel. Reinterpreting Lucas in terms of ages of information technology, we see that oral civilizations produce the epic, scribal culture the romance, and print culture the novel. What will be the major narrative form of digital information technology, if any? Or, put another way, will the major form be hyperfiction—or something else, perhaps poetry? I believe it's obviously too soon to make any sure predictions. One can, however, make a few observations, the first of which is that the first decades of digital literature consist largely of movemented or moving text, hyperpoetry, and fusions of word and image. The expected explosion of hyperfiction does yet not seem to have taken place and, moreover, much hypertext fiction, including some of the best, exhibits a minimum hypertextuality. As the examples of two well-regarded works—Shelley Jackson's *Patchwork Girl* and Caitlin Fisher's *Waves of Girls*—reveals, much of the limited hypertextuality in these and similar works takes the form of an organizational superstructure, a top-level branching structure that leads to multiple, relatively isolated linear narratives. Looking back at the brief history of hyperfiction, one is surprised to note how few works have accepted the challenge of Michael Joyce's *afternoon* to create branching story lines. Joyce's linking produces what we may term *branching narrators,* and one would expect that more writers would have tried varieties in chronology, setting, character, and so on. Of course, an advocate of the view that hypermedia is chiefly a poetic form can point to the other part of *afternoon* in which links produce a poemlike collage of texts from Creeley and Basho. Even one of the most successful pioneers of hypertext with multiple narrative lines moves in the direction of poetry when he begins to explore the medium.

One can hazard a few explanations for the turns that digi-lit has taken. One possible explanation would lie in the simple fact that it's too soon to take stock of this new literary form. If it took a hundred years to invent the title page and other distinguishing features of the print codex, such as pagination and the alphabetized index, at this moment we might find ourselves too early in the learning curve for any assessments. Another related possibility is that writers are immersing themselves in various capacities of digital text—including blending of word and image, animating words, and exploring the ludic or the gamelike possibilities—because they delight in the new possibilities of text. Again, this could just be a stage in the development of a new literary form. A third explanation might center on the claim that human beings in all times and cultures, including our own, depend on linear narrative. "We tell ourselves stories," Joan Didion points out, "in order to live." Perhaps linear narrative has too much human importance to abandon.

A final possibility: hypertext as a creative medium is not fundamentally a narrative form; hypertext, this argument goes, is an information technology unsuited to telling stories—just as orality, so McLuhan argued in *The Gutenberg Galaxy,* makes precise logical argumentation unlikely because it cannot be remembered and repeated. William Ivins similarly pointed out many decades ago that a scribal culture, which has no means of accurately reproducing and hence communicating color and form, does not permit the development of many forms of modern science, such as zoology and botany. What, then, would be the message of hypertext as medium? What features of text does it privilege and thereby make likely? Since the link characterizes hypertext, and links are reified associations, a poetic mode or form seems especially suited to hypertext. Looking at a range of digital works, we see that much hyperfiction actually takes the form of hyperpoetry.

Coover himself has expressed the idea that hypertext might turn out to be more a poetic than a narrative form, and many of the webs at which we have already looked, particularly those by Guyer and Joyce, suggest that such might be the case. They reveal, as Jean Clement has argued, "a shift from narration to poetry in fiction hypertexts." According to Clement, "hypertexts produce— at the level of narrative syntax—the same 'upheaval' as poems produce at the level of phrastic syntax." In other words, the way links reconfigure narrative leads to a defamiliarization that parallels the effects of characteristically poetic departures from word order, common usage, and the like. Clement continues: "Hypertexts free narrative sequences from their subjection to the syntax of conventional narration to insert them into the multidimension[al] space of a totally new and open structure, as poems free words from their linkage to the straightness of the syntagmatic axis to put them in a network of thematic, phonetic, metaphoric (and so on) connections which create a multi-isotopic configuration" (71). The explanation may be even simpler: the link, the element that hypertext adds to writing, bridges gaps between text, bits of text, and thereby produces effects similar to analogy, metaphor, and other forms of thought, other figures, that we take to define poetry and poetic thought.

Yet another ground for believing that hypertext might privilege poetry, particularly its lyric forms, appears in Jerome J. McGann's argument that "the object of poetry is to display the textual condition. Poetry is language that calls attention to itself, that takes its own textual activities as its ground subject." He emphasizes that such a claim does not assume "poetic texts lack polemical, moral, or ideological materials and functions. The practice of language takes place within those domains. But poetical texts operate to display their own practices, to put them forward as the subject of attention" (*Textual Condition*, 10–11), or, as McGann explains later, "The object of the poetical text is to thicken the medium as much as possible—literally, to put the resources of the medium on full display, to exhibit the processes of self-reflection and self-generation which texts set in motion, which they *are*" (14). McGann refers to the bibliographical and linguistic resources of written and printed language, but hypertext adds a new element—the link—to the mix. Since the link and various associated functions, such as lists of destination lexias, serve as the defining resources of hypertext, one expects to find them foregrounded in literary webs, and such is in fact the case. Of course, at this point in the development of this new medium, one cannot tell whether the sheer novelty of the medium motivates foregrounding these writing resources, and later hyperwriters will not do so, though I must admit that I find it difficult imagining literary hypertext that does not in this poetic manner make the most of its unique features.

Certainly, as we have already seen, poetry appears throughout the docuverse, often in unexpected places. That is, we encounter poetry not only in the form of scholarly hypertext editions, such as Peter Robinson's Chaucer project, or in translations into the hypertextual, such as Espen Aarseth's Hyper-Card version of Raymond Queneau's *Cent Mille Milliards de Poèmes* or Jon Lanestedt's and my *In Memoriam Web*. It also appears brushing up against other forms and modes, sometimes merely as a defining allusion within the midst of a prose fiction (*afternoon* and Joshua Rappaport's *Hero's Face*) and other times as part of a prose mystory (Taro Ikai's *Electronic Zen*). Karen Lee's *Lexical Lattice* demonstrates that poetry comes braided together with theory and literary history. Poetry also shows up in the most unexpected places within hypertext webs, none more so than in Stuart Moulthrop's *Victory Garden,* where at least one of his link menus forms a sonnet!

You will have noticed, I suspect, that more of the examples of both digital text and hypermedia textuality and rhetoric summoned in earlier chapters came from poetry than from fiction. The dozens of poems on the Brazilian and Spanish CD-ROMs show poets themselves exploring the possibilities of both animated text and hypermedia. Almost all the works on the New River website take the form of hyperpoetry: David Herrstrom's "City of Angels and Anguish" and "To Find the White Cat," Stephanie Strickland's *Vniverse,* Christie Sanford's "Light-Water: A Mosaic," and Robert Kendall's "A Study in Conveyance." Similarly, much of the Eastgate offering is hyperpoetry, including works in Storyspace (Ed Falco's *Sea Island,* Richard Gess's *Mahasukha Halo,* Kathryn Kramer's *In Small and Large Pieces,* and Kathy Mac's *Unnatural Habitats*) and those in other environments (Jim Rosenberg's *Intergrams,* Robert Kendall's *A Life Set for Two,* and Judy Malloy and Cathy Marshall's *Forward Anywhere: Notes on an Exchange between Intersecting Lives*). The enormous amount of hyperpoetry in the Electronic Literature Directory—twenty-nine screens of around ten poems each—suggests the amount of it being written in hypertext.

Hypertext poets have created their work in a wide variety of software environments, although HTML and Flash have recently become the most popular. A number of them have used Storyspace, but William Dickey, one of the first (possibly the very first) hypertext poets, used HyperCard, some poets in France

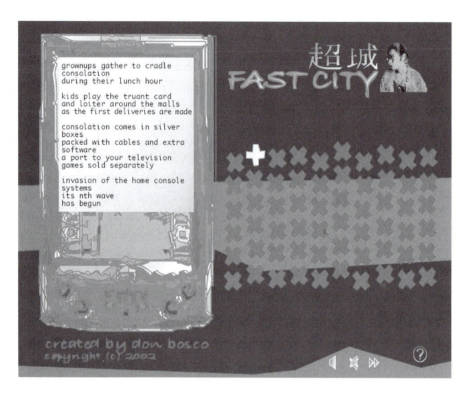

Figure 33. A Spatial Hypertext that Readers Perform: Don Bosco's *Fast City.* (Courtesy of Eastgate Systems.)

have used the help system for Windows, and Robert Kendall and Ian M. Lyons write in Visual Basic. Many of those working in new media, as I suggested above, tend to produce genres that seem more like poetry than narrative.

Take, for example, *Fast City,* Don Bosco's witty, combinatorial mosaic portrait of modern urban life, which he composed in Flash (Figure 33). *Fast City* appears in the online journal *Tekka,* and to read it the user downloads either the Windows or Macintosh version, double clicks on its pink and gray icon, and watches the opening animation. First, to the sound of a drumbeat and sounds of urban dissonance, giant Chinese characters and then the words "Fast City," which I assume refers to Singapore, overwhelm a black and gray background, shrinking in size and settling at upper-screen right, next to the image of a woman. Then the image of a blue PDA appears on the left and increases in size until it occupies a little less than half of the screen, after which six rows of ten "X's" snap into place accompanied by sound. Bosco's mosaic of modern urban lives takes the form of sixty lexias whose texts appear on the screen of the PDA. One can "play" the PDA, creating an assemblage of texts, and one can also create one's own dub music by manipulating the buttons. Mousing over each of the sixty X's at the right of the PDA produces a snippet of urban sound—barking dogs, sirens, for example—and brings up an associated text, each a mini-portrait of life in the *Fast City*—the out-of-work surveillance expert looking at the classifieds, urbanites assembling a home entertainment center, kids playing video games, rush hour traffic jams, dance-clubbers immersing themselves in the bangbangbang of "guitaristic assaults," the fashion model on the runway, and the soldier back for R&R: lives and deaths, all permeated by modern media:

the news channel is attractive
for adults going nowhere fast
playing out international
economics
with all the fury of mtv
creating fantastic paradigms
of market behaviour
businesscasters in their
powersuits
mouthing miraculous updates
every minute, second, split
second
inside of a nano-second.

As a Singaporean author and an inhabitant of the world's most globalized (and technologized) nation-state, Bosco effortlessly moves between Asian, American, and European scenes. As he writes, "Each simple lexia" is "a potent meme engine inserting its own unique values . . . into the flow of everyday media." Since the reader can see only one text at a time, the succession of them as one explores each X-shaped button forms more of a montage than a collage: the lexias appear sequentially rather than across the screen, and the method here is associative rather than narrative. After mousing over the various buttons for a while, readers begin to remember the texts and sounds that some of them produce, and then they can return to their favorites or construct some sort of order. Like a musician sight-reading a score, instrument in hand, the reader becomes a performer.

Gray Matters, a three-dimensional hyperpoem created in Pad ++ by a team at the NYU Media Lab, uses "15 images from *Gray's Anatomy* [to] form a patchwork body. Embedded within each image are a text from Henry Gray, Kirstin Kantner, Chris Spain, and Noah Wardrip-Fruin" (*Gray Matters* site).

Figure 34. Text in a Virtual Reality Environment: *Screen* by Noah Wardrip-Fruin, Andrew McClain, Shawn Greenlee, Robert Coover, and Josh Carroll. This photograph of a session in the Brown University Cave environment shows a reader after the text has begun to cascade off the walls. In the original photograph the letters appear in light blue and green while the heap directly in front of the reader is white. (Used by permission of Noah Wardrip-Fruin.)

Using a mouse the reader flies into images of body parts taken from the famous anatomy book, and as one approaches more closely, texts appear.

The illusion of immersing oneself in the poem leads to Wardrip-Fruin's later projects in the Virtual Reality Cave at Brown University, where he worked with Andrew McClain, Shawn Greenlee, Robert Coover, and Josh Carroll to create poetry in a more complete immersive environment.[18] In their *Screen,* the reader-viewer (immersee?) dons headtracker goggles and a VR glove, which allows the system to track her position and movement, and then enters the three-sided space whose white walls three projectors cover with text rather than images or, rather, with texts *as* images (Figure 34). As the introduction to the *Iowa Review* Web interview points out, "This experience in virtual reality is very different from the Holodeck vision of total immersion in a make-believe world. Screen does not attempt to replicate a real-world environment, but instead immerses the user in a reflexive literary representation, one in which words and narrative remain predominant." The reader experiences the text as existing in three dimensions rather than on a flat surface. As the piece begins, a voice—a second "reader"—begins to read: "In a world of illusions, we hold ourselves in place by memories." The texts hovering in space on three sides variously relate the memories of a man and a woman who feel them weaken, fade, vanish.

She uncurls her arm,
reaches back to lay her hand across
his thigh, to welcome him home,
but touches only a ridge of sheet,
sun warmed, empty.

After the voice has finished reading, the words, first on one wall then another, begin to fall. Using the VR glove, the reader can at first grab the words and replace them, but if more than one word is falling, words that she catches do not return to their original position, so despite her best efforts, text becomes corrupted. Then, as the words begin to cascade ever more quickly off the walls, she moves even faster but misses more and more of them until finally, realizing that she cannot stay time, she stops moving her hand and stands motionless. *Screen* thus includes text, movement, sound, and the reader's own actions, which in one sense are ultimately useless and in another absolutely necessary to read the poem successfully. I do not know if we can legitimately term *Screen,* which includes elements of narrative, lyric, animation, interactivity, and immersive VR, hypertext, though it does show the poetic possibilities of New Media.

In conclusion, as the example of *afternoon* and *Waves of Girls* demonstrates, hypertext fiction that compels the interest of readers is clearly possible. We also find a number of examples of fiction in hypermedia environments, like *Patchwork Girl,* that has a little hypertext branching in the main narratives and uses hypertext chiefly to create a contents page. Spatial hypertext has also played a role in hyperfictions, such as *Adam's Bookstore, Patchwork Girl,* and *Quibbling.* It has also been used to generate combinatorial fiction like Tom McHarg's just as it has been used in combinatorial poetry like Aarseth's translation of Queneau. Although I suspect that poetry will probably dominate hypermedia, hypertext fiction still seems to have great promise. If, as Didion says, we need stories to live, authors will always find themselves tempted to tell stories in any and all media.

Reconfiguring Literary Education

Like many other observers of the relations between information technology and education, Jean-François Lyotard, writing as early as 1979, perceived that "the miniaturization and commercialization of machines is already changing the way in which learning is acquired, classified, made available, and exploited. It is reasonable to suppose that the proliferation of information-processing machines is having, and will continue to have, as much of an effect on the circulation of learning as did advances in human circulation (transportation systems) and later, in the circulation of sounds and visual images (the media)" (*Postmodern Condition,* 4). One chief effect of electronic hypertext lies in the way it challenges now conventional assumptions about teachers, learners, and the institutions they inhabit. It changes the roles of teacher and student in much the same way it changes those of writer and reader. Its emphasis on the active, empowered reader, which fundamentally calls into question general assumptions about reading, writing, and texts, similarly calls into question our assumptions about literary education and its institutions that so depend on these texts. Gary Marchioni who created evaluation procedures for the pioneering Project Perseus, reminds us that "each time a new technology is applied to teaching and learning, questions about fundamental principles and methods arise" ("Evaluating Hypermedia-Based Learning," 10.1). Hypertext, by holding out the possibility of newly empowered, self-directed students, demands that we confront an entire range of questions about our conceptions of literary education.

Hypertext systems promise—or threaten—to have major effects on literary education, and the nature of hypertext's potential effect on human thought appears in descriptions of it from its earliest days. Writing of Bush, Engle-bart, Nelson, and other pioneers of hypertext, John L. Leggett and members of his team at the Hypermedia Lab at the University of Texas point out that "the revolutionary content of their ideas was, and continues to be, the extent to which these systems engage the user as an active participant in interactions with information" ("Hypertext for Learning," 2.1), Students making use of hypertext systems participate actively in two related ways: they act as reader-authors both by choosing individual paths through linked primary and secondary texts and by adding texts and links to the docuverse.[1]

Now that more than a decade and a half has passed since I began teaching with hypertext—and a decade since I completed the typescript of the first version of this book—I can see that hypertext has been used in four ways. One cannot accurately term them stages since several coincided with each other, and all continue in use today on the Web. Starting with Intermedia, read-only hypermedia helped students acquire both information and habits of thinking critically in terms of multiple approaches or causes. These first two uses or results represent the effects of employing an information medium based on connections to help students develop the habit of making connections. Next, almost immediately we discovered that Intermedia, which provided a participatory reading-and-writing environment, empowered students by placing them within—rather than outside—the world of research and scholarly debate. Finally, writing hypermedia enabled students to explore and create new modes of discourse appropriate for the kind of reading and writing we shall do increasingly in e-space, the writing necessary for the twenty-first century. In the first version of this book, by necessity I cited materials chiefly available only on Intermedia at Brown. Because many of these webs have now moved into other systems, I shall use examples easily accessible to readers, choosing, whenever possible, either materials published in Story-space or other environments or available on the Web.

All these effects or applications encourage and even demand an active student. The ways in which hypertext does so leads writers on the medium like David H. Jonassen and R. Scott Grabinger to urge that "hypermedia learning systems will place more responsibility on the learner for accessing, sequencing and deriving meaning from the information." Unlike users of "most information systems, hypermedia users must be mentally active while interacting with the information" ("Problems and Issues," 4).

From this emphasis on the active reader follows a conception of an active, constructivist learner and an assumption that, in the words of Philippe C. Duchastel, "hypermedia systems should be viewed not principally as teaching tools, but rather as learning tools" (139). As Terry Mayes, Mike Kibby, and Tony Anderson from the Edinburgh Centre for the Study of Human-Computer Interaction urge, systems of computer-assisted learning "based on hypertext are rightly called *learning systems,* rather than *teaching systems.* Nevertheless, they do embody a theory of, at least an approach to, instruction. They provide an environment in which *exploratory* or *discovery* learning may flourish. By requiring learners to move towards nonlineal thinking, they may also stimulate processes of integration and contextualization in a way not achievable by linear presentation techniques" (229). Mays and his collaborators therefore claim:

> At the heart of understanding interactive learning systems is the question of how deliberate, explicit learning differs from implicit, incidental learning. Explicit learning involves the conscious evaluation of hypotheses and the application of rules. Implicit learning is more mysterious: it seems almost like a process of osmosis and becomes increasingly important as tasks or material to be mastered becomes more complex. Much of the learning that occurs with computer systems seems implicit. (228)

Rand J. Spiro, working with different teams of collaborators, has developed one of the most convincing paradigms yet offered for educational hypertext and the kind of learning it attempts to support. Drawing on Ludwig Wittgenstein's *Philosophical Investigations,* Spiro and his collaborators propose that the best way to approach complex educational problems—what he terms "ill-structured knowledge domains"—is to approach them as if they were unknown landscapes: "The best way to [come to] understand a given landscape is to explore it from many directions, to traverse it first this way and then that (preferably with a guide to highlight significant features). Our instructional system for presenting complexly ill-structured 'topical landscape' is analogous to physical landscape exploration, with different routes of traversing study-sites (cases) that are each analyzed from a number of thematic perspectives" ("Knowledge Acquisition," 187). Concerned with developing efficient methods of nurturing the diagnostic skills of medical students, Spiro's team of researchers involve themselves in knowledge domains that present problems similar to those found in the humanistic disciplines. Like individual literary texts, patients offer the physician ambiguous complexes of signs whose interpretation demands the ability to handle diachronic and synchronic approaches. Young medical doctors, who must learn how to "take a history," confront symptoms that often point to multiple possibilities. They must therefore learn how to relate particular symptoms to a variety of different conditions and diseases. Since patients may suffer from a combination of several conditions at once, say, asthma, gall bladder trouble, and high blood pressure, physicians have to learn how to connect a single symptom to more than one explanatory system.

Spiro's explanation of his exploration-of-landscape paradigm provides an excellent description of educational hypertext:

> The notion of "criss-crossing" from case to case in many directions, with many thematic dimensions serving as routes of traversal, is central to our theory. The treatment of an irregular and complex topic *cannot be forced in any single direction* without curtailing that potential for transfer. If the topic can be applied in many different ways, none of which follow in rule-bound manner from the others, then limiting oneself in acquisition to, say, a single point of view or a single system of classification, will produce a relatively *closed* system instead of one that is open to context-dependent variability. By criss-crossing the complex topical landscape, the twin goals of highlighting multifacetedness and establishing multiple connections are attained. Also, awareness of variability and irregularity is heightened, alternative routes of traversal of the topic's complexities are illustrated, multiple entry routes for later information retrieval are established, and the general skill of working around that particular landscape (domain-dependent skill) is developed. *Information that will need to be used in a lot of different ways needs to be taught in lots of different ways.* (187–88)

In such complex domains, "single (or even small numbers of) connecting threads" do not run "continuously through large numbers of successive cases." Instead, they are joined by "'woven' interconnectedness. In this view, strength of connection derives from partial overlapping of many different strands of connectedness across cases rather than from any single strand running through large numbers of the cases" (193).

Reconfiguring the Instructor

Educational hypertext redefines the role of instructors by transferring some of their power and authority to students. This technology has the potential to make the teacher more a coach than a lecturer, and more an older, more experienced partner in a collaboration than an authenticated leader. Needless to say, not all my colleagues respond to such possibilities with cries of glee and hymns of joy.

Before some of my readers pack their bags for the trip to Utopia and others decide that educational computing is just as dangerous as they thought all along, I must point out that hypertext systems have a great deal to offer instructors in all kinds of institutions of higher education. To begin with, a hypermedia corpus of multidisciplinary materials provides a far more efficient means of developing, preserving, and obtaining access to course materials than has existed before. One of the greatest problems in course development lies in the fact that it takes such a long time and that the materials developed, however pioneering or brilliant, rarely transfer to another teacher's course because they rarely match that other teacher's needs exactly. Similarly, teachers often expend time and energy developing materials potentially useful in more than one course that they teach but do not use the materials because the time necessary for adaptation is lacking. These two problems, which all teachers face, derive from the classic, fundamental problem with hierarchical data structures that was Vannevar Bush's point of departure when he proposed the memex. A hypertext corpus, which is a descendant of the memex, allows a more efficient means of preserving the products of past endeavors because it requires so much less effort to select and reorganize them. It also encourages integrating all one's teaching, so that one's efforts function synergistically. A hypermedia corpus, such as a website, has the potential to preserve and make easily available one's past efforts as well as those of others.

Hypertext obviously provides us with a far more convenient and efficient means than has previously existed of teaching courses in a single discipline that need the support of other disciplines. As I discovered in my encounter with the nuclear arms materials, which I discussed in chapter 4, this educational technology permits teachers to teach in the virtual presence of other teachers and other subsections of their own discipline or other closely related disciplines. Thus, someone teaching a plant-cell biology course can draw upon the materials created by courses in very closely related fields, such as animal-cell biology, as well as slightly more distant ones, such as chemistry and biochemistry. Similarly, someone teaching an English course that concentrates on literary technique of the nineteenth-century novel can nonetheless draw upon relevant materials in political, social, urban, technological, and religious history. All of us try to allude to such aspects of context, but the limitations of time and the need to cover the central concerns of the course often leave students with a decontextualized, distorted view.

Inevitably, the Web and other forms of hypertext give us a far more efficient means than has previously existed of teaching interdisciplinary courses, of doing, that is, which almost by definition "shouldn't be done." (When most departmental and university administrators are not applying for funding from external agencies, they use the term *interdisciplinary* to mean little more than "that which should not be done" or "that for which there is no money." After all, putting together biology and chemistry to study the chemistry of organisms is not interdisciplinary; it is the subject of a separate discipline called biochemistry.) Interdisciplinary teaching no longer has its earlier glamour for several reasons. First, some have found that the need to deal with several disciplines has meant that some or all end up being treated superficially or only from the point of view of another discipline. Second, such teaching requires faculty and administration to make often extraordinarily heavy commitments, particularly when such courses involve teams of two or more instructors. Then, when members of the original team take a leave or cover an essential course for their department, the interdisciplinary course comes to a halt. In contrast to previous educational technology, hypertext offers instructors the continual virtual presence of teachers from other disciplines.

All the qualities of connectivity, preservation, and accessibility that make hypertext an enormously valuable teaching resource also make it equally valuable as a scholarly tool. The medium's integrative quality, when combined with its ease of use, offers a means of efficiently integrating one's scholarly work and work-in-progress with one's teaching. In particular, one can link portions of data on which one is working, whether they take the form of primary texts, statistics, chemical analyses, or visual materials, and integrate these into courses. Such methods, which we have already tested in undergraduate and graduate courses at Brown, allow faculty to explore their own primary interests while showing students how a particular discipline arrives at the materials, the "truths of the discipline," it presents to students as worthy of their knowledge. Materials on anti-Catholicism and anti-Irish prejudice in Victorian Britain created by Anthony S. Wohl, like some of my recently published work on Graham Swift and sections of this book, represent such integration of the instructor's scholarship and teaching. Such use of one's own work for teaching, which one can use to emphasize the more problematic aspects of a field, accustoms students to the notion that for the researcher and theorist many key problems and ideas remain in flux.

Hypermedia linking, which integrates scholarship and teaching and one discipline with others, also permits the faculty member to introduce beginners to the way advanced students in a field think and work while it gives beginners access to materials at a variety of levels of difficulty. Such materials, which the instructor can make easily available to all or only advanced students, again permit a more efficient means than do textbooks of introducing students to the actual work of a discipline, which is often characterized by competing schools of thought. Because hypertext interlinks and interweaves a variety of materials at differing levels of difficulty and expertise, it encourages both exploration and self-paced instruction. The presence of such materials permits faculty members to accommodate the slower as well as the faster, or more committed, learners in the same class. I've often been asked if the very abundance of materials, which I've just praised, does not confuse and intimidate students. I don't believe that it does, chiefly because students, like other computer users, generally employ these kinds of electronic resources for a specific task. Many doubters, it is clear, envisaged students wandering aimlessly about enormous information spaces, but the key here is that students, who have limited amounts of time to prepare for each assignment, tend to act rationally and first concentrate on obviously relevant materials.

A reasonably well-organized website will permit students to find what they need in short order, though it may also encourage them to explore— something that assignments can encourage. *The Victorian Web,* which includes about twenty thousand documents, some of which are entire book chapters, obviously offers a daunting amount of information, but students don't immediately encounter all that information. Instead, upon arriving at the site, users find an overview that offers twenty categories. Students wishing to read about Dickens's *Great Expectations* see the icon labeled "Authors," click on it, receiving a list of writers, and then choose the Dickens link. Upon opening that author's overview, they go to "Works" and thence to *Great Expectations.* Even here, students do not find themselves in some inchoate information space, since they have a local sitemap that makes it easy for them to locate essays, reading questions, and other materials that meet their needs. A student looking for information about the novelist's relation to Carlyle can go to the "Literary Relations" section while one wanting discussions of Dickens's early life chooses "Biography."

Reconfiguring the Student

For students hypertext promises new, increasingly reader-centered encounters with text. In the first place, experiencing a text as part of a network of navigable relations provides a means of gaining quick and easy access to a far wider range of background and contextual materials than has ever been possible with conventional educational technology. Students in schools with adequate libraries have always had the materials available, but availability and accessibility are not the same thing. Until students know how to formulate questions, particularly about the relation of primary materials to other phenomena, they are unlikely to perceive a need to investigate context, much less know how to go about using library resources to do so.

Even more important than having a means of acquiring factual material is having a means of learning what to do with such material when one has it in hand. Critical thinking relies on relating many things to one another. Since the essence of hypertext lies in its making connections, it provides an efficient means of accustoming students to making connections among materials they encounter. A major component of critical thinking consists in the habit of seeking the way various causes impinge upon a single phenomenon or event and then evaluating their relative importance, and well-designed hypertext encourages this habit.

Hypertext also offers a means for a novice reader to learn the habit of multisequential reading necessary for both student anthologies and scholarly apparatuses. Hypertext, which has been defined as text designed to be read nonsequentially or in a nonlinear mode, efficiently models the kind of text characteristic of scholarly and scientific writing. These forms of writing require readers to leave the main text and venture out to consider footnotes, evidence of statistics and other authorities, and the like. Our experience at Brown University in the late 1980s suggests that using hypertext teaches students to read in this advanced manner. This effect on reading, which first appears in students' better use of anthologies and standard textbooks, exemplifies the way that hypertext and appropriate materials together can quickly get students up to speed.

In addition, a corpus of hypertext documents intrinsically joins materials students encounter in separate parts of a single course and in other courses and disciplines. Hypertext, in other words, provides a means of integrating the subject materials of a single course with other courses. Students, particularly novice students, continually encounter problems created by necessary academic specialization and separation of single disciplines into individual courses. In the course of arguing for the historic contextualization of literary works, Brook Thomas describes this all too familiar problem:

The notion of a piece of literature as an organic, autonomous whole that combats the fragmentation of the modern world can easily lead to teaching practices that contribute to the fragmentation our students experience in their lives; a fragmentation confirmed in their educational experience. At the same time sophomores take a general studies literature course, they might also take economics, biology, math, and accounting. There is nothing, not even the literature course, that connects the different knowledge they gain from these different courses . . . Furthermore, because each work students read in a literature course is an organic whole that stands on its own, there is really no reason why they should relate one work to another in the same course. As they read one work, then another, then another, each separate and unique, each reading can too easily contribute to their sense of education as a set of fragmented, unrelated experiences in which wholeness and unity are to be found only in temporary, self-enclosed moments. (229)

Experience of teaching with various kinds of hypertext demonstrates that its intrinsic capacity to join varying materials creates a learning environment in which materials supporting separate courses exist in closer relationship to one another than is possible with conventional educational technology. As students read through materials for one English course, they encounter those supporting others and thereby perceive relationships among courses and disciplines.

Learning the Culture of a Discipline

Hypertext also offers a means of experiencing the way a subject expert makes connections and formulates inquiries. One of the great strengths of hypertext lies in its capacity to use linking to model the kinds of connections that experts in a particular field make. By exploring such links, students benefit from the experience of experts in a field without being confined by them, as students would be in a workbook or book approach.

Hypertext thus provides novices with a means of quickly and easily learning the culture of a discipline. From the fact that hypertext materials provide the student with a means of experiencing the way an expert works in an individual discipline it follows that such a body of electronically linked material also provides the student with an efficient means of learning the vocabulary, strategies, and other aspects of a discipline that constitute its particular culture.

The capacity of hypertext to inculcate the novice with the culture of a specific discipline and subject might suggest that this new information medium has an almost totalitarian capacity to model encounters with texts. The intrinsically antihierarchical nature of hypertext, however, undercuts such possibilities and makes it a means of efficiently adapting the materials to individual needs. A body of hypertext materials functions as a customized electronic library that makes available materials as they are needed and not, as lectures and other forms of scheduled presentation of necessity must often do, just when the schedule permits.

The infinitely adaptable nature of this information technology also provides students a way of working up to their abilities by providing access to sophisticated, advanced materials. Considered as an educational medium, hypertext also permits the student to encounter a range of materials that vary in terms of difficulty, because authors no longer have to pitch their materials to single levels of expertise and difficulty. Students, even novice students, who wish to explore individual topics in more depth therefore have the opportunity of following their curiosity and inclination as far as they wish. At the same time, more advanced students always have available more basic materials for easy review when necessary.

The reader-centered, reader-controlled characteristics of hypertext mean that it offers student-readers a way of shaping and hence controlling major portions of what they read. Since readers shape what they read according to their own needs, they explore at their own rate and according to their own interests. In addition, the ease of using hypertext means that any student can contribute documents and links to the system. Students can thus experience the way contributions in various fields are made.

Finally, hypertext produces an additional form of discussion and a new means of contributing to class discussions that assists many students. Jolene Galegher and Robert Kraut, like most students of cooperative work, point out that "one of the failures of group discussion is the social influence that inhibits the quantity of original ideas that the members would have generated had they been working in isolation." In this context, hypermedia exemplifies those "permissive technologies" that "allow current practices to be extended into new realms in which they had previously been impracticable" (9). This feature of hypertext doubly permits students to contribute to the activity of a class: they can contribute materials in writing if they find group discussions difficult, and other students can cite and discuss their hypertext contributions. By giving an additional means of expression to those people shy or hesitant about speaking up in a group, electronic conferencing, hypertext, and other similar media shift the balance of exchange from speaking to writing, thus addressing Derrida's calls to avoid phonocentricism in that eccentric, unexpected, very literal manner that, as we have seen before, characterizes such hypertext instantiations of theory.

Nontraditional Students: Distant Learners and Readers outside Educational Institutions

The combination of the reader's control and the virtual presence of a large number of authors makes an efficient means of learning at a distance. The very qualities that make hypertext an efficient means of supporting interdisciplinary learning also permit students to work without having to be in residence at a geographical or spatial site. In other words, the adaptable virtual presence of hypermedia contributors serves both the distant, unconventional learner and the college student in a more conventional setting. For those interested in the efficient and just distribution of costly educational resources, hypertext offers students at one institution a way to share resources at another. Hypermedia provides an efficient means for students anywhere potentially to benefit from materials created at any participating institution.

The very strengths of hypertext that make it work so well in conventional educational settings also makes it the perfect means of informing, assisting, and inspiring the unconventional student. Because it encourages students to choose their own reading paths, hypertext provides the individualistic learner with the perfect means for exploration and enrichment of particular areas of study. By permitting one to move from relatively familiar areas to less familiar ones, a hypertext corpus encourages the autodidact, the resumed education student, and the student with little access to instructors to get in the habit of making precisely those kinds of connections that constitute such an important part of the liberally educated mind today so necessary in government and business. At the same time the manner in which hypertext places the distant learner in the virtual presence of many instructors both disperses the resources they have created in a particularly effective manner and allows the individual access to some of the major benefits of an institutional affiliation without the cost to either party in terms of time and money.

The World Wide Web has enabled the rapid development of a widespread, if all but unnoticed, form of distant learning. With good reason, attention has been paid to the use of the Web for distant learning courses offered by both conventional tertiary institutions and those, such as the Open Universities of the United Kingdom and Catalonia, dedicated to distant learning. Meanwhile, secondary and tertiary students have quietly used the Web with or without the knowledge of their instructors. I first became aware of this phenomenon after I began to receive e-mail thanking me for my *Victorian* and *Postcolonial* sites from undergraduates, postgraduate students, instructors, and even provosts of institutions from Europe, Asia, Australia, and the Americas. At the same time, contributions to the sites by students as well as faculty from all these areas continue to arrive. These distant students—distant, that is, to the web servers in New York and Singapore and to the university where I am paid to teach—came to the sites by different routes: some have been assigned to use them by their instructors, others have followed recommendations by the ministries of education in France and Sweden, National Endowment for the Humanities (NEH), the BBC, or groups representing individual disciplines, such as history, art history, and the sciences; yet others discovered them by using popular search engines. This kind of use of these sites, which now receive 15 million hits/month, suggests several things. First of all, while experts in distance education have been understandably concentrating the use of digital technologies by institutions like the Open University and the U.S. Military, students and faculty have quietly begun to use the Web to supplement educational resources at their own institutions.[2] Second, skeptics like Vincent Mosco have denigrated as mere hype—contributions to the "cyberspace myth"—predictions about the future of higher education made by cyber enthusiasts, such as William J. Mitchell in *City of Bits* (65–70). Many claims about the effect of the Internet, however, appear to becoming true outside the purview of the elite institutions, which thereby lose an important opportunity to influence the course of higher education. Universities that support important scholarly and educational sites, like the University of Virginia, create cultural influence and academic reputation far from their physical campuses.

Two of the most exciting and objectively verifiable effects of using educational hypertext systems involve the way they change the limiting effects of time. The modularization that John G. Blair has described as characteristic of American (as opposed to European) higher education appears in the concepts of credit hours, implicitly equivalent courses, and transcripts.[3] It also appears, one may add, in the precise, necessarily rigid scheduling of the syllabus for the individual course, which embodies what Joseph E. McGrath describes as a naively atomistic Newtonian conception of time:

Two of the assumptions of the Newtonian conception of time, which dominates our culture and organizations within it, are (a) an atomistic assumption that time is infinitely divisible, and (b) a homogeneity assumption that all the "atoms" of time are homogeneous, that any one moment is indistinguishable from and interchangeable with any other. But these assumptions do not hold in our experience . . . Ten 1-minute work periods, scattered throughout the day, are not of equivalent productivity value to one 10-minute period of work from 9:15 to 9:25 a.m. Nor is the day before Christmas equivalent to February 17th for most retailers. A piece of time derives its epochal meaning, and its temporal value, partly in terms of what activities can (or must) be done in it. (38)

The division—segregation, really—of individual weeks into isolated units to which we have all become accustomed has the unfortunate effect of habituating students to consider in isolation the texts and topics encountered during these units. The unfortunate effects of precise scheduling, which coverage requires, only became apparent to me after teaching with hypertext. Here, as in other cases, one of the chief values of teaching with a hypertext system has proved to be the light it unexpectedly has cast on otherwise unexamined, conventional assumptions about education.

One ethnographical team devoted three years to studying the effects of hypermedia in teaching.[4] The first effect comes from the experience of Peter Heywood, associate professor of biology, who used an Intermedia component in his upper-class course in plant-cell biology. The term paper for his course, which he intended to be a means of introducing students to both the literature of the field and the way it is written, required that students include all materials on their particular topic that had seen publication up to the week before papers were handed in. This demanding assignment required that Heywood devote a great deal of time to assisting individual students with their papers and their bibliographies, and one of the chief attractions of the Intermedia component to him lay in its potential to make such information more accessible. Using hypermedia greatly surprised Heywood by producing a completely unexpected effect. In the previous seventeen years that he had taught this course, he had discovered that many term papers came in after the deadline, some long after, and that virtually all papers concerned topics covered in the first three weeks of the course. The first year that students used the Intermedia component, all thirty-four papers came in on time. Moreover, their topics were equally distributed throughout the fourteen weeks of the semester. Heywood explains this dramatic improvement in student performance as a result of the way hypertext linking permits students to perceive connections among materials covered at different times during the semester. Although all other components of the course remained the same, the capabilities of hypermedia permitted students to follow links to topics covered later in the course and thereby encounter attractive problems for independent work. For example, while reading materials about the cell membrane in the first weeks of the course, students could follow links that brought them to related materials not covered until week eight, when the course examined genetics, or until the last weeks of the course, when it concerned ecological questions or matters of bioengineering. Many who enrolled in this upper-division course had already taken other advanced courses in genetics, biochemistry, or similarly related subjects. From the very beginning of the semester, linking permitted these students to integrate materials encountered early in this course with those previously encountered in other classes.

Educational hypertext in this way serves what McGrath describes as one of those "technological tools . . . designed in part to ease the constraints of the time/activity match in relation to communication in groups. For example, certain forms of computer conference arrangements permit so-called asynchronous communication among group members" (39). As the example from Heywood's course shows, hypertext systems also support this "asynchronous communication" between students and chronologically ordered modular components of the course.

The way that hypertext frees learners from constraints of scheduling without destroying the structure and coherence of a course appears in more impressionistic observations reported by members of both biology and English courses. One of Heywood's students described working with hypertext as providing something like the experience of studying for a final examination every week, by which, he explained that he meant that each week, as students encountered a new topic, they discovered they were rearranging and reintegrating the materials they had already learned, an experience that previously they had encountered only during preparations for major examinations. English students similarly contrasted their integrative experience of course readings with those of acquaintances in sections of the survey course that had not used hypermedia. The English students, for example, expressed surprise that whereas they placed each new poem or novel within the context of those read previously as a matter of course—considering, say, the relation of *Great Expectations* to "Tintern Abbey" and "The Vanity of Human Wishes" as well as to *Pride and Prejudice* and *Gulliver's Travels*—their friends in other sections assumed that, once a week was over, one should set aside the reading for that week until the final exam. In fact, students in other sections apparently expressed surprise that my students wanted to make all these connections.

A second form of asynchronous communication involves the creation by hypertext of a course memory that reaches beyond a single semester. Galegher and Kraut propose that "technologies that allow users to observe each other's contributions (such as computer conferences and hypermedia systems) may provide a system for sustaining group memory independent of the presence of specific individuals in an organization" (15). The contributions of individual student (and faculty) reader-authors, which automatically turn Intermedia and its Web-based descendants into fully collaborative learning environments, remain on the system for future students to read, quote, and argue against. Students in my literature courses encounter essays and reading questions by at least a dozen groups of students from earlier years plus by students at other universities whose work they or their instructors submitted. Coming upon materials created by other students, some of whom one may know or whose name one recognizes, serves to convince them that they are in a very different, more active kind of learning situation. As we shall also observe when we return to this subject in discussing the political implications of such educational media, this technology of memory produces effects quite unusual in a university setting.

Reconfiguring Assignments and Methods of Evaluation

To take advantage of hypertext's potential educational effects, instructors must decide what role it will play and must consciously teach with it. One must make clear to students both the goals of the course and the role of the hypertext system, generally a website, in meeting them. Peter Whalley correctly points out that "the most successful uses of hypertext will involve learners and lead them to adopt the most appropriate learning strategy for their task. They must . . . allow the learner to develop higher level skills, rather than simply become the passive recipients of a slick new technology" (68). Instructors therefore must create assignments that emphasize precisely those qualities and features of hypertext that furnish the greatest educational advantages. I have elsewhere described in detail such an initial assignment and will summarize it below before providing the example of a more complex exercise.

Whether it is true that readers retain less of the information they encounter while reading text on a screen than while reading a printed page, electronically linked text and printed text have different advantages. One should therefore prepare an initial assignment that provides the student with experience of its advantages—the advantages of connectivity. Obviously, instructors wishing to introduce students to the capacity hypertext gives them to choose their own reading paths and hence construct their own document must employ assignments that encourage students to do so.

The first hypertext assignment I formerly used derives from one first developed for Intermedia and then modified for Storyspace and later the World Wide Web. This assignment instructed students to follow specific links, report what they found, and offer both suggestions for additional links and sample passages from a work being discussed in class. In recent years students have become so Web savvy that I no longer need this kind of basic introductory assignment; in fact, students occasionally enter my classes having used *The Victorian Web* in courses in other departments. Since I employ a corpus of linked documents to accustom students to discovering or constructing contexts for individual blocks of text or data, my assignments require multiple answers to the same question or multiple parts to the same answer. If one wishes to accustom students to the fact that complex phenomena involve complex causation, one must arrange assignments in such a way as to make students summon different kinds of information to explain the phenomena they encounter. Since my courses have increasingly taken advantage of hypertext's capacity to promote collaborative learning, my assignments, from the beginning of the course, produce material that becomes part of the website.

Instructors employing educational hypertext must also rethink examinations and other forms of evaluation. If the Web's greatest educational strength as well as its most characteristic feature is its connectivity, then tests and other evaluative exercises must measure the results of using that connectivity to develop the ability to make connections. Independent of educational use of hypertext, dissatisfaction with American secondary school students' ability to think critically has recently led to a new willingness to try evaluative methods that emphasize conceptual skills—chiefly making connections—rather than those that stress simple data acquisition.

Taking advantage of the full potential of hypertext obviously forces instructors to rethink the goals and methods of education. If one wishes to develop student skills in critical thinking, then one might have to make one's goal elegance of approach rather than quantitative answers. Particularly when dealing with beginning students, instructors will have to emphasize that several correct answers may exist for a single problem and that such multiplicity of answers does not indicate that the assigned problem is subjective or that any answer will do. If, for example, one asks students to provide a context in contemporary philosophy or religion for a literary technique or historical event, one can expect to receive a broad range of correct solutions.

A Hypertext Exercise

Several of the courses that I teach with the Web employ the following exercise, which may take the form of either an in-class exercise or a take-home exam that students have a week or more to complete. The exercise consists of a series of passages from the assigned readings that students have to identify and then relate to a single work in brief essays; in the past, these exercises have used Wordsworth's "Tintern Abbey," Dickens's *Great Expectations,* and Austen's *Pride and Prejudice* as the central texts; those for courses in Victorian literature have similarly employed Charlotte Brontë's *Jane Eyre,* Elizabeth Barrett Browning's *Aurora Leigh,* Thomas Carlyle's "Signs of the Times," and other works. The instructions for the exercise asking students to relate passages to a specific text directs them thus:

Begin each essay by identifying the full name, exact title, and date of the passage, after which you should explain at least three ways in which the passage relates (whatever you take that term to mean) to the poem. One of these connections should concern theme, a second should concern technique, and a third some aspect of the religious, philosophical, historical, or scientific context . . . Not all the relations you discover or create will turn out to be obvious ones, such as matters of influence or analogous ideas and techniques. Some may take the form of contrasts or oppositions that tell us something interesting about the authors, literary forms, or times in which these works appeared.

To emphasize that demonstrating skill at formulating possible explanations and hypothesizing significant relations counts as much as factual knowledge alone, the directions explain that some subjects, "particularly matters of context, may require you to use materials" in whatever website or book they use "to formulate an hypothesis," and the assignment goes on to warn that in many cases, the hypermedia materials, like the library, "provide the materials to create an answer but not answers themselves."

Using this exercise in six iterations of the survey course as well as in various other upper-level courses convinces me that it provides a useful and accurate means of evaluation that has several additional beneficial effects. Although the exercise does not directly ask for specific factual information other than titles, authors, and dates, students soon recognize that without such information they cannot effectively demonstrate connections between or among texts. In comparing a passage from Pope's "Essay on Man" with "Tintern Abbey," for example, they soon realize that only specific examples and specific comments on those examples produce effective discussion. Gary Marchioni points out that "hypermedia is an enabling technology rather than a directive one, offering high levels of user control. Learners can construct their own knowledge by browsing hyperdocuments according to the associations in their own *cognitive structures*. As with access, however, control requires responsibility and decision making" ("Evaluating Hypermedia-Based Learning," 356). By making students choose which literary techniques, themes, or aspects of context they wish to relate, the exercise emphasizes the major role of student choice.

This assignment itself also proves an effective educational tool because while attempting to carry it out many students realize that they have difficulty handling matters of context, which at the beginning they often confuse with the theme or main idea of a passage. Discussions of context require one to posit a connection between one phenomenon, say, the imagery in a poem, and some other, often more general, phenomenon, such as conceptions of the human mind, gender roles, or religious belief contemporaneous with that imagery. Perceiving possible connections and then arguing for their validity is a high-level intellectual skill. Since students are permitted and in fact encouraged to redo these exercises as many times as they wish, these exercises simultaneously furnish students the opportunity to make conceptual breakthroughs and teachers the opportunity to encourage and then measure them.

Two additional advantages of this exercise for the courses in which it appears involve writing. Since both the survey and the more advanced courses are intended to be intensive writing courses, the opportunity to do a large amount of writing (and rewriting) supports one of their goals, although obviously that might prove a hindrance in other kinds of courses, particularly those with large enrollments. Second, the several short essays that the structure of the assignment requires seem to accomplish more than a single long essay. At the same time that students find writing many short essays easier than constructing a single much longer one, they cover far more material than they could with a more conventional assignment and they cover different approaches, each demanding the kind of materials generally available only in a hypermedia corpus.

Another advantage of this exercise, which I find well suited to courses with hypertext supplements, lies in the fact that, particularly in its take-home version, it demonstrates the usefulness of the website at the same time that it draws on skills encouraged by using it. The hypertext materials show students possible connections they might wish to make and furnish information so they can make their own connections. Our hypermedia corpora, which have taken the form of websites for more than a decade, also permit them to range back and forth throughout the course, thereby effecting their own syntheses of the materials.

A final utility of this exercise lies in the fact that by encouraging the students to take a more active, collaborative approach to learning, it thereby creates exemplary materials for students to read. I used to require students to hand in both paper copies of their papers and HTML versions on disks, which I provided along with templates, but as computer displays dramatically improved and reading on them became more pleasant, I have the essays submitted by e-mail; this practice saves paper, and it provides me with the opportunity to make interlinear comments easily and return the essay with my comments to the student; this way I no longer devote an hour writing comments on a term paper with serious problems only to discover, as one often does, that the author of the essay never bothers to pick it up, and it remains in a box outside one's office until, a year or so later, it ends up in the trash.

During the past few years, I have used what has turned out to be a particularly effective assignment involving student contributions to websites. These weekly exercises take the form of question sets. As the syllabi for several of my courses explain, these weekly reading and discussion questions have three parts: "(a) a substantial passage of 1–3 paragraphs from the assigned readings (please include page numbers and give your question set a title); (b) a graceful and effective introduction to the passage that suggests why the reader wants to read it; and (c) 4–5 questions, chiefly concerning matters of technique or relations to previous readings, for which you do not have to have answers." The class uses e-mail to submit these exercises, which provide the basis for class discussion, to me no later than 6:00 p.m. the day before we begin talking about the reading or painting. In addition to the minor goal of providing a way for students to bootstrap a new section of the website, I find the assignment has several beneficial effects for both instructor and student, the first of which is that it encourages students to come to class having read the assignment with care. In addition, it encourages active discussions in which all members of the class participate, even those too shy to do so willingly without such a prop, and because presentations begin with references to specific texts, the discussions are much more substantial than they usually are.

This assignment also encourages students at all levels—and I've used it for freshmen and for graduate students—to develop important academic skills, which include learning how to use textual data to support an argument, choosing appropriate passages, and, a technique almost all students need to learn, developing effective ways to introduce quoted material. The assignment also has the major advantage for the instructor of permitting him or her to work with very brief, but typical, passages of a student's writing each week, thereby providing a convenient and yet effective way to correct common mistakes and encourage new skills. Beginning students tend to write fairly brief question sets, and I find that I devote most of my comments to their writing; juniors, seniors, and graduate students tend to send in substantial brief essays, even though the assignment does not require them to do so, and most of my comments involve matters of rhetoric and interpretation. (Anyone who wants to take a look at these question sets can find typical examples in the *Victorian Web*'s sections on Elizabeth Barrett Browning, Thomas Carlyle, J. E. Millais, and A. C. Swinburne; just click on the icon labeled "Leading Questions" in each author's overview.)

Each week, when the question sets arrive, I respond to them by return e-mail, often adding interlinear comments, after which I place the reading question in a previously prepared template, and upload it to the relevant section of the website. Starting with the second week of the course, I teach the class HTML tags that create paragraphs, indented passages, and various forms of emphasis, so after the first few weeks the students become HTML "experts" and I have to do little formatting. When and if students ask, I also teach them how to make links, because some of them wish to refer to discussion questions from previous weeks or other material on the site. These question sets remain online after the course ends, and the fact that students who come after them might read their work, like the fact the site is public, provides student-authors with a crucial sense that they are writing for an audience and that they are engaged in collaborative learning.

All texts on a hypertext system potentially support, comment on, and collaborate with one another. Once placed within a hypertext environment, a document created by a student no longer exists alone. It always exists in relation to other documents in a way that a book or printed document never does and never can. From this follow two corollaries. First, any document placed on a networked system that supports electronically linked materials potentially exists in collaboration with any and all other documents on that system. Second, any document electronically linked to any other document collaborates with it.[5]

To create a document or a link in hypertext is to collaborate with all those who have used it previously and will use it in the future. The essential connectivity of the medium encourages and demands collaboration. By making each document in the docuverse exist as part of a larger structure, hypertext places each document in what one can term the *virtual presence* of all previously created documents and their creators. This electronically created virtual presence transforms individual documents created in an assembly-line mode into ones that could have been produced by several people working at the same time. In addition, by permitting individual documents to contribute to this electronically related overarching structure, hypertext also makes each contribution a matter of versioning. In so doing, it provides a model of scholarly work in the humanities that better records what actually takes place in such disciplines than does traditional book technology.

The same factors—connectivity, virtual presence, and shifting of the balance between writer and reader—that prompt major, perhaps radical, shifts in teaching, learning, and the organization of both activities inevitably have the potential to affect the related notions of canon and curriculum. For a work to enter the literary canon—or, more properly, to be entered into the canon—gains it certain obvious privileges. That the passive grammatical construction more accurately describes the manner in which books, paintings, and other cultural texts receive that not-so-mysterious stamp of cultural approval reminds us that those in positions of power decide what enters this select inner circle. The gatekeepers of the fortress of high culture include influential critics, museum directors and their boards of trustees, and a far more lowly combine of scholars and teachers. One of the chief institutions of the literary canon is the middlebrow anthology, that hanger-on of high culture that in the Victorian period took the form of pop anthologies like *Golden Treasury* and today exists principally in the form of major college anthologies. In America, to be in the Norton or the Oxford anthology is to have achieved, not greatness, but what is more important, certainly—status. And that is why, of course, it matters that so few women have managed to gain entrance to such anthologies.

The notion of a literary canon descends from that of the biblical one, in which, as Gerald L. Bruns explains, canonization functions as "a category of power":

What is important is not only the formation, collection, and fixing of the sacred texts, but also their application to particular situations. A text, after all, is canonical, not in virtue of being final and correct and part of an official library, but because it becomes *binding* upon a group of people. The whole point of canonization is to underwrite the authority of a text, not merely with respect to its origin as against competitors in the field . . . but with respect to the present and future in which it will reign or govern as a binding text . . . From a hermeneutic standpoint . . . the theme of canonization is *power.* (81, 67)

One sees the kind of privileges and power belonging to canonization in the conception that something is a work of art; the classification of some object or event as a work of art enters it into a form of the canon. Such categorization means that the work receives certain values, meanings, and modes of being perceived. A work of art, as some modern aestheticians have pointed out, is functionally what someone somewhere takes to be a work of art. Saying it's so makes it so. If one says the found object is a work of art, then it is; and having become such (however temporarily), it gains a certain status, the most important factor of which is simply that it is looked at in a certain way: taken as a work of art, it is contemplated aesthetically, regarded as the occasion for aesthetic pleasure or, possibly, for aesthetic outrage. It enters, one might say, the canon of art; and the contemporary existence in the Western world of galleries permits it to inhabit, for a time, a physical space that is taken by the acculturated to signify, "I am a work of art. I'm not (simply) an object for holding open a door. Look at me carefully." If that object is sold, bartered, or given *as a work of art* to one who recognizes the game or accedes in the demand to play her or his role in it, then it brings with it the capacity to generate that special space around it that signals it to be an object of special notice and a special way of noticing.

In precisely the same way, calling something a work of literature invokes a congeries of social, political, economic, and educational practices. If one states that a particular text is a work of literature, then for one it is, and one reads it and relates it to other texts in certain definite ways. As Terry Eagleton correctly observes, "anything can be literature, and anything which is regarded as unalterably and unquestionably literature—Shakespeare, for example—can cease to be literature. Any belief that the study of literature is the study of a stable, well-definable entity, as entomology is the study of insects, can be abandoned as a chimera . . . Literature, in the sense of a set of works of assured and unalterable value, distinguished by certain shared inherent properties, does not exist" (*Literary Theory,* 10–11). The concept of literature (or literariness) therefore provides the fundamental and most extended form of canonization, and classifying a text as a work of literature is a matter of social and political practice.

I first became aware of the implications of this fact a bit more than several decades ago when I was reading the sermons of the Evangelical Anglican, Henry Melvill, in an attempt to understand Victorian hermeneutic practice. Upon encountering works by a man who was the favorite preacher of John Ruskin, Robert Browning, W. E. Gladstone, and many of their contemporaries, I realized that his sermons shared literary qualities found in writings by Ruskin, Carlyle, Arnold, and Newman. At first Melvill interested me solely as an influence on Ruskin and as a means of charting the sage's changing religious beliefs. In several studies I drew upon his extraordinarily popular sermons as extraliterary sources or as indications of standard Victorian interpretative practice. If I were to write my study of Ruskin now, three decades later, I would treat Melvill's sermons also as works of literature, in part because contemporaries did so and in part because classifying them as literature would foreground certain intertextual relations that might otherwise remain invisible. At the time, however, I never considered discussing Melvill's sermons as literary texts rather than as historical sources, and when I mentioned to colleagues that his works seemed in some ways superior to Newman's, none of us considered the implications of that remark for a concept of literature. Remarks by colleagues, even those who specialized in Victorian literature, made clear that paying close attention to such texts was in some way eccentric and betokened a capacity to endure reading large amounts of necessarily boring "background material." When I taught a course in Anglo-American nonfiction some fifteen years after first discovering Melvill, I assigned one of his sermons, "The Death of Moses," for students to read in the company of works by Thomas Carlyle and Henry David Thoreau. Reading Melvill's sermon for an official course given under the auspices of the department of English, they assumed that it was a work of literature and treated it as such. Considering "The Death of Moses," which has probably never before appeared in an English course, as a work of "real" literature, my students, it became clear, assumed that Melvill's writing possessed a certain canonical status.

The varieties of status that belonging to the canon confers—social, political, economic, aesthetic—cannot easily be extricated one from the others. Belonging to the canon is a guarantee of quality, and that guarantee of high aesthetic quality serves as a promise, a contract, that announces to the viewer, "Here is something to be enjoyed as an aesthetic object. Complex, difficult, privileged, the object before you has been winnowed by the sensitive few and the not-so-sensitive many, and it will *repay* your attention. You will receive a frisson; at least you're supposed to, and if you don't, well, perhaps there's something wrong with your apparatus." Such an announcement of status by the poem, painting, building, sonata, or dance that has appeared ensconced within a canon serves, as I have indicated, a powerful separating purpose: it immediately stands forth, different, better, to be valued, loved, enjoyed. It is the wheat winnowed from the chaff, the rare survivor, and has all the privileges of such survival.

Anyone who has studied literature in a secondary school or university in the Western world knows what that means. It means that the works in the canon get read, read by neophyte students and expert teachers. It also means that to read these privileged works is a privilege and a sign of privilege. It is also a sign that one has been canonized oneself—beautified by the experience of being introduced to beauty, admitted to the ranks of those of the inner circle who are acquainted with the canon and can judge what belongs and does not. Becoming acquainted with the canon, with those works at the center, allows (indeed, forces) one to move to the center or, if not absolutely to the center, at least much closer to it than one had been before.

This canon, it turns out, appears far more limited to the neophyte reader than to the instructor, for few of the former read beyond the reading list of the course, few know that one *can* read beyond, believing that what lies beyond is by definition dull, darkened, dreary. One can look at this power, this territoriality of the canonized work, in two ways. Gaining entrance clearly allows a work to be enjoyed; failing to do so thrusts it into the limbo of the unnoticed, unread, unenjoyed, unexisting. Canonization, in other words, permits the member of the canon to enter the gaze and to exist. Like the painting accepted as a painting and not, say, a mere decorative object or even paint spill, it receives a conceptual frame; and although one can remark upon the obvious fact that frames confine and separate, it is precisely such appearance within the frame that guarantees its aesthetic contemplation— its capacity to make the viewer respect it.

The very narrowness of the frame and the very confinement within such a small gallery of framed objects produces yet another effect, for the framed object, the member of the canon, gains an intensification not only from its segregation but also because, residing in comparative isolation, it gains splendor. Canonization both permits a work to be seen and, since there are so relatively few objects thus privileged, canonization intensifies the gaze; potentially distracting objects are removed from the spectator's view, and those that are left benefit from receiving exclusive attention.

Within academia, however, to come under the gaze, works must be teachable. They must conform to whichever currently fashionable pedagogy allows the teacher to discuss this painting or that poem. In narrating the formation of the modernist canon, Hugh Kenner explains that "when Pound was working in his normal way, by lapidary *statement,* New Critics could find nothing whatever to say about him. Since 'Being-able-to-say-about' is a pedagogic criterion, he was largely absent from a canon pedagogues were defining. So was Williams, and wholly. What can Wit, Tension, Irony enable you so say about The Red Wheelbarrow?"[6] Very little, one answers, and the same is true for the poetry of Swinburne, which has many similarities to that of Stevens but which remained unteachable for many trained in New Criticism. In painting the situation is much the same: critics of purely formalist training and persuasion had nothing to say about the complex semiotics of Pre-Raphaelite painting. To them it didn't really seem to be art.

Thematic as well as formal filters render individual texts teachable. As

Sandra M. Gilbert and Susan Gubar, Ellen Moers, Elaine Showalter, and many others have repeatedly demonstrated, people who for one reason or other do not find interesting a particular topic—say, the works, fates, and subjectivities of women—do not see them and have little to say about them. They remove them from view. If belonging to the canon brings a text to notice, thrusts it into view, falling out of the circle of light or being absent or exiled from it keeps a text out of view. The work is in effect excommunicated. For, as in the Church's excommunication, one is not permitted to partake of the divine refreshing acts of communion with the divinity, one is divorced from sacramental life, from participation in the eternal, and one is also kept from communicating with others. One is exiled from community. Likewise, one of the most savage results of not belonging to the canon is that these works do not communicate with one another. A work outside the canon is forgotten, unnoticed, and if a canonical author is under discussion, any links between the uncanonical work and the canonical tend not to be noticed.

I write *tend* because under certain conditions, and with certain gazes, they can be at the other end of the connections. But within the currently dominant information technology, that of print, such connections and such linkages to the canonical require almost heroic and certainly specialized efforts. The average intelligent educated reader, in other words, is not expected to be able to make such connections with the noncanonical work. For him or her they do not exist. The connections are made among specialized works and by those readers—professionalized by the profession of scholarship—whose job it is to explore the reader's equivalent of darkest Africa of the nineteenth- and early-twentieth-century Western imagination—the darkest stacks of the library where reside the unimportant, unnoticed books, those one is supposed not to know, not even to have seen. The situation, not so strangely, resembles that of the unknown dark continent, which certainly was not dark or unknown to itself or to its inhabitants but only to Europeans, who labeled it so because to them, from their vantage point, it was out of view and perception. They did so for obviously political—indeed, obviously colonialist—reasons, and one may inquire if this segregation, this placement at a distance, accurately figures the political economy of works canonized and uncanonized.

Like the colonial power, say, France, Germany, or England, the canonical work acts as a center—the center of the perceptual field, the center of values, the center of interest, the center, in short, of a web of meaningful interrelations. The noncanonical works act as colonies or as countries that are unknown and out of sight and mind. That is why feminists object to the omission or excision of female works from the canon, for by not appearing within the canon works by women do not . . . appear. One solution to this more or less systematic dis-appearance of women's works is to expand the canon.

A second approach to the decanonization of works is the creation of an alternate tradition, an alternate canon. Toril Moi points to the major problems implicit in the idea of a feminist canon of great works (though she does not point to the possibility of reading without a canon) when she argues that all ideas of a canon derive from the humanist belief that literature is "an excellent instrument of education" and that the student becomes a better person by reading great works. "The great author is great because he (occasionally even she) has managed to convey an authentic vision of life." Furthermore, argues Moi—and thus incriminates all canons and all bodies of special works with the same brush—"the literary canon of 'great literature' ensures that it is this 'representative experience' (one selected by male bourgeois critics) that is transmitted to future generations, rather than those deviant, unrepresentative experiences discoverable in much female, ethnic, and working-class writing. Anglo-American feminist criticism has waged war on this self-sufficient canonization of middle-class male values. But they have rarely challenged the very notion of such a canon" (78). Arguing that Showalter aims to create a "separate canon of women's writing, not to abolish all canons," she points out that "a new canon would not be intrinsically less oppressive than the old" (78).

Unfortunately, one cannot proclaim the end of canons, or do away with them since they cannot be ended by proclamation. "To teach, to prescribe a curriculum, to assign one book for a class as opposed to another," Reed Way Dasenbrock points out, "is ineluctably to call certain texts central, to create a canon, to create a hierarchy" ("What to Teach When the Canon Closes Down," 67). Rather, we must learn to live with them, appreciate them, benefit from them, but, above all, remain suspicious of them. Grandiose announcements that one is doing away with The Canon fall into two categories: announcements, doomed to failure, that one is no longer going to speak in prose, and censorship that in totalitarian fashion tells others what they cannot read. Doing away with the canon leaves one not with freedom but with hundreds of thousands of undiscriminated and hence unnoticeable works, with works we cannot see or notice or read. Better to recognize a canon, or numerous versions of one, and argue against it, revise it, add to it.

Having thus far paraphrased—but I hope not parodied—now-popular notions of the positive and negative effects of a literary canon, I have to express some reservations. I have little doubt that a canon focuses attention, provides status, and screens noncanonical works from the attention of most people. That seems fairly clear. But I do not believe the one canon about which I know very much, that for English and American literature, has ever been terribly rigid. The entire notion of world literature, great touchstones, and studying English academically has a comparatively brief history. Victorian literature, that area of literature to which I devote most of my attention, certainly shows astonishing changes of reputations. When I first encountered the Victorians in undergraduate courses some thirty years ago, Tennyson, Browning, and Arnold claimed positions as the only major poets of the age, and Hopkins, when he was considered, appeared as a protomodernist. In the following decades, Swinburne and the Pre-Raphaelites, particularly Christina Rossetti and her brother Dante Gabriel Rossetti, have seemed more important, as has Elizabeth Barrett Browning, who had a major reputation during her own lifetime. Arnold, meanwhile, has faded rather badly. Looking at older anthologies, one realizes that some of the poets whose reputations have of late so taken a turn for the better had fairly strong reputations in the 1930s and 1940s but disappeared into a shade cast by modernism and the New Criticism.

Such evidence, which reminds us how ideological and critical fashions influence what we read as students and what we have our students read now, suggests, perhaps surprisingly, that the literary canon, such as it is, changes with astonishing speed. Viewing it over a scholarly or critical career, only the historically myopic could claim that the academic canon long resists the pressures of contemporary interests. No matter how rigid and restrictive it may be at any one moment, it has shown itself characterized by impermanence, even transience, and by openness to current academic fashion, over a university "generation," a far shorter span of time, the lag seems intolerably long. What good does it do an individual student to know that students will be able to study, say, a particular Nigerian writer a few years after *they* graduate?

Nonetheless, the canon, particularly that most important part of it represented by what educational institutions offer students in secondary school and college courses, takes a certain amount of time to respond. One factor in such resistance to change derives from interest and conviction, though as we have seen, such conviction can change surprisingly quickly in the right circumstances—right for change, not necessarily right according to any other standard. Another factor, which every teacher encounters, derives from book technology, in particular from the need to capitalize a fixed number of copies of a particular work. Revising, making additions, taking into account new works requires substantial expenditure of time and money, and the need to sell as many copies as possible to cover publication costs means that one must pitch any particular textbook, anthology, or edition toward the largest possible number of potential purchasers.

As Richard Ohmann has so chillingly demonstrated in "The Shaping of a Canon: U.S. Fiction, 1960–1975," the constraints of the marketplace have even more direct control of recent fiction, both bestsellers and those few books that make their way into the college curriculum. The combination of monopoly capitalism and a centralized cultural establishment, entrenched in a very few New York–based periodicals has meant that for a contemporary novel to "lodge itself in our culture as precanonical—as 'literature,'" however briefly, it had to be "selected, in turn, by an agent, an editor, a publicity department, a review editor (especially the one at the *Sunday New York Times*), the New York metropolitan book buyers whose patronage [is] necessary to commercial success, critics writing for gatekeeper intellectual journals, academic critics, and college teachers" (381). Once published, "the single most important boost" for a novel is a "prominent review in the *Sunday New York Times*," which, Ohmann's statistics suggest, heavily favors the largest advertisers, particularly Random House (380).

Historians of print technology have long argued that the cost of book technology necessitates standardization, and although education benefits in many ways from such standardization, it is also inevitably harmed by it as well. Most of the great books courses, which had so much to offer within all their limitations, require some fixed text or set of texts.[7]

Although hypertext can hardly provide a universal panacea for all the ills of American education, it does allow one to individualize any corpus of materials by allowing reader and writer to connect them to other contexts. In fact, the connectivity, virtual presence, and shifting of the balance between writer and reader that permit interdisciplinary team teaching to do away with this kind of time lag at the same time permit one to preserve the best parts of book technology and its associated culture. Let me give an example of what I mean. Suppose, as is the case, that I am teaching a survey course in English literature, and I wish to include works by women. A few years ago, if one turned to the Oxford or Norton anthologies, one received the impression that someone had quite consciously excluded the presence of women from them—and therefore from most beginning undergraduates' sense of literature. One could of course complain, and in fact many did. After a number of years, say, seven or eight, a few suitable texts began to appear in these anthologies, though Norton also took the route of publishing an anthology of women's literature in English. This new presence of women is certainly better than the former nonpresence of women, but it takes and is taking a long time. What is worse, many of the texts that appear at last in these anthologies may well not be those one would have chosen.

Let us consider a second problem I have encountered in introducing new materials into my teaching, one less likely to find redress anywhere as quickly as has the first. I refer to the difficulty of introducing authors of non-English ethnic backgrounds who write in English. This problem, which precisely typifies the difficulties of redefining the canon and the curriculum alike, arises because a good many of Britain's major authors during the past century have not been English.[8] In England, where the inhabitants distinguish quite carefully among English, Welsh, Scots, and Irish, the major figures since the rise of modernism have not necessarily been English: Conrad was Polish; James, American; Thomas, Welsh; and Joyce and Yeats, Irish. Generally, anthologies work in these figures without placing too much emphasis on their non-Englishness, which shows a nice capacity to accommodate oneself to the realities of literary production. Of course, such accommodation has taken a rather long time to materialize.

Today the situation has become far more complex, and in Great Britain's postcolonial era, if one wishes to suggest the nature of writing in English—which is how I define English literature—one must include both writers of Commonwealth and ex-Commonwealth countries and also those with a wide range of ethnic origins who live in the United Kingdom and write in English. Surveying leading novelists writing in English in Britain, one comes upon important English men and women, of course, like Graham Swift, Jane Gardam, and Penelope Lively; but such a survey almost immediately brings up the matter of national origins. After all, among the novelists who have won prestigious prizes of late one must include Salman Rushdie (India and Pakistan), Kazuo Ishiguro (Japan), and Timothy Mo (Hong Kong), and if one includes novels in English written by authors occasionally resident in Britain, one must include the works of Chinua Achebe and Nobel Prize winner Wole Soyinka (Nigeria), and of Anita Desai (India). And then there are all the Canadian, Australian, not to mention American novelists who play important roles on the contemporary scene. The contemporary English novel, in other words, is and is not particularly English. It is English in that it is written in English, published in England, and widely read in England and the rest of Britain; it is non-English insofar as its authors do not have English ethnic origins or even live in England.

The canon, such as it is, has rather easily accommodated itself to such facts, and while the academic world churns away, attacking or defending the supposedly fearsome restrictions of the canon and the virtual impossibility of changing it, contemporary writers, their publishers, and readers have made much of the discussion moot, if not downright comical. The problem faced by the teacher of literature, then, is how in the case of contemporary English literature to accommodate the curriculum to a changing canon. Of course, one can include entire novels in a course on fiction, but that means that the new does not enter the curriculum very far. In practice, the academic version of the expanded canon of contemporary literature will almost certainly take the form of African American literature, which now appears in separate courses and is experienced as essentially unconnected to the central, main, defining works.

Hypertext offers one solution to the problem of accommodating the curriculum to a changing canon. In my section of the standard survey course, which is a prerequisite for majoring in English at Brown University, I included works by Derek Walcott (Jamaica) and Wole Soyinka. How can hypertext aid in conveying to students the ongoing redefinition, or rather self-redefinition, of English literature? First of all, since Soyinka writes poems alluding to *Ulysses* and *Gulliver's Travels,* one can easily create electronic links from materials on Joyce and Swift to Soyinka, thus effortlessly integrating the poems of this Nigerian author into the literary world of these Irish writers.

Since hypertext linking also encourages students to violate the rigid structure of the standard week-by-week curriculum, it allows them to encounter examples of Soyinka's work or questions about its relation to earlier writers in the course of reading those writers earlier in the curricular schedule. By allowing students to range throughout the semester, hypertext permits them to see various kinds of connections, not only historical ones of positive and negative influence but equally interesting ones involving analogy. In so doing, this kind of educational technology effortlessly inserts new work within the total context.

Such contextualization, which is a major strength of hypermedia, has an additional advantage for the educator. One of the great difficulties of introducing someone like Soyinka into an English literature course, particularly one that emphasizes contextualization, involves the time and energy—not to mention additional training required—to add the necessary contextual information. Our hypertext component, for example, already contains materials on British and continental history, religion, politics, technology, philosophy, and the like. Although Soyinka writes in English, received his undergraduate degree from Leeds, and wrote some of his work in England, he combines English and African contexts, and therefore to create for him a context analogous to that which one has created for Jonathan Swift and Robert Browning, one has to provide materials on colonial and postcolonial African history, politics, economics, geography, and religion. Since Soyinka combines English literary forms with Yoruban myth, one must provide information about that body of thought and encourage students to link it to Western and non-Western religions.

Such an enterprise, which encourages student participation, draws upon all the capacities of hypertext for team teaching, interdisciplinary approaches, and collaborative work and also inevitably redefines the educational process, particularly the process by which teaching materials, so called, develop. In particular, because hypertext corpora are inevitably open-ended, they are inevitably incomplete. They resist closure, which is one way of stating they never die; and they also resist appearing to be authoritative: they can provide information beyond a student's or teacher's wildest expectations, but they can never make that body of information appear to be the last and final word.

Creating the New

Discursive Writing

Since writing the first version of *Hypertext* my interest in the educational applications of this information technology has increasingly shifted from read-only informational hypertext to those forms created by one or more students. Although I continue to use hypertext in the ways described in the preceding sections for courses on both literature and critical theory, students in the theory-related courses have begun to invent the ways of writing hypermedia at which we have already looked in chapter 5. Equally important, they have also in often brilliant and unexpected ways tested my proposal that hypertext offers a rare laboratory in which to test the ideas of poststructuralist theory.

As part of my courses in hypertext and critical theory, I developed the electronic versions of this volume in Intermedia and other systems described in chapter 5 as an example of translating a print book into hypertext. As I've already explained in chapter 5, students radically reconfigured the original in several ways, since they read *Hypertext* as wreaders—as active, even aggressive readers who can and do add links, comments, and their own subwebs to the larger web into which the print version has transformed itself.

Although students continued to make similar contributions while working with the Storyspace version of *Hypertext,* they also began to create their own self-contained sets of interlinked lexias. In thus moving from Intermedia, a truly real-time or synchronous collaborative environment, to Storyspace, students repeated many of the changes, advantages, and disadvantages that occurred when my institution switched most of its word-processing activities from centralized networked mainframe computing to stand-alone personal computers: the personal computer brought with it both greater convenience and resultant wider usage but also a marked loss in certain forms of computer literacy based on networked computing. Many more people used computers, though often inefficiently as little more than typewriters, but comparatively few took advantage of electronic mail, bulletin boards, and discussion groups. Without access to the kind of networked textuality provided by Intermedia, students found synchronous collaboration more difficult to carry out. Fortunately, when moving from Intermedia to Storyspace a great deal was also gained.

The sophistication and intellectual accomplishments exemplified by these first student webs compensated in many ways for the loss of an immensely powerful, if occasionally unstable, networked environment. The very first webs demonstrated more clearly than could any theoretical argument that writing in this medium creates new genres and new expectations. As one looks at these projects, it is clear that new kinds of academic writing were taking form. A few of them, like David Stevenson's *Freud Web,* whose dozen and a half lexias offer an introduction to Freud's theories, represent attempts to create hypertext versions of the standard academic term paper. Intrigued by the possibilities of hypertext, which he had encountered in English 32, Stevenson asked permission to create his term paper in Intermedia, and not surprisingly he followed the approaches used in developing the lexias he had seen in *Context32;* that is, like the developers of the course materials, he wrote each of his substantial discussions of free association, libido, and the like, and to these he added a bibliography, chronology, bibliography, and various graphic presentations of Freud's model of the mind.

The Freud Web, like a number of others, moved first to Storyspace and then Stevenson himself created an HTML version, making it one of the very first set of humanities materials available anywhere on the World Wide Web (it resided on a server belonging to the High Energy Theory Group in Brown's physics department). In its earliest version Stevenson interlinked it to the text of Rudyard Kipling's "Mary Postgate," a narrative of psychosexual violence that he believed Freud's theories would illuminate. Looking in retrospect at this pioneering student web, one sees how it combines two different kinds of writing. *The Freud Web* itself contains only materials written by a single author, but he then pushed the resulting web up against a literary text, thereby creating a hybrid form of writing in which the intellectual connections and interpretations consist only in links.

In contrast to Stevenson's approach of linking *from outside,* Steve Boyan's adaptation of Edgar Lee Master's *Spoon River Anthology,* like the *In Memoriam Web,* uses paths or trails of links through an existing text to permit reading the poem more easily in ways that the print version already encourages or even demands. Its added interpretative link paths serve as readings, or rather as records of readings, that, if we wish, we can make into our own.

Most student academic webs, however, rely less on either central print texts or on ways of writing associated with them. Once students began to use Storyspace, a hypertext environment that easily imports text, I began to notice something that I have since realized characterizes hyperwriting—its tendency, already observed in the discussion of hypertext as collage, to take the form of appropriation and abrupt juxtaposition. For example, Tom Meyer's *Plateaus* appropriates and interlinks a broad variety of materials to explain the relation of Deleuze and Guattari's thought to hypertext. In addition to Meyer's substantial discussion-lexias and material garnered from the Internet, his Storyspace web has folders containing multiple documents from *A Thousand Plateaus,* the Cabbala, Calvino's *Invisible Cities,* Burrough's *Naked Lunch,* and *The Satyricon.*

In addition to this tendency to exploit electronic collage for purposes of interpretive juxtaposition and comparison, the student webs share other qualities, one of which involves joining what one might consider academic and so-called creative writing; that is, poetry and fiction. Again, this tendency appeared in some of the earliest Storyspace webs. *Adam's Bookstore* by Adam Wenger developed in two stages, the first as hyperfiction and the second as a laboratory for theory. As a midterm exercise Wenger created a Borgesian tale that readers can enter and leave at any point, something enforced by the fact he provided no title screen and arranged his lexias as a circle in the Storyspace view (see Figure 32). For his final project, Wenger, who was highly skeptical of Barthes's approach in *S/Z,* applied the theorist's five codes to his own work, producing a very heavily linked web. Of the fifty-one lexias and 354 links that constitute *Adam's Bookstore,* approximately half consist of the original story, and if one clicks on the hot text in a specific lexia, one receives a list of six, eight, or even more links, the first several constructing the narrative, those that appear farther down in the list constructing Wenger's Barthean reading.

Although few webs thus self-consciously apply critical theory to the student-author's own texts, a large number move effortlessly between theory and fiction or poetry. Karen Kim's *Lexical Lattice,* Shelley Jackson's *Patchwork Girl,* and Michael DiBianco's *Memory, Inc.* (created in HTML and now part of the *Cyberspace Web*) all thus interweave substantial lexias containing text similar to standard academic discourse with fiction or poetry.

Lars Hubrich's *In Search of the Author, or Standing up Godot,* first created in Storyspace and then recreated by him for the Web, exemplifies the playful

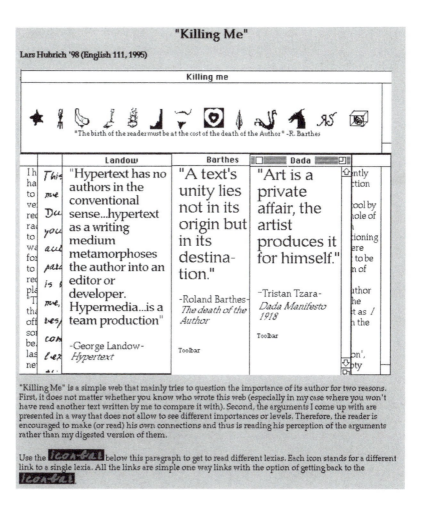

Figure 35. Lars Hubrich's *Killing Me*. This web, which represents another translation of Storyspace web into HTML, playfully explores the notions of authorship in e-space, braiding together texts from Foucault, Barthes, and Landow with the student-author's own challenges to the reader.

examination of central critical issues that often characterize this writing (Figure 35). An introductory title screen explains that readers can choose to begin with either of its two separate parts or subwebs, one of which, "Killing Me," wryly meditates on the ways hypertext reconfigures our conceptions of authorship. In the World Wide Web version, "Killing Me" begins with a screen shot of the Storyspace original, showing six-layered, overlapping lexias on this theme, three of which are entirely visible—those from Barthes's "The Death of the Author," Tristran Tzara's *Dada Manifesto,* and my *Hypertext.*

Beneath the defining image provided by this screenshot, Hubrich places a brief introduction that explains several ways in which his web reveals the problematic nature of conventional understandings of authorship, after which the text directs readers to use an immediately following set of thirteen cryptic magenta-and-yellow icons that stretch across the screen. Clicking on them brings the reader to individual statements about issues of authorship, intellectual property, and our assumptions about them. Thus, in addition to the three statements one can read in the screenshot, one comes upon additional passages from Barthes, Emile Benveniste's definition of the self, and Michel Foucault's "What Is an Author?" as well as questions to the reader about Hubrich's educational background and a humorous example of the way people use the author function in making aesthetic choice and evaluation:

I have a friend who hates U2. One day, I went to his house to find him very excited. He had just recorded a song from the radio and wanted to play it to me. He said that this was the best song he heard for months and that he had to find out which band recorded it. What he finally played to me was U2's "The Fly." When I told him that, he frowned and shut off the music mumbling something like "That can't be . . . sounded much better last time I heard it." He never again mentioned U2 to me. ("U2")

Other brief lexias challenge our habit of reading a coherent authorial self out of a text. In fact, the very first lexia readers are likely to encounter—that obtained by clicking on the icon at the extreme left—reads: "I paid someone to do this midterm assignment for me. I really had no time at all to get it done. Therefore, everything you are going to read and what you already read has been written by someone else." Another announces: "I don't know if you care, but you are misinterpreting this web. I never meant what you think this web is about." And yet another entitled "Handwriting" takes the form of an image of what appears to be a handwritten three-paragraph statement, which begins: "This was written by me and Marcel Duchamp. Who do you think holds the authorship of this paragraph?" Duchamp, "for it is his handwriting," or Hubrich, who wrote the lexia? Using a computer font named "Duchamp" based on the artist's handwriting, Hubrich created his lexia in Storyspace and then made an image of it for the HTML translation. With effective playfulness he uses it to question our assumptions about authorship on several levels. As he explains in his introductory lexia, he has arranged his materials nonhierarchically in a way that makes his text multivocal. "Therefore, the reader is encouraged to make (or read) his own connections and thus is reading his perception of the arguments rather than my digested version of them."

A great many of the other several hundred student projects in Storyspace and HTML that students have created take the form of similar experiments, for they use hypertext to test the theories of Barthes, Derrida, and others. Borges often appears as the Vergilian guide to these electronic explorations. Karen Kim thus created a hypertext version of Borges's "Grains of Sand" to the individual lexias of which she linked analyses in the manner of Barthes's *S/Z*, and other students have taken similar approaches to works of Carroll, Lorca, Maupassant, and Proust. Derrida, Bakhtin, Baudrillard, Haraway, and theorists not mentioned in the print version of *Hypertext* also appear within such laboratory-for-theory webs.

Many student-created webs exemplify that new form of discourse proposed in Gregory Ulmer's *Teletheory* (where, however, he presents it in the context of video and film; he has since discovered hypertext and become a major innovator using it, particularly in the form of the World Wide Web, to teach large classes at the University of Florida in writing and literature). This genre, which Ulmer terms "mystory," combines autobiography, public history, and popular myth and culture. As Ulmer explains, his proposed new mode of writing "brings into relationship the three levels of sense—common, explanatory, and expert—operating in the circulation of culture from 'low' to 'high' and back again," and thereby offers a means of

researching the equivalencies among the discourses of science, popular culture, everyday life, and private experience. A mystory is always specific to its composer, constituting a kind of personal periodic table of cognitive elements, representing one individual's intensive reserve. The best response to reading a mystory would be a desire to compose another one, for myself . . . mystory assumes that one's thinking begins not from the generalized classifications of subject formation, but from the specific experiences historically situated, and that one always thinks by means of and through these specifics, even if that thinking is directed against the institutions of one's own formation. (vii–viii)

Although Ulmer presented his Derridean notions of the new writing in *Teletheory,* a work subtitled *Grammatology in the Age of Video,* it turns out to describe not so much—or at least not only—the kind of textuality one finds in the analogue media of film and video but that emanating from (or instantiated by) digital word and digital image. As we have several times observed, hypertext, a border- and genre-crossing mode of writing, inevitably stitches together lexias written "in" different modes, tones, genres, and so on. Ulmerian mystory provides us with a first, possibly preliminary, model of how to write hypermedia.

One of the most interesting of such mystories is Taro Ikai's *Electronic Zen,* which uses hypertext linking to allow the reader to travel among lexias relating his experiences as a security guard in Tokyo, work with a Zen master, and Japanese poetry. Following directions and clicking on the introduction, one encounters two possible routes—"water" or "chef"—and following the first, one encounters four lexias that, taken together, produce the following:

[1] Water flows, unceasingly.

[2] It never stops. Not for a second.

[3] To hear it makes me think that I can hear the sound of time trickling down like water.

[4] Look without your eyes, straight—at all that has life, and simply to obey them.

Following the link from this last lexia opens an image of the night sky from which one can take a dozen different paths, some of which cycle back through the sky. At first, like the lexia entitled "chef," some on this path appear to contrast sharply with the tone and subject of the Zen materials, but increasingly as one encounters and reencounters them, these supposedly disparate subjects begin to interpenetrate and interilluminate one another, drawing closer together, as it were: the hard-working short-order cook turns out to fulfill the Nun Aoyama's injunction "Don't think about yourself" while the words of a half-witted co-worker, obsessed with the weather, blend eerily with those of his Zen master.

Some mysteries, to be sure, may well be fictional through and through; that is, like *Jane Eyre* and *Great Expectations,* they imitate or simulate autobiographies, and however much autobiographical material may permeate them, they nonetheless take the form of autobiographies of fictional characters. Helene Zumas's *Semio-Surf,* at which we have already looked when discussing the rhetoric of writing hypermedia for the World Wide Web, exemplifies such a possibly fictional mystory, and so do several other works submitted as course projects.

In contrast, Jeffrey Pack's *Growing Up Digerate,* which now forms a part or subweb of the *Cyberspace, Virtual Reality, and Critical Theory Web,* combines theory, here chiefly relating to cyberspace, and autobiography of someone who grew up "'digitally literate,' that is, having a familiarity with computers." As Pack's "Introduction" points out,

Most autobiographies start at birth, or with a short prelude describing how one's parents met. For this web, however, such things aren't very important. A birthdate (February 14, 1977) may prove useful if you're the sort of person who likes to do the math and figure out how old I was when various things happened, but isn't very necessary since I *am* that sort of person and will probably do it for you if I feel it's important. Where this story *really* begins is in 1983, when our family purchased its first computer.

At this point one can follow links within the text from the phrases "William Gibson's *Neuromancer*" or "its first computer," or one can use Pack's footer links to open an index that lists alphabetically approximately forty items ranging from America On-Line and Apple IIe through MS-DOS and MUD to World Wide Web and Zork. One can read this mystory more or less linearly, or one can go to the index or cycle through it by means of its many links— the lexia entitled "MS-DOS," for example, has seven in addition to the four footer links—and in doing so one receives both a personal history of computer literacy and a personal history.

As these few examples show, hypertext is here and undergraduate students are already mapping out the new forms of discourse that this combined information technology promises. After giving readings of these and similar webs at conferences and workshops, I am often asked how I go about evaluating them, and I respond that I combine the requirements of the old and the new; that is, accuracy, quality of research, writing at the level of the individual sentence and paragraph, and rhetorical effectiveness still count for a great deal, but webs also have to show visual literacy, skillful linking, clear and effective organization, and the like.

From Intermedia to the Web— Losses and Gains

After changes in Apple Computers' hardware and software effectively ended the Intermedia project, my students and I used various other hypertext systems, each of which has its own distinctive strengths and disadvantages and its corresponding educational effects. Experiences with these different systems revealed several important points of interest to anyone working with educational hypermedia, the first of which is that the apparently most minute technological change, such as system speed or screen size, can have unexpected, broad effects on reading, writing, and learning with hypertext.

Storyspace, which works on both Macintosh and Windows machines, does not have Intermedia's UNIX-based system of varying permissions (which permits an instructor to fix or freeze a document while allowing students to link to it), and it also does not have either Intermedia's structured graphics editor or its ability to permit individual documents to participate in multiple webs. On the other hand, it has a range of valuable qualities, not the least of which is that it will work on any Macintosh or Windows machine; unfortunately, moving webs between environments is not entirely automatic. Importing text and images, making links between words and phrases, full-text searching, and organizing documents are all very easy, and although this system does not have Intermedia's Web View, the Storyspace Roadmap (see Figure 17), which one can call up by pressing a simple key combination, provides a partial analogue to this invaluable feature by furnishing a reading history and list of link destinations for each individual document.

Perhaps most important, the simple fact that Storyspace runs on any PC created novel portability for all the webs originally created for Intermedia. Since students were able to copy any web from a server situated in the Computing and Information Technology building but accessible from various parts of the university, including some residence halls, they both read and wrote webs anywhere they had access to a Macintosh. (Since Storyspace permits one to copy linked sets of lexias easily, one can create comments at home and later paste them into the master or server version of any web to share with others.) The ease and convenience of working with what J. David Bolter called the "poor man's Intermedia" led, particularly in my hypertext and literary theory courses, to students creating their own considerable hypertext webs, some of them quite massive.

Storyspace has proved itself extremely useful but reduced the ability of students to read spontaneously as wreaders. Yes, students can add links but, without Intermedia's UNIX-based system of hierarchical privileges, they can only do so if they meet in lab with the person having the course password. In retrospect, one can see that the convenience of using a hypertext system based on the standard operating system meant that we gained greater ease of use, particularly when importing materials, and far greater accessibility. In return, we lost a real-time—as opposed to asynchronous—collaborative work environment. Nonetheless, having experience of working with Intermedia, I could easily develop strategies to ensure that students created valuable collaborative webs and even added their own links. Collaborative authoring, however, proved, and remains, a much easier matter than does adding links. One of the first exercises using Intermedia and Storyspace involved students creating their own links in the course webs, but the absence of a convenient way of doing so meant that students did not tend to think of working in this manner unless an assignment called for them to do so.

Using Storyspace also affected the kinds of visual materials students created, paradoxically reducing the visual literacy of student work in the purely literary courses while radically increasing it in those on digital culture and critical theory. The reason for the first change lies in the fact that whereas

Intermedia had a simple graphics editor that permitted student-wreaders to create diagrams, concept maps, and overviews within the system, Storyspace and World Wide Web viewers do not, thus requiring them to use Photoshop or similar graphics software. The immediate effect of the switch from Intermedia to Storyspace, therefore, was that students in my survey and Victorian courses stopped producing interesting concept maps that I discussed above. At the same time, because students experienced in using Photoshop, Illustrator, and image scanning programs found adding visual materials to their webs very easy to do in Storyspace—one simply copies images from a graphics program and pastes them directly into a Storyspace lexia—they began to use images (and video and sound) much more often.

Another anticipated difference appeared in the way students manipulate the Storyspace view to convey information. Although both Intermedia and Storyspace share what at first appear near-identical folder structures, authors can arrange the individual items in the Storyspace view to create patterns and hence display a web's organization. Experimenting with this feature, students quickly began to use it as a visual element in their writing (for examples, see Figures 14 and 26).

Using the World Wide Web again confronts the teacher with another set of advantages and disadvantages. Most obviously, available resources and potential collaborations are truly worldwide rather than being limited to a single class or campus, and students find creating basic HTML documents very easy to do, particularly if instructors provide simple templates. Although images consume time and resources, they are easy to employ, and sheer visual literacy has risen greatly with the Web. Assuming that students have access to a server on which they can place their own documents (and a considerable number of students in my hypermedia classes do), the World Wide Web once again grants student-collaborators the power to create their own documents and sets of links.

On the other hand, as we have already seen, HTML viewers come with a heavy cost as well. HTML produces a relatively flat version of hypertext, and students used to working with two features shared by Intermedia and Storyspace—one-to-many linking and various aspects of the multiwindow feature—often complain bitterly how confining and disorienting they find the Web to be. As I have already explained in chapter 5, course templates, identifying headers, and sets of linked footer icons solve many of the potential problems of navigation and orientation in HTML-based systems. One-to-many linking, which I take to be one of the defining qualities of a true hypertext system and one of its educationally most valuable, proves harder to replace or find an equivalent. The laborious task of creating and then maintaining sub-overviews for each item in an author and text overview solves the problem of using effective overview and crossroads documents on the Web, but the common Intermedia, Storyspace, and Microcosm practice of attaching several links to a word or phrase in a text document—particularly useful because multiple links produce a valuable preview function in the form of automatically generated menus—simply disappears.

Answered Prayers, or the

Academic Politics of Resistance

After a lecture I had delivered at an Ivy League campus on the role of hypertext in literary education, a distinguished historical scholar worried aloud in conversation with me that the medium might serve primarily to indoctrinate students into poststructuralism and Marxist theory. After another talk at a large state university in the Deep South, a younger academic, concerned with critical theory and the teaching of writing, argued (on the basis of my use of hypermedia in a historical survey) that it would necessarily enforce historical approaches and prevent the theorizing of literature. Such responses have proved typical of a sizable minority of those to whom I and others who work with this new medium have introduced educational and other applications of hypertext. Many with whom I have spoken have shown interest and enthusiasm, of course, and some of those concerned with critical theory as a major professional interest have responded with valuable suggestions and advice, even when remaining guardedly skeptical. For a sizable minority, however, hypertext represented and still represents the unknown, and one is not surprised to find that they project their fears on it, as people do on any unknown Other.

Not all observers find themselves troubled by the entrance of this latest educational technology into the portals of academe. Jean-François Lyotard, for example, argued as early as 1979 that "it is only in the context of the grand narratives of legitimation—the life of the spirit and/or the emancipation of humanity—that the partial replacement of teachers by machines may seem inadequate or even intolerable" (*Postmodern Condition,* 51). Since he has abandoned these "grand narratives," he does not resist technology that might threaten them. The historical record reveals, however, that university teachers have fiercely resisted all educational technology and associated educational practice at least since the late Middle Ages. Those who feel threatened by hypertext and associated technologies might do well to remember that, as Paul Saenger points out, when the introduction of spacing between words made reading to oneself possible, in "fourteenth-century universities, private silent reading [was] forbidden in the classroom" ("Books of Hours," 155). One can easily imagine the objections to the new technology and its associated prac-tice, since those objections have not changed very much in the past seven cen-turies: "Students, if left to their own devices, will construe the texts incorrectly. Everyone knows that permitting them such control over their own education before they are ready for it is not good for them. They don't yet know enough to make such decisions. And besides, what is to become of us if they use this insidious technology by themselves? What are we to *do*?" Similarly, when books appeared, many faculty members feared these dangerous new teach-ing machines, which clearly ceded much of the instructor's knowledge and power to the student. The mass production and wide distribution made pos-sible by printing, which threatened to swamp ancient authority in a flood of modern mediocrity, also permitted people to teach themselves outside insti-tutional control. Therefore, well into the eighteenth century, undergraduates in European universities had access to the library only a few hours per week.

Hypertext systems, just like printed books, dramatically change the roles of student, teacher, assignment, evaluation, reading list, relations among individual instructors, courses, departments, and disciplines. No wonder so many faculty find so many "reasons" not to look at hypertext. Perhaps scari-est of all for the teacher, hypertext answers teachers' sincere prayers for active, independent-minded students who take more responsibility for their education and are not afraid to challenge and disagree. The problem with answered prayers is that one may get that for which one asked, and then . . . What more terrifying for professors of English, who have for decades called for creativity, independent-mindedness, and *all those other good things,* to receive them from their students! Complaining, hoping, even struggling heroically, perhaps, to awaken their students, they have nonetheless accom-modated themselves to present-day education and its institutions, which include the rituals of lecture, class discussions, and examinations through which they themselves have passed and which (they are the evidence) have some good effects on some students.

What Chance Has Hypertext in Education?

My experience of teaching with hypertext since 1987 con-vinces me that the Web materials currently available have enormous potential to improve teaching and learning. Skep-tical as I first was when I became involved with the Interme-dia hypertext experiment, I discovered two years later that the hypertext com-ponent of my courses allowed me to accomplish far more with them than ever before possible. In the decade since I began to work with educational hypermedia, I have observed increasingly computer-literate students either demand hypertext materials or, now that the World Wide Web has arrived, go in search of them independent of their instructors' suggestions, wishes, or even knowledge. One of my favorite stories in this regard involves a student in one of the earliest classes to use Intermedia who took a visiting year at Columbia University. After the opening meeting of a course on James Joyce, he perplexed the lecturer by asking, "Where is your Intermedia web on *Ulysses*?" Now they just ask for the URL of the course website. Students, in other words, increasingly drive the use of hypertext—just as they did the use of silent reading in the late Middle Ages. As we have seen in the discussion of distance learning, students at traditional tertiary institutions already use the Web for their courses—whether or not their instructors tell them to do so.

Nonetheless, even with the enormous impetus provided by the World Wide Web I do not expect to see dramatic changes in educational practice for some time to come, in large part because of the combination of technological conservatism and general lack of concern with pedagogy that characterizes the faculty at most institutions of higher learning, particularly at those that have pretensions to prestige. There is, however, occasion to hope as I first wrote in 1991, because as one of those attending a 1988 conference on educational hypermedia at Dartmouth commented: "It took only twenty-five years for the overhead projector to make it from the bowling alley to the classroom. I'm optimistic about academic computing; I've begun to see computers in bowling alleys."[9] Academic computing in the form of course and university websites, e-mail, and online course information has in fact arrived, and there is much about which to be pleased, particularly concerning administrative matters like submitting and advertising course descriptions, obtaining classlists, ordering books, scheduling meetings, and so on. Teaching has yet to take full advantage of the Web, in part because many educational technologists and faculty users still think in terms of the book.

Getting the Paradigm Right

Although the World Wide Web has obviously already had a major effect on colleges, universities, and other cultural institutions, it has not realized many of the more utopian visions of hypertext. A small part of the reason involves the already-discussed limitations of the Web as form of hypertext. At this point, however, the limitations of our mentalware are clearly more to blame than the limitations of our software. Too many of us—and I include teachers, educational technologists, webmasters, and software developers—remain so deep inside the culture of the book that we automatically conceive of digital media in terms of the printed book. We base our ideas about the nature of teaching, the purpose of documents, and their relation to courses, disciplines, and universities on the mistaken assumption that electronic documents are essentially the same as printed ones. They're not.

Digital media, hypertext, and networked computing, like other innovations, at first tend to be (mis)understood in terms of older technologies. We often approach an innovation, particularly an innovative technology, in terms of an analogy or paradigm that at first seems appropriate but later turns out to block much of the power of the innovation. Thinking about two very different things only in terms of their points of convergence promotes the assumption that they are in fact more alike than they really are. Such assumptions bring much comfort, for they remove much that is most threatening about the new. But thus emphasizing continuity, however comforting, can blind us to the possibilities of beneficial innovation. Yes, it is easier to understand an automobile as a horseless carriage or a personal computer as a convenient form of typewriter. But our tendency to put new wine in old bottles, so common in early stages of technological innovation, can come at a high cost: it can render points of beneficial difference almost impossible to discern and encourage us to conceptualize new phenomena in inappropriate ways. Thus, thinking of an automobile as a horseless carriage not only emphasizes what is missing (a horse) but also fails to take into account the way speed greatly changes the vehicle's relation to many aspects of self and society. Similarly, thinking of a computer, as so many users do, as a fancy typewriter that easily makes corrections prevents taking advantage of the labor-saving possibilities of the digital text, such as its configurability by styles or the ways it permits seamless movement between paper documents and those moved about by e-mail.

Working with the right paradigm—that is, conceiving digital media in the correct terms—is essential if one is to take advantage of their special strengths. The paradigm, in other words, is more important than the purchase. Unfortunately, many computer users still think primarily in terms of the book. Examining educational institutions, including my own, reveals both that they commonly use the print paradigm in inappropriate applications and they often fail to take advantage of the particular strengths of the digital technology in which they have invested so heavily.

There is nothing strange about such resistance during a transitional period, and examples lie close at hand. During the first decades of e-mail, many potential users in business, education, and government preferred faxes, which still rely on the physical page, despite the fact that for those with Internet access, the telephone charges associated with faxing documents made them much more expensive for the individual person or administrative unit.

I'm sure we all recall hearing fax users boast how they'd entered the electronic age while refusing to use e-mail. The fax machine still has its function, though it is a diminishing one, since facsimiles can now be sent to one's e-mail account, and some business, educational, and governmental organizations permit and even encourage one to e-mail important correspondence as long as it is accompanied by a digital image of one's signature—an ingenious compromise that will no longer be necessary once secure electronic signatures become widely used.

The inappropriate use of the printed page as a basic model in an electronic environment appears again in the widespread misuse of PDF (Portable Document Format) files on websites. A PDF version of text documents has the great value of preserving the exact appearance of the original document. It has the strengths—but also the limitations—of print. Within an electronic environment, a PDF presentation of a document represents a refusal to employ any of the advantages of digital technology other than its ability to send copies quickly and cheaply over a network. It permits, however, neither searching nor linking, thereby creating an annoyingly inefficient means of conveying information. At my university, for example, some departments place their course listings on the institution's website as PDF files while others present them in HTML. If a faculty member or student wishes to look at English department course offerings or find important information, such as the section number of an individual course (necessary when placing book orders with the university store), she must download a PDF file on her computer's desktop, open it with the proper software, and read through page after page of text. Then, she either stores the PDF file or discards it. Instead of using PDF files, some departments and administrative units present the information in HTML, which is searchable and which also permits linking from course titles to brief descriptions, reading lists, and information about the course instructor. The use of PDF here instead of HTML represents a refusal to take advantage of the institution's expensive electronic infrastructure. It exemplifies, in other words, what happens when one thinks in terms of print or the printed book when using digital media. (New versions of Adobe Acrobat do in fact allow both text searching and linking, but my point still holds: many organizations eagerly choose to use earlier versions of PDF documents despite—or perhaps because—they sacrificed the advantages of digital text by closely imitating those of print.)

In contrast to portions of the institution that chose the PDF route, the university library has a beautiful and superbly conceived site that efficiently permits users to obtain information about its collections and also employ a large number of digital reference materials, such as encyclopedias, diction-aries, bibliographies, and collections of scholarly articles. Ironically, that part of the institution specifically dedicated to storing, preserving, and dispensing books and other physical information media, such as microfilm and micro-fiche, has made an especially well thought-out application of digital media.

Thus far I have discussed only colleges and universities as educational institutions, but museums are also educational institutions, and their use of digital resources often shows dependence on the print paradigm. Two examples from small museum websites demonstrate the high cost in lost opportunities when one fails to conceive an innovative technology in its own terms. In the first, a small historical museum in a region of the United States once dominated by the logging industry created a website as part of its man-date to play a greater role in the cultural life of its community. Featuring an exhibition of what life was like in old logging days, it encouraged visitors to record their handwritten comments in a guestbook. Visitors responded by writing that their fathers had worked in the logging industry, or they remem-bered it as part of their own childhoods. What's wrong here? Having conceived its website as a printed book, the museum has blinded itself to the possibili-ties of the new technology. In particular, by assuming that a website is *essen-tially* a book, its creators suppressed various innovative capacities that would have well served their project and mission. Working with the flawed assump-tion that the website is fundamentally a particular kind of book—in this case, a print exhibition catalogue placed in the gallery with a guest book next to it for handwritten comments—the website's developers made several unfortu-nate corollary assumptions. They took it for granted, for example, that the printed book's separation of author and audience is the right way to concep-tualize the relationship between website and user. But is it? Since one of the purposes of this site involves building a sense of community and creating a community memory, why not take advantage of visitors' comments by adding them to the website, inviting people to expand upon them, provide family information, photographs, and the like? Why not use the fundamental charac-teristics of linked digital information resources (hypertext) to "grow the site"? Why not use a dynamic site to create or enhance a sense of community among its constituents? A dynamic, fluid textuality, such as that found on websites, can change and easily adapt to its users, taking advantage of the modularity and capacity for change of digital text. But it cannot do so if its developers and home institution only think of it as a book—wonderful as books are.

Another example: a small anthropological museum at a Midwestern American university created a website with elegant graphic design obviously intended both to draw visitors and allow those who cannot come in person to enjoy some of its treasures. Individual screens present images of North American Indian artifacts together with basic information about them. So far so good. Unfortunately, that's all there is, for the entire site is nothing more than a direct electronic presentation of a museum catalogue. Putting print-derived text and images online, however, requires doing much more than formatting them in HTML, and a description of the site shows why. Since the designers began with the idea that a website is little more than an electronified book, they also assumed that readers would begin at the opening screen and make their way through one of several tightly limited paths. Making a common error, they failed to permit readers to return easily to the opening screen or sitemap, much less provide similar access to sitemaps for subcategories (departments) in the museum. Such book-blindered design, all too common on websites, also shows the designers never took into account that many web-readers will not arrive at the front entrance of the museum but led there by Internet search tools, arrive by falling through the roof and landing in the middle of a strange gallery. At a very minimum, they need to know where they are, and where they can go next.

The failings I've described thus far exemplify what happens when one assumes that a website, a nonphysical, electronic form, has the attributes of a book, which has physical form. A related, though less obvious set of problems has broader cultural, educational, and political implications. In the first kind of mistake, one uses a potential innovation inefficiently; in the following, one suppresses it entirely. Looking at the elegant graphic design of the site, one realizes the brief texts describing the represented objects in the collection contain no links. Conceptualized as old-fashioned catalogue entries, these passages fail to take advantage of the innovative capacity of links, which can provide basic glossary items that help younger users or those unfamiliar with the topic under discussion. Links can also lead interested readers to more advanced materials. Links, which can produce a kind of customizable text, have the power to turn such a site into a fully functioning educational resource. Used in this way, they serve to enrich and deepen the site as an introduction to the entire museum itself. Furthermore, linking to documents about materials outside the museum can also reconfigure the site's relation to its intellectual community. Here, however, relying on the book as thought-form or conceptual model prevented such innovation. Although this anthropology museum site exists at a university that also has a department of anthropology, it makes no attempt to connect the two. The site contains no documents by members of that department, nor does it list relevant courses available at the university.

What could one do differently with such a resource if one understands that a website can be more than a booklike static introduction to a museum collection? Since links cross borders and reconfigure our senses of the relationships, why not use them to reconfigure the relations of museum and university? The site could include relevant departmental research, all or parts of previously published papers, bibliographies, research guides, exemplary work by undergraduate and graduate students, even material from other universities' collections, and so on. Once one conceives a website from the vantage point of innovation—asking what's different about this new information technology and what we can do with such differences—one can conceptualize it as a network within other networks, and not simply (and misleadingly) as a book. At the same time, we must not exaggerate the differences between electronic hypertext and print text, for the many continuities between them require that we not only pay close attention to the relative strengths of each but that we also take from the printed page as much can help us in the new media. Hypertext is still text.

The same approaches to incorporating websites into institutional practice that we saw in museums occur in many colleges and universities. Increasingly, educational institutions around the world have turned to WebCT, Singapore's Interactive Virtual Learning Environment (IVLE), and similar software packages, which provide many valuable features for administering and teaching a course. These include not only electronic class lists and spread sheets for grades joined to a central database but also other features more directly related to teaching than to administration, such as collections of texts and images, discussion lists, multiple-choice tests, and course websites. What's wrong with a course-based website? Those that move beyond serving as mere replacements for paper syllabi often show intelligent planning and valuable resources, but most that I've seen have three major shortcomings. In the first place, they do not always include student work, and when they do they almost never keep these materials online past the end of the class, thereby sacrificing any opportunity for developing a critical mass of student-created materials that later students could use and that could convince students that they actively contribute to the educational process. Erasing these materials at semester's end also destroys the possibility of creating a valuable course memory.

Second, course websites are almost always closed to the public, thereby losing opportunities to obtain contributions of useful materials and to show off what the class is doing. At one institution at which I've taught, class websites proved valuable informational and recruiting tools. Many universities prevent public access to course websites either because they contain materials for they have paid a subscription or, in other cases, because instructors have illegally used copyrighted materials without permission. Still, that's no reason to close off the entire site.

Third, a course website is almost always associated with a particular instructor, and only very rarely will others in the same department use it, which means in practice that the time and energy that go into creating a website can never be shared or leveraged. For that reason I would urge departments and institutions to create broader sites like *The Victorian Web* and *The Postcolonial Web* that many courses and instructors can use. For example, a website with materials about eighteenth-century Anglo-European culture could be of use not only to courses in English, German, French, Italian, Spanish, and other literatures but also to those in nonliterature departments, such as art history, religious studies, social and political history, music, philosophy, and so on. Websites, like all hypertext, are fundamentally modular. Therefore, such sites can grow slowly and only in certain areas. I find that people are far more willing to contribute a small module to an ongoing enterprise, such as this kind of a website, than to begin creating a body of materials by themselves.

The next step away from course-based websites involves creating departmental or institutional ones that include more than administrative information. When I served as the founding dean of the University Scholars Program (USP), an interdisciplinary honors college at the National University of Singapore, we created the usual institutional site with information about admissions and sections for all disciplines and courses. We also used hypertext as an institutional paradigm.[10] Courses ranging from writing, ethics, and the culture of Islam to physics and statistics raised questions with students about connections of their classes to other fields. The site for Science, Technology, and Society in addition to including lists of courses offered also had brief introductions to the relations of technology, particularly information technology, to literature, computer science, ethics, and so on. Since Singapore is a multiethnic, multiracial society, many courses tried to reflect that fact, incorporating materials from two or more cultures. Thus an introduction to political theory included both European political thinkers and those from Islamic countries, a course in the history of cosmologies included India, China, the Middle East, and Europe. Like the sitemaps in *The Victorian Web*, the USP site was designed to suggest to students the many different ways of approaching a subject, not all of which any one person could cover.

The Politics of Hypertext:
Who Controls the Text?

Can Hypertext Empower

Anyone? Does Hypertext

Have a Political Logic?

After all my claims for reader empowerment in educational applications of hypermedia, the question remains, in what senses can hypertext or any other information technology empower anyone? In what sense and to what extent can we claim that they democratize, or even tend to democratize, the people and societies that use them? Certainly, despite all the enthusiasm for hypermedia, the Web, and the Internet, skeptical voices have made themselves heard early and often. As early as 1989—that is, before the appearance of the World Wide Web—Norman Meyrowitz, a leader of the team that developed Intermedia, delivered a keynote address at a major computer conference entitled "Hypertext—Does It Reduce Cholesterol, Too?" Several obvious reasons for skepticism come to mind when encountering statements by those whom Vincent Mosco calls the "visionaries promising an electronic utopia" (118), part of which is supposedly created by hypermedia and the Internet.[1] Most obviously, as Mosco, Tom Standage, and others have pointed out, during the past two centuries enthusiasts have proclaimed that every new technology would transform the world for the better. Electricity, telegraphy, photography, cinema, microfilm, video, cable television, computing, space satellites, and the Internet all were supposed to bring the world peace, prosperity, and freedom. All these technologies, as it turned out, did have major economic and social effects, though they hardly engendered any of the promised utopias.

Others who doubt claims that hypermedia or other digital technologies can produce beneficial political effects fall into several categories. First, there are those like Meyrowitz, myself, and other experienced workers with hypertext who reject outlandish claims for it. Then, there are those like Mitchell who recognize the "no free lunch" factor—in other words, that every advantage of a new information medium brings with it a possible disadvantage. For example, many have proclaimed, quite correctly, the enormous benefits of the Internet, but connectivity has its downside, too, for as Mitchell warns after 9/11, "in a networked, electronically interconnected world, there is no fundamental difference between addresses and targets" (*Me++*, 5). "The densely, globally networked world," Mitchell further explains, "is emphatically not (as early cyberspace utopians had sometimes imagined) inherently one of self-regulating, libertarian harmony. The proliferation and geographic distribution of access points—the very essence of the benefits of networks—also multiplies and distributes opportunities to create threats to the safety and well-being of those who have come to rely upon network capabilities" (*Me++*, 179). Moreover, even those who remain convinced of the positive personal and political benefits of the Net, such as the cyber-activist Geert Lovink, have learned from experience that discussion lists and other electronic forums require some form of central control—editors, webmasters, moderators, gatekeepers.[2] Multiuser digital environments, like all human enterprises, turn out to need organizational techniques found outside cyberspace: "After a brief period of excitement, the newly founded web sites, lists, servers, media labs, etc. have to find ways to deal with growth, economic issues, internal hierarchies, ever-changing standards, ongoing convergence problems between platforms, and incompatible software while establishing a form of cybernetic normalcy in the process" (*Dark Fiber*, 4).

Another reason for skepticism about the possibility of hypermedia, the Internet, and other digital technologies fulfilling predictions that they would democratize institutions and empower users derive from justifiable fears that, as Lovink puts it, "the Internet, bit by bit, is being closed down, sealed off by filters, firewalls, and security laws, in a joint operation by corporations and government in order to create a 'secure' and 'safe' information environment, free of dissents and irritants to capital flows" (*Dark Fiber*, 11–12). Lovink, one of many who believe that "it is time to say goodbye to the short summer of the internet" (19), fears that governmental and commercial interference with the Internet threatens to choke off its potential for political good:

The primary values of the early Internet, with its Usenet, virtual communities and focus on the fight against censorship are under threat. The consensus myth of an egalitarian, chaotic system, ruled by self-governing users with the help of artificial life and friendly bots, is now crushed by the take-over of telecom giants, venture capital and banks and the sharp rise in regulatory efforts by governments. ("Information Warfare: From Propaganda Critique to Culture Jamming," 309)

China's decade-long efforts to censor the Internet, the U.S. government's tracking Internet users, and Microsoft's continuing attempts to control the consumer and business market exemplify narrowing the possibilities of Internet freedoms. As a 1996 article in the *Wall Street Journal* explains, China "is determined to do what conventional wisdom suggests is impossible: Join the information age while restricting access to information" (Kahn, Chen, and Brauchli, A1), and its authorities hope to do so by creating an electronic Wall of China, a heavily filtered and censored "'intranet' or Internet-lite" (A4) with a "monolithic Internet backbone, centrally administered, that minimizes the threat of the Internet's amoeba-like structure" and thereby control the "two things China's authoritarian government most dreads, political dissent and pornography" (A4). Eight years later, the struggle to control the Internet continues, with the government sending mixed signals. On the one hand, a court "recently announced that an Internet democracy advocate charged with subversion would get a suspended sentence instead of a long prison term." On the other, the government relies on Internet controls and surveillance of users. According to Howard French's article in the *New York Times*, "Internet café users in China have long been subject to an extraordinary range of controls. They include cameras placed discreetly throughout the establishments to monitor and identify users and Web masters, and Internet café managers who keep an eye on user activity, whether electronically or by patrolling the premises." In addition, approximately thirty thousand Internet police play "a cat-and-mouse game with equally determined Web surfers, blocking access to sites that the government considers politically offensive, monitoring users who visit other politically sensitive sites and killing off discussion threads on Internet bulletin boards." Web users who try to reach censored sites "receive messages announcing a page is no longer accessible, or their computer screen may simply go blank, or they may be redirected to unrelated sites." Furthermore, to join a discussion on politically sensitive topics, users must identify themselves by their real names, e-mail addresses, and even phone numbers. The government appears alarmed by the sudden popularity of blogs, in large part because as Xiao Qiang, director of the China Internet Project at Berkeley, explains, "'the volume of online information is increasing vastly, and there's nothing the government can do about that. You can monitor hundreds of bulletin boards, but controlling hundreds of thousands of bloggers is very different.'"

According to researchers at Berkeley, Cambridge, Harvard, and Toronto, the Chinese government may well have found a way to control this vast amount of information using a variety of filtering software. One method uses filtering technology that in effect disables features of the search engine Google by tapping "into snapshots of web pages stored on Google's servers—which are based outside China" that formerly provided "a common way for Chinese to view sites that were otherwise blocked" (Hutzler, B1). According to researchers at Berkeley, a second part of "the Great Firewall of China" takes the form of "a list of banned words and phrases that a Chinese company embeds in desktops to filter messaging among PCs and cellphones. Among the more than 1,000 taboo terms: 'democracy,' 'sex,' and 'Hu Jintao,' China's president" (B1). The new filtering technologies comb the Internet and make sure that e-mails objectionable to the government become lost "in Chinese cyberspace and never reach their destinations, and requests to search engines, which provide lists of Web sites based on words, can go unanswered" (B1).

The government of Singapore, "one of the world's most enthusiastic users of the Internet," also wishes to take advantage of the new technology while simultaneously silencing any liberatory message that might be in the medium, for as Dan McDermott wrote on March 6, 1996,

Chill winds blew through Singapore cyberspace yesterday, as the government announced sweeping plans to filter what the average Singaporean can see and say on the Internet. Joining several other governments in seeking to filter the rivers of words and pictures pouring onto the Internet, Singapore said it will hold both content providers and access providers responsible for keeping pornographic and politically objectionable material out of the country's 100,000 Internet accounts. (A1)

Singapore has already blocked computer sites objectionable to the government, shutting down access both to Playboy Enterprises Inc. homepage and to that of the Socratic Circle, "an informal discussion group that . . . briefly held some animated political discussions last year" (A1). In the intervening years, the Singaporean government has somewhat loosened its grip, announcing, for example, that Internet providers would no longer be held responsible for content placed on their servers by users—a crucial point in a country with especially stringent libel laws.

For many observers, increasing commercial control of the Internet arouses more concern than government surveillance and censorship. According to Mosco, the most worrisome changes includes "three interrelated trends: the digitization and commodification of communication, corporate integration and concentration in the communication industry, and the deregulation of the industry" (143). Lovink, who took part in Amsterdam's early experiments in using the Internet to empower citizens—see his "The Digital City—Metaphor and Community"—provides an example of what happens when several large corporations try to control the technology. He was one of many observers who noted that in the 1990s Microsoft and other large corporate interests tried to undermine the fundamental user-centered nature of the World Wide Web by turning it into another broadcast medium: "Due to the commercialization of the net, big publishing houses, cable giants, telecoms and software companies have moved in and are now pushing the web in the direction of old-style broadcasting technologies. *Wired* calls this the revenge of 'TV'" ("A Push Media Critique," 130). Lovinck wrote this in 1997, when Microsoft tried to direct users of its Internet Explorer in discrete channels through which information could be "pushed." The attempt was a complete failure, for users preferred to search the Web, however inefficiently, and choose their own links and paths. The notorious failure of push media, I would argue, demonstrates that users believe that user-centered hypermedia best serves their needs—and that networked computer environments do in fact empower users to act as more than mere consumers. In his "Insider's Guide to Tactical Media," Lovink argues that information technologies do in fact have an ideological bias: "Being a 'difference engine' on the level of representation may put out a lot of useful public content, but it does not touch on the 'media question.' What is of interest are the ideological structures written into the software and network architecture. It is not just enough to subvert or abuse this powerful structure" (263). The reactions by users to commercial attempts to turn the World Wide Web into another form of broadcast media—that is, *the choices they made*—demonstrate that users experienced the Web as having a bias toward reader empowerment.

The most extreme doubters include those like Espen Aarseth, who denies the possibility that hypertext in any way empowers or liberates its users. Let us look first at Aarseth, whom we may accurately describe as the leading antihypertext theorist. In his otherwise valuable *Cybertext* (1997), which advances pioneering ideas about computer games, he mounts a fierce attack on most earlier writings on hypertext, particularly on those that invoke poststructuralist theory to explain digital media or that claim that the new media in any way empowers users. Although he himself freely draws on Eco, Genette, and Barthes at points in his book, he charges that people who use critical theory to explain hypertext, or to point out parallels between them, are using an "imperialist pretext" to "colonize" another field (83). Similarly, to call paper-based texts that partially anticipate electronic hypertext protohypertexts is an "imperialist classification" (75). Strong words, but not surprising when one considers the context in which they appear. Most of *Cybertext* concerns computer games and other forms of digital media, such as MUDs and MOOs, which Aarseth earnestly wishes to establish as important cultural forms. For some reason, he seems to believe—or at least *writes* as if he believes—that to clear intellectual space for games, MUDs, and other forms of what he terms "cybertext," he must trash hypertext, denying that it has any positive qualities. Certainly, at the time he began his project, hypertext and hypermedia were the forms of digital text that received most attention, particularly in Norway, which has important hypermedia theorists and practitioners like Gunnar Liestøl.

Reading *Cybertext,* I was struck by how little actual hypertext Aarseth seems to have read, and how few hypertext systems he seems to have used. Michael Joyce's pioneering *afternoon,* which is one of the few hypertexts he mentions, is the only one he discusses at any length, and his remarks closely follow the writings of Jane Yellowlees Douglas and Stuart Moulthrop. His specific comments show little understanding or experience of either *afternoon* or its software environment, Storyspace. According to Aarseth, in Storyspace "readers could follow *only* the sequences laid down by the writer. Hyperfictions written in *Storyspace,* like *Afternoon,* do not allow its readers free browsing, unlike any codex fiction in existence" (77). His general remarks about Storyspace, which apply only to the page reader version used by Joyce, are wrong on several counts. As we've already observed, Storyspace hypertexts, like *Patchwork Girl* and *Quibbling,* which employ the Storyspace reader that includes the software's characteristic graphic map, permit readers to browse freely and indeed encourage readers to do so. Other Storyspace webs, such as the *In Memoriam Web, Breath of Sighs,* and *Hero's Face,* were published in the Demo version of the authoring environment, which not only features "free browsing" but also includes a search tool.

Another fundamental flaw in Aarseth's chapter on hypertext, perhaps suggested by his almost total reliance on a single work of hyperfiction, is that he almost completely neglects informational and educational hypertext and examines no specific examples of them. This turns out to be a major weakness in his evaluation of reader empowerment, because these kinds of hypertext offer much clearer positive choices, in part because many of them have vastly more lexias than any existing hyperfiction, and this factor provides a greater range of *informed* choices. One of the few places he mentions such forms of hypertext comes when he responds to my claim that "hypertext blurs the boundary between author and reader" ("Hypertext, Metatext," 70):

"First by permitting various paths through a group of documents (one can no longer write 'one document or text'), it makes readers, rather than writers, control the materials they read and the order in which they read them. Second, true hypertext, such as the Intermedia system developed at Brown University, permits readers to become authors by adding electronic links between materials created by others and also by creating materials themselves." Aarseth comments: "Landow's project at Brown is one of institutional reform, and even if he bestows the role of reformer on the technology—in this case, hypertext—it really belongs to him" (171). I certainly agree that my teaching with hypermedia from Intermedia to the present involves institutional as well as pedagogical reform, but one hardly needs Aarseth to point that out, since the eighty-page chapter in which the quotation appears continually emphasizes that to activate the educational potential of hypertext, instructors must rethink and reconfigure the subjects, procedures, assignments, and evaluation methods they use. Here as throughout much of his critique of statements about the potentially positive nature or effects of hypertext, Aarseth falsely claims that someone else has identified hypertext as the sole and sufficient cause of something when they have in fact actually argued that it provides an enabling, though not by itself a sufficient, condition.

One of the most disturbing aspects of Aarseth's critique involves his lack of any comment on the following sentence: "Second, true hypertext, such as the Intermedia system developed at Brown University, permits readers to become authors by adding electronic links between materials created by others and also by creating materials themselves." I find such omissions very odd since I am claiming that read-write hypertext systems like Intermedia fulfill his own definition of an author. As he points out, "the politics of the author-reader relationship, ultimately, is not a choice between various pairs of media or forms of textuality, but instead is whether the user has the ability to transform the text into something that the instigator of the text could not foresee or plan for" (164). After all his denials that choosing links in any way affects the relation between author and reader, he doesn't even acknowledge the claims of a hypermedia system in which readers clearly can act as authors.

Aarseth also claims that "it seems somewhat self-contradictory to claim, as Landow does, that hypertext blurs the distinction between reader and author while at the same time permitting the former to become the latter" (173). His difficulties in comprehending my ideas of hypertext readers here come from his habit of thinking in terms of binary oppositions. In fact, I argue that the reader who chooses among links or takes advantage of Storyspace's spatial-hypertext capabilities shares *some* of the power of the author.

We both agree, following Foucault, that authorship "is a social category and not a technological one" (172)—the point, after all, of much of my chapter 4—and I claim that different forms of information technology plus their social and political contexts have produced different notions of authorship.

Perhaps our fundamental points of disagreement appear when he asserts that "there is no evidence that the electronic and printed texts have clearly divergent attributes" (70) while also denying that hypermedia in any way empowers anyone or could tend toward democratization. Here in response are three real-life examples of the ways hypertext and the Internet empower users by democratizing access to information. Note that I do not claim that any of these examples shows that the Internet and associated technologies by themselves produce political democracy. I also do not claim that they in any way make the reader of Internet materials an author of them. In the first case, a young man encounters some pills that resemble an antihistamine for which he has a prescription, but he cannot definitely identify them for he no longer has on hand the bottle in which they came. Each pill, however, has three numbers impressed in the surface, so he types just the three numbers into Google, a popular Web search tool, and the first hit brings one to a pharmaceutical manufacturer's site for a drug, which identifies it as a very different medication—an antibiotic, in fact—and he does not take the wrong medication. Even if one had a *Physicians Desk Reference* (*PDR*) at hand, one could not search by number but would have to compare the pill to images of various medications. In a similar situation involving unidentified medications for elderly people, consulting a pharmacist did not help, since the pharmacist claimed the *PDR* contained so many pills, he did not have the time to try to identify any. In other words, information one would have to expend many hours or even days to acquire was located on the Web in less than fifteen seconds.

This user's experience, I submit, provides clear answers to two questions. First, does the Internet democratize information and empower users? I believe the obvious answer is, yes, for as this example demonstrates, digital information technology provides to people other than physicians and pharmacists information when they need it. In fact, anyone with a computer connected to the Internet, even if located far from a hospital, physician's office, or medical library, has access to the needed information. Second, does the quality—in this case, chiefly time—needed to access information prove an important enough factor to distinguish media from one another? Again, I would say, yes, for even though the same information (the identity of the pill) is available in both print and digital media, the vastly different experience of using them constitutes a major, distinguishing characteristic that separates different information technologies.

The previous example is, properly speaking, more a matter of information retrieval on the Internet than pure hypertext, but since Google ranks search results in part by numbers of links to each site, even this example involves hypertext. The next example does so directly. It also involves medication: some time after an elderly male relative begins taking a newly prescribed medication, he becomes less energetic and clear-headed. Since his children do not know much about the medication, they Google its name and immediately (within less than a second) they locate a list of sites, and picking the highest-ranked one they open a web document devoted to this drug, which contains links to its possible side effects, contraindications, statistics, and so on. Determining that the medication could not possibly cause the new symptoms, the family at least knows that there is no need to contact a physician.

The next example concerns access to sources about international news events. While in Singapore when NATO began bombing Kosovo as a means of halting the ethnic cleansing in the former Yugoslavia, I wanted to know what *The Providence Journal,* my local newspaper at home, had to say, and so I opened my web browser and typed in http://www.projo.com. When the homepage for the online version of the newspaper appeared, I discovered links to a story about the Kosovo situation. Following the link I was surprised to find not just the usual news from a wire service like the Associated Press but also a dozen links to statements by the governments of concerned countries, including Serbia, Russia, and the United States, thus presenting opposing interpretations of events. Following another link labeled something like "NATO starts bombing campaign," I found myself in a document containing links to information about the airplanes and weaponry of all parties. Additional links led to information about the number, cost, development problems, and legislative history of each airplane plus similarly detailed information about every kind of bomb, rocket, or other munitions employed. Other links brought me to the website of an Italian airbase, which contained many photos of actions taking place there, including images of ground crews from one European NATO nation arming the planes of another. Finally, I came upon information about when each group of planes had arrived at the base or was scheduled to do so as well as additional links to information about individual pilots and aircrews. By this point, I wondered why governments employed spies if all this information could be found so easily on the Net.

In 1993, shortly after the appearance of the World Wide Web, I gave a talk at the Niemann Foundation at Harvard about the implications of the new technology for journalism. I speculated that if we could apply the hypertext paradigm to news, we might have news media in which individual readers could pursue subjects that interested them as long as they had time and patience to do so.[3] I hardly expected to encounter such a rich example half a dozen years later! Let me emphasize how this hypertext version of the news differs from both the television and newspaper ones. To procure the same information I obtained on the web, I would need to expend several days and possibly weeks using a conveniently located information resource, such as a large library, and I would also have to supplement my search with telephone calls to various embassies, including those located abroad. Compare this to what we may term the *Aarseth principle*—the idea that if two different information technologies contain essentially the same information, no difference exists between them. As in the searches for information about medications, this hypertextual presentation of news about a current event clearly empowers the user—if by "empower" we mean, as I do here, "provides information that would be difficult if not impossible to obtain otherwise." Again, Aarseth finds absolutely no difference between information technologies that contain the same information, but to do so he has to ignore the vastly different ways each one is experienced. Using the words Mitchell uses to distinguish between digital and printed text, I emphasize "that is relevant at the level of everyday experience." Whether or not they have a place in Aarseth's theories, the democratization effects of hypermedia and the Internet are relevant at the level of everyday experience.

The Marginalization of Technology and the Mystification of Literature

Discussions of the politics of hypertext have to mention its power, at least at the present time, to make many critical theorists, particularly Marxists, very uncomfortable. Alvin Kernan wryly observes, "That the primary modes of production affect consciousness and shape the superstructure of culture is, not since Marx, exactly news, but . . . both Whiggish theories of progress and Marxist historical dialectic have failed to satisfy the need to understand the technologically generated changes or to provide much real help in deciding what might be useful and meaningful responses to such radical change" (3). Anyone who encounters the statements of Frederic Jameson and other critical theorists about the essential or basic lack of importance of technology, particularly information technology, to ideology and thought in general recognizes that these authors conspicuously marginalize technology. As Terry Eagleton's fine discussions of general and literary modes of production demonstrate, contemporary Marxist theory has drawn upon the kind of materials Kernan, McLuhan, and other students of information technology have made available.[4] For this reason, when other Marxists like Jameson claim that examining the effects of technology on culture inevitably produces technological determinism, one should suspect that such a claim derives more from widespread humanist technophobia than from anything in Marxist thought itself. Jameson's statements about technological determinism bear directly on the reception of ideas of hypertext within certain portions of the academic world for which it has most to offer but which, history suggests, seem most likely to resist its empowerment. This rejection of a powerful analytic tool lying ready to hand appears particularly odd given that, as Michael Ryan observes, "technology—form-giving labor—is, according to Marx, the 'nature' of human activity, thereby putting into question the distinction between nature and culture, at least as it pertains to human life."[5]

In *Marxism and Form,* Jameson reveals both the pattern and the reason for an apparently illogical resistance to work that could easily support his own. There he argues that

however materialistic such an approach to history may seem, nothing is farther from Marxism than the stress on invention and technique as the primary cause of historical change. Indeed, it seems to me that such theories (of the kind which regard the steam engine as the cause of the Industrial Revolution, and which have been rehearsed yet again, in streamlined modernistic form, in the works of Marshall McLuhan) function as a substitute for Marxist historiography in the way they offer a feeling of concreteness comparable to economic subject matter, at the same time that they dispense with any consideration of the human factors of classes and of the social organization of production. (74)

One must admire Jameson's forthrightness here in admitting that his parodied theories of McLuhan and other students of the relations of technology and human culture potentially "function as a substitute for Marxist historiography," but the evidence I have presented in previous pages makes clear that Eisenstein, McArthur, Chartier, Kernan, and many other recent students of information technology often focus precisely on "the human factors of classes and of the social organization of production." In fact, these historians of information technology and associated reading practices offer abundant material that has potential to support Marxist analyses.

Jameson attacks McLuhan again a decade later in *The Political Unconscious.* There he holds that an old-fashioned, naive conception of causality, which he "assumed to have been outmoded by the indeterminancy principle of modern physics," appears in what he calls "that technological determinism of which MacLuhanism [*sic*] remains the most interesting contemporary expression, but of which certain more properly Marxist studies like Walter Benjamin's ambiguous Baudelaire are also variants." In response to the fact that Marxism itself includes "models which have so often been denounced as mechanical or mechanistic," Jameson gingerly accepts such models, though his phrasing suggests extraordinary reluctance: "I would want to argue that the category of mechanical effectivity retains a purely local validity in cultural analysis where it can be shown that billiard-ball causality remains one of the (nonsynchronous) laws of our particular fallen social reality. It does little good, in other words, to banish 'extrinsic' categories from our thinking, when they continue to have a hold on the objective realities about which we plan to think." He then offers as an example the "unquestioned causal relationship" between changes in "the 'inner form' of the novel itself" (25) and the late-nineteenth-century shift from triple-decker to single-volume format. I find this entire passage very confusing, in part because in it Jameson seems to end by accepting what he had begun by denying—or at least he accepts what those like McLuhan have stated rather than what he apparently assumes them to have argued. His willingness to accept that "mechanical effectivity retains a purely local validity in cultural analysis" seems to do no more than describe what Eisenstein, Chartier, and others do. The tentativeness of his acceptance also creates problems. I do not understand why Jameson writes, "I would want to argue," as if the matter were as yet only a distant possibility, when the end of this sentence and those that follow show that he definitely makes that argument. Finally, I find troubling the conspicuous muddle of his apparently generous admission that "it does little good . . . to banish 'extrinsic' cate-gories from our thinking, when they continue to have a hold on the objective realities about which we plan to think." Such extrinsic categories might turn out to match "the objective realities about which we plan to think," or again, these objective realities might turn out to support the hypothesis contained in extrinsic categories, but it only mystifies things to describe categories as having "a hold on . . . objective realities."

Such prose from Jameson, who often writes with clarity about particu-larly difficult matters, suggests that this mystification and muddle derives from his need to exclude technology and its history from Marxist analyses. We have seen how hard Jameson works to exclude technological factors from consideration, and we have also observed that they not only offer no threat to Jamesonian Marxism but even have potential to support it.[6] Jameson's exclusions, I suggest, therefore have little to do with Marxism. Instead, they exemplify the humanist's common technophobia, which derives from that "venerable tradition of proud ignorance of matters material, mechanical, or commercial" Elizabeth Eisenstein observes in students of literature and history (706).

Such resistance to the history of technology does not appear only in Marxists, though in them, as I have suggested, the exclusion strikes one as particularly odd. While reading Annette Lavers's biography of Roland Barthes, I encountered another typical instance of the humanist's curious, if characteristic, reticence to grant any importance to technology, however defined, as if so doing would remove status and power: "The contemporary expansion of linguistics into cybernetics, computers, and machine translation," she tells us, "probably played its part in Barthes's evolution on this subject; but the true reason is no doubt to be found in the metaphysical change in outlook which resulted in his new literary doctrine" (138). After pointing to Barthes's obvious intellectual participation in some of the leading currents of his own culture (or strands that weave his own cultural context), she next takes back what she has granted. Although her first clause announces that computing and associated technologies "probably" played a part in "Barthes's evolution on this subject," she immediately takes back that "probably" by stating unequivocally that "the true reason"—the other factors were apparently false reasons, now properly marginalized—"without a doubt" lies in Barthes's "metaphysical change." One might have expected to encounter a phrase like "the most important reason," but Lavers instead suddenly changes direction and brings up matters of truth and falsity and of doubt and certainty.

Two things about Lavers's discomfort deserve mention. First, when confronted with the possibility that technology may play a contributing role in some aspect of culture, Lavers, like Jameson and so many other humanists, resorts to devices of mystification, which suggests that such matters intrude in some crucial way upon matters of power and status. Second, her mystification consists of reducing complexity to simplicity, multivocality to univocality. Her original statement proposes that several possible contributing factors shaped Barthes's "evolution," but once we traverse the semicolon, the possibilities, or rather probabilities, that she herself has just proposed instantly vanish into error, and a "metaphysical change in outline" in all its vagueness becomes the sole causation.

One wonders why critical theorists thus marginalize technology, which, like poetry and political action, is a production of society and individual imagination. Since marginalization results from one group's placement of itself at a center, one must next ask which group places itself at the center of power and understanding, and the answer must be one that feels itself threatened by the importance of technology. Ryan asks, "What is the operation of exclusion in a philosophy that permits one group, or value, or idea to be kept out so that another can be safeguarded internally and turned into a norm?" (*Marxism and Deconstruction,* 3). One such operation that I have frequently encountered after talks on educational hypertext takes the form of a statement something like "I am a Luddite" or "What you say is very interesting, but I can't use (or teach with) computers, because I'm a Luddite." (Can you imagine the following? "I can't use lead pencils—ballpoint pens—typewriters—printed books—photocopies—library catalogues because I'm a Luddite.") All the self-proclaimed Luddites in academe turn out to oppose only the newest machines, not machines in general and certainly not machines that obviate human drudgery. Such proclamations of Ludditism come permeated by irony, since literary scholars as a group entirely depend on the technologies of writing and printing. The first of these technologies, writing, began as the hieratic possession of the politically powerful, and the second provides one of the first instances of production-line interchangeable parts used in heavily capitalized production. Scholars and theorists today can hardly be Luddites, though they can be suspicious of the latest form of information technology, one whose advent threatens, or which they believe threatens, their power and position. In fact, the self-presentation of knowledge workers as machine-breakers defending their chance to survive in conditions of soul-destroying labor in bare, subsistence conditions tells us a lot about the resistance. Such mystification simultaneously romanticizes the humanists' resistance while presenting their anxieties in a grotesquely inappropriate way. In other words, the self-presentation of the modern literary scholar or critical theorist as Luddite romanticizes, in other words, an unwillingness to perceive actual conditions of his or her own production.

Perhaps my favorite anecdote and possibly one that makes a particularly significant contribution to our understanding of resistance is this: after a lecture on hypertext and critical theory at one institution, a young European-trained faculty member who identified his specialty as critical theory candidly admitted, "I've never felt old-fashioned before." As the latest of the newfound, new-fangled developments, hypertext, like other forms of New Media, in general has the (apparent) power to make those who position themselves as the advocates of the new appear to themselves and others as old-fashioned.

The Politics of Particular Technologies

Discussions of hypertext all raise political questions—questions of power, status, and institutional change. All these changes have political contexts and political implications. Considerations of hypertext, like all considerations of critical theory and literature, have to take into account what Jameson terms the basic "recognition that there is nothing that is not social and historical—indeed, that everything is 'in the last analysis' political" (*Political Unconscious*, 20). A fully implemented embodiment of a networked hypertext system such as I have described obviously creates empowered readers, ones who have more power relative both to the texts they read and to the authors of these texts. The reader-author as student similarly has more power relative to the teacher and the institution. This pattern of relative empowerment, which we must examine with more care and some skepticism, appears to support the notion that the logic of information technologies, which tends toward increasing dissemination of knowledge, implies increasing democratization and decentralization of power.

Technology always empowers someone. It empowers those who possess it, those who make use of it, and those who have access to it. From the very beginnings of hypertext (which I locate in Vannevar Bush's proposals for the memex), its advocates have stressed that it grants new power to people. Writers on hypertext almost always continue to associate it with individual freedom and empowerment. "After all," claim the authors of a study concerning what one can learn about learning from the medium, "the essence of hypertext is that users are entirely free to follow links wherever they please" (Mayes, Kibby, and Anderson, 228). Although Bush chiefly considered the memex's ability to assist the researcher or knowledge worker in coping with large amounts of information, he still conceived the issue in terms of ways to empower individual thinkers in relation to systems of information and decision. The inventors of computer hypertext have explicitly discussed it in terms of empowerment of a more general class of reader-authors. Douglas Englebart, for example, who invented the first actual working hypertext environment, called his system Augment; and Ted Nelson, who sees Xanadu as the embodiment of the 1960s New Left thought, calls on us to "imagine a new accessibility and excitement that can unseat the video narcosis that now sits on our land like a fog. Imagine a new libertarian literature with alternative explanations so that anyone can choose the pathway or approach that best suits him or her; with ideas accessible and interesting to everyone, so that a new richness and freedom can come to the human experience; imagine a rebirth of literacy" (*Computer Lib*, 1/4).[7]

Like other technologies, those centering on information serve as artificial, human-made means of amplifying some physical or mental capacity. Jean-François Lyotard describes computing and other forms of information technology in terms usually assigned to wooden legs and artificial arms: "Technical devices originated as prosthetic aids for the human organs or as physiological systems whose function it is to receive data or condition the context. They follow a principle, and it is the principle of optimal performance: maximizing output (the information or modifications obtained) and minimizing input (the energy expended in the process)" (*Postmodern Condition*, 44). According to *The American Heritage Dictionary*, the term *prosthesis* has the two closely related meanings of an "artificial replacement of a limb, tooth, or other part of the body" and "an artificial device used in such replacement." Interestingly, prosthesis has an early association with language and information, since it derives from the late Latin word meaning "addition of a letter or syllable," which in turn comes from the Greek for "attachment" or "addition, from *prostithenai*, to put, add: *pros-*, in addition + *tithenai*, to place, to put." Whereas its late Latin form implies little more than an addition following the rules of linguistic combination, its modern application suggests a supplement required by some catastrophic occurrence that reduced the individual requiring the prosthesis to a condition of severe need, as in the case of a person who has lost a limb in war, in an automobile accident, or from bone cancer or, conversely, of a person suffering as a result of a "birth defect." In each case the individual using the prosthesis requires an artificial supplement to restore some capacity or power.

Lyotard's not uncommon use of this term to describe all technology suggests a powerful complex of emotional and political justifications for technology and its promises of empowerment. Transferring the term *prosthesis* from the field of rehabilitation (itself an intriguing term) gathers a fascinating, appalling congeries of emotion and need that accurately conveys the attitudes contemporary academics and intellectuals in the humanities hold toward technology. Resentment of the device one needs, resentment at one's own need and guilt, and a Romantic dislike of the artificiality of the device that answers one's needs mark most humanists' attitudes toward technology, and these same factors appear in the traditional view of the single most important technology we possess—writing. These attitudes result, as Derrida has shown, in a millennia-long elevation of speech above writing, its supposedly unnatural supplement.

Walter J. Ong, who reminds us that writing is technology, exemplifies the comparatively rare scholar in the humanities who considers its artificiality as something in its favor: "To say that writing is artificial is not to condemn it but to praise it. Like other artificial creations and indeed more than any other, it is utterly invaluable and indeed essential for the realization of fuller, interior, human potentials . . . Alienation from a natural milieu can be good for us and indeed is in many ways essential for full human life. To live and to understand fully, we need not only proximity but also distance" (*Orality and Literacy,* 82). Like McLuhan, Ong claims that "technologies are not mere exterior aids but also interior transformations of consciousness" (82), and he therefore holds that writing created human nature, thought, and culture as we know them. Writing empowers people by enabling them to do things otherwise impossible—permitting them not just to send letters to distant places or to create records that preserve some information from the ravages of time but to think in ways otherwise impossible.

Abstractly sequential, classificatory, explanatory examination of phenomena or of stated truths is impossible without writing and reading . . . In the total absence of any writing, there is nothing outside the thinker, no text, to enable him or her to reproduce the same line of thought again or even to verify whether he or she has done so . . . In an oral culture, to think through something in non-formulaic, non-patterned, non-mnemonic terms, even if it were possible, would be a waste of time, for such thought, once worked through, could never be recovered with any effectiveness, as it could be with the aid of writing. It would not be abiding knowledge but simply a passing thought. (8–9, 34–35)

Technology always empowers someone, some group in society, and it does so at a certain cost. The question must always be, therefore, what group or groups does it empower? Lynn White showed in *Medieval Technology and Social Change* that the introduction from the Far East of three inventions provided the technological basis of feudalism: the horse collar and the metal plow produced far higher yields than had scratch plowing on small patches of land, and these two new devices produced food surpluses that encouraged landowners to amass large tracts of land. The stirrup, which seems to have come from India, permitted a heavily armored warrior to fight from horseback; specifically it permitted him to swing a heavy sword or battle axe, or to attack with a lance, without falling off his mount. The economic power created by people employing the horse collar and the metal plow provided wealth to pay for the expensive weaponry, which in turn defended the farmers. According to White, these forms of farming and military technology provided crucial, though not necessarily defining, components of feudalism. Whom did this technology empower? Those who ultimately became knights and landowners in an increasingly hierarchical society obviously obtained more power, as did the Church, which benefited from increasing surplus wealth. Those who made and sold the technology also obtained a degree of status, power, and wealth. What about the farm worker? Those freemen in a tribal society who lost their land and became serfs obviously lost power. But were any serfs better off, either safer or better fed, than before feudalism, as apologists for the Middle Ages used to argue? I do not know how one could answer such questions, though one component of an answer is certain: even if one had far more detailed evidence about living conditions of the poor than we do, no answer will come forth garbed in neutrality, because one cannot even begin to consider one's answer without first deciding what kind of weight to assign to matters such as the relative value of nutrition, safety, health, power, and status both in our own and in an alien culture. Another thing is clear as well: the introduction of new technology into a culture cuts at least two ways.

Like other forms of technology, those involving information have shown a double-edged effect, though in the long run—sometimes the run has been very long indeed—the result has always been to democratize information and power. Writing and reading, which first belonged to a tiny elite, appears in the ancient Middle East as an arcane skill that supports the power of the state by recording taxes, property, and similar information. Writing, which can thus conserve or preserve, has other political effects, Ong tells us, and "shortly after it first appeared, it served to freeze legal codes in early Sumeria" (*Orality and Literacy*, 41). Only careful examinations of the historical evidence can suggest which groups within society gained and which lost from such recording. In a particular society within a particular battle of forces, only nobility or nobility and priesthood could have gained, whereas in other situations the common person could have benefited from stability and clear laws.

Another political implication inheres in the fact that a "chirographic (writing) culture and even more a typographic (print) culture can distance and in a way denature even the human, itemizing such things as the names of leaders and political divisions in an abstract, neutral list entirely devoid of a human action context. An oral culture has no vehicle so neutral as a list" (42). The introduction of writing into a culture effects many changes, and all of them involve questions of power and status. When it first appeared in the ancient world, writing made its possessors unique. Furthermore, if writing changes the way people think as radically as McLuhan, Ong, and others have claimed, then writing drove a sharp wedge between the literate and the illiterate, encouraged a sharp division between these two groups that would rapidly become classes or castes, and greatly increased the power and prestige of the lettered. In the millennia that it took for writing to diffuse through large proportions of entire societies, however, writing shifted the balance from the state to the individual, from the nobility to the polis.

Writing, like other technologies, possesses a logic, but it can produce different, even contrary, effects in various social, political, and economic contexts. Marshall McLuhan points to its multiple, often opposing effects when he remarked that "if rigorous centralism is a main feature of literacy and print, no less so is the eager assertion of individual rights" (*Gutenberg Galaxy,* 220). Historians have long recognized the contradictory roles played by print in the Reformation and in the savage religious wars that followed. "In view of the carnage which ensued," Elizabeth Eisenstein observes, "it is difficult to imagine how anyone could regard the more efficient duplication of religious texts as an unmixed blessing. Heralded on all sides as a 'peaceful art,' Gutenberg's invention probably contributed more to destroying Christian concord and inflaming religious warfare than any of the so-called arts of war ever did" (319).[8] One reason for these conflicts, Eisenstein suggests, lies in the fact that when fixed in print—put down, that is, in black and white, "positions once taken were more difficult to reverse. Battles of books prolonged polarization, and pamphlet wars quickened the process" (326).

I contend that the history of information technology from writing to hypertext reveals an increasing democratization or dissemination of power. Writing begins this process, for by exteriorizing memory it converts knowledge from possession of one to the possession of more than one. As Ryan correctly argues, "writing can belong to anyone; it puts an end to the ownership or self-identical property that speech signaled" (*Marxism and Deconstruction,* 29). The democratic thrust of information technologies derives from their diffusing information and the power that such diffusion can produce.[9] Such empowerment has always marked applications of new information technology to education. As Eisenstein points out, for example, Renaissance treatises, such as those for music, radically reconfigured the cultural construction of learning by freeing the reader from a subordinate relation to a particular person: "The chance to master new skills without undergoing a formal apprenticeship or schooling also encouraged a new sense of independence on the part of many who became self-taught. Even though the new so-called 'silent instructors' did no more than duplicate lessons already being taught in classrooms and shops, they did cut the bonds of subordination which kept pupils and apprentices under the tutelage of a given master" (244). Eisenstein cites Newton as an example of someone who used books obtained at "local book fairs and libraries" to teach himself mathematics with little or no outside help (245). First with writing, then with print, and now with hypertext, one observes increasing synergy produced when readers widely separated in space and time build upon one another's ideas.

Tom McArthur's history of reference materials provides another reminder that all developments and inflections of such technology serve the interests of particular classes or groups. The early-seventeenth-century "compilers of the hard-word dictionaries" did not in the manner of modern lexicographers set out to record usage. Instead, they achieved great commercial success by "transferring the word-store of Latin wholesale into their own language . . . They sought (in the spirit of both the Renaissance and Reformation) to broaden the base of the educated Elect. Their works were for the nonscholarly, for the wives of the gentry and the bourgeoisie, for merchants and artisans and other aspirants to elegance, education, and power" (87). These dictionaries served, in other words, to diffuse status and power, and the members of the middle classes who created them for other members of their classes self-consciously followed identifiable political aims.

The dictionary created by the French Academy, McArthur reminds us, also embodies a lexicographical program that had clear and immediate political implications. Claude Favre de Vaugelas, the amateur grammarian who directed the work of the Academy, sought "to regulate the French language in terms of aristocratic good taste" as a means of making French the "social, political and scientific successor to Latin" (93). This dictionary is one of the most obvious instances of the way print technology sponsors nationalism, the vernacular, and relative democratization. It standardizes the language in ways that empower particular classes and geographical areas, inevitably at the expense of others. Nonetheless, it also permits the eventual homogenization of language and a corollary, if long-in-coming, possibility of democratization.

By the end of the eighteenth century, Kernan argues, print technology had produced many social and political changes that transformed the face of the literary world. "An older system of polite or courtly letters—primarily oral, aristocratic, amateur, authoritarian, court-centered—was swept away at this time and gradually replaced by a new print-based, market-centered, democratic literary system" (4). Furthermore, by changing the standard literary roles of scholar, teacher, and writer, print "noticeably increased the importance and the number of critics, editors, bibliographers, and literary historians" at the same time that it increasingly freed writers from patronage and state censorship. Print simultaneously transformed the audience from a few readers of manuscripts to a larger number "who bought books to read in the privacy of their homes." Copyright law, which dates from this period, also redefined the role of the author by making "the author the owner of his own writing" (4–5).

Like earlier technologies of information and cultural memory, electronic computing has obvious political implications. As Gregory Ulmer argued during a recent conference on electronic literacy, artificial intelligence projects, which use computers either to model the human mind or to make decisions that people would make, necessarily embody a particular ideology and a particular conception of humanity.[10] What, then, are the political implications of hypertext and hypertext systems?

I propose to begin examining that question by looking at the political implications of events described in a scenario that opened an article on hypertext in literary education that I published several years ago. It is 8:00 p.m., and, after having helped put the children to bed, Professor Jones settles into her favorite chair and reaches for her copy of Milton's *Paradise Lost* to prepare for tomorrow's class. A scholar who specializes in the poetry of Milton's time, she returns to the poem as one turns to meet an old friend. Reading the poem's opening pages, she once again encounters allusions to the Old Testament, and because she knows how seventeenth-century Christians commonly read these passages, she perceives connections both to a passage in Genesis and to its radical Christian transformations. Furthermore, her previous acquaintance with Milton allows her to recall other passages later in *Paradise Lost* that refer to this and related parts of the Bible. At the same time, she recognizes that the poem's opening lines pay homage to Homer, Vergil, Dante, and Spenser and simultaneously issue them a challenge.

Meanwhile John H. Smith, one of the most conscientious students in Professor Jones's survey of English literature, begins to prepare for class. What kind of a poem, what kind of text, does he encounter? Whereas Professor Jones experiences the great seventeenth-century epic situated within a field of relations and connections, her student encounters a far barer, less connected, reduced poem, most of whose allusions go unrecognized and almost all of whose challenges pass by unperceived. An unusually mature student, he pauses in his reading to check the footnotes for the meaning of unfamiliar words and allusions, a few of which he finds explained. Suppose one could find a way to allow Smith to experience some of the connections obvious to Professor Jones. Suppose he could touch the opening lines of *Paradise Lost,* for instance, and the relevant passages from Homer, Vergil, and the Bible would appear, or that he could touch another line and immediately encounter a list of other mentions of the same idea or image later in the poem or elsewhere in Milton's writing—or, for that matter, interpretations and critical judgments made since the poem's first publication—and that he could then call up any or all of them.

This scenario originally ended with my remark that hypertext allows students to do "all these things." Now I would like to ask what such a scenario implies about the political relations that obtain between teachers and students, readers and authors. These issues, which writers on hypertext have long discussed, also arose in questions I encountered when delivering invited talks on my experiences in teaching with hypertext. One of the administrators at my own university, for example, asked a question I at first thought rather curious but have since encountered frequently enough to realize is quite typical for those first encountering the medium. After I had shown some of the ways that hypertext enabled students to follow far more connections than ever before possible between texts and context, she asked if I was not worried because it limited the students too much, because it restricted them only to what was available on the system. My first response then as now was to remark that as long as I used print technology and the limited resources of a very poor university library, no administrator or member of the faculty ever worried that I found myself unable to suggest more than a very limited number of connections, say, five or six, in a normal class discussion; now that I can suggest six or ten times that number, thus permitting students a far richer, less controlled experience of text, helpful educators suddenly begin to worry that I am "limiting" students by allowing them access to some potentially totalitarian system.

One part of the reason for this reaction to educational hypertext lies in a healthy skepticism. Another appears in the way we often judge new approaches to pedagogy as simultaneously ineffective, even educationally useless, and yet overpoweringly and dangerously influential. Nonetheless, the skeptical administrator raised important questions, for she is correct that the information available limits the freedom of students and general readers alike. At the early, still experimental stage in the development of hypertext when I was asked that question, one had to pay great attention to ensuring a multiplicity of viewpoints and kinds of information. For this reason I emphasize creating multiple overviews and sets of links for various document sets, and I also believe that one must produce educational materials collaboratively whenever possible; as I have suggested, such collaboration is very easy to carry out between individual instructors in the same department as well as between those in different disciplines and different institutions. Now that the World Wide Web includes so much information, finding the multiple points of view and learning how to evaluate them become crucial.

Several key features of hypertext systems intrinsically promote a new kind of academic freedom and empowerment. Reader-controlled texts permit students to choose their own way. The political and educational necessity for this feature provides one reason why hypertext systems must always contain both bidirectional links and efficient navigational devices; otherwise developers can destroy the educational value of hypertext with instructional systems that alienate and disorient readers by forcing them down a predetermined path as if they were rats in a maze. A second feature of hypertext that has crucial political implications appears in the sheer quantity of information the reader encounters, since that quantity simultaneously protects readers against constraint and requires them to read actively, to make choices. A third liberating and empowering quality of (read-write) hypertext appears in the fact that the reader also writes and links, for this power, which removes much of the gap in conventional status relations between reader and author, permits readers to read actively in a much more powerful way—by annotating documents, arguing with them, leaving their own traces. As long as any reader has the power to enter the system and leave his or her mark, neither the tyranny of the center nor that of the majority can impose itself. The very open-endedness of the text also promotes empowering the reader.

The Political Vision of Hypertext; or, The Message in the Medium

Does hypertext as medium have a political message? Does it have a particular bias? As the capacity of hypertext systems to be infinitely recenterable suggests, they have the corollary characteristic of being antihierarchical and democratic in several different ways. To start, as the pioneering authors of "Reading and Writing the Electronic Book" point out, in such systems, "ideally, authors and readers should have the same set of integrated tools that allow them to browse through other material during the document preparation process and to add annotations and original links as they progress through an information web. In effect, the boundary between author and reader should largely disappear." One sign of the disappearance of boundaries between author and reader consists in its being the reader, not the author, who largely determines how the reader moves through the system, for the reader can determine the order and principle of investigation. Hypertext has the potential, thus far only partially realized, to be a democratic or multicentered system in yet another way: as readers contribute their comments and individual documents, the sharp division between author and reader that characterizes page-bound text begins to blur and threatens to vanish, with several interesting implications: first, by contributing to the system, users accept some responsibility for materials anyone can read; and second, students thus establish a community of learning, demonstrating to themselves that a large part of any investigation rests on the work of others.

Writing about electronic information technology in general rather than about hypertext in particular, McLuhan proposed: "The 'simultaneous field' of electronic information structures, today reconstitutes the conditions and need for dialogue and participation, rather than specialism and private initiative in all levels of social experience" (*Gutenberg Galaxy*, 141). McLuhan's point that electronic media privilege collaborative, cooperative practice, which receives particular support from hypertext, suggests that such media embody and possibly support a certain political system or construction of relations of power and status. J. Hillis Miller similarly argues that "one important aspect of these new technologies of expression and research is political. These technologies are inherently democratic and transnational. They will help create new and hitherto unimagined forms of democracy, political involvement, obligation, and power" ("Literary Theory," 20). Writing in the spring of 1989, Miller commented: "Far from being necessarily the instruments of thought control, as Orwell in 1984 foresaw, the new regime of telecommunications seems to be inherently democratic. It has helped bring down dictator after dictator in the past few months" (21).

Hypermedia seems to embody a truly decentered, or multiply centered, politics that seems the political equivalent of Richard Rorty's edifying philosophy whose purpose is "to keep the conversation going rather than to find objective truth . . . The danger which edifying discourse tries to avert is that some given vocabulary, some way in which people might come to think of themselves, will deceive them into thinking that from now on all discourse could be, or should be, normal discourse. The resulting freezing-over of culture would be, in the eyes of edifying philosophers, the dehumanization of human beings" (377).[11] Like Bakhtin and Derrida, Rorty presents his views as an explicit reaction against totalitarian centrism. Bakhtin had the example of Marxist-Leninism, particularly during the Stalin years, whereas Derrida and Rorty react against Plato and his heirs in a manner reminiscent of Karl Popper in *The Open Society and Its Enemies*.[12] Hypertext is potentially the technological embodiment of such a reaction and such a politics.

Gregory Ulmer comments that "the use of communications technology is a concretization of certain metaphysical assumptions, consequently that it is by changing these assumptions (for example, our notion of identity) that we will transform our communicational activities" (*Applied Grammatology*, 147). We may add that the use of communications technology is also a concretization of certain political assumptions. In particular, hypertext embodies assumptions of the necessity for nonhierarchical, multicentered, open-ended forms of politics and government.

Hypertext and Postcolonial Literature, Criticism, and Theory

Postcolonial literatures, criticism, and theory have numerous important relations both to hypertext as a medium and to hypertext as a theoretical paradigm. These connections range from the cultural applications of this new computing technology to the use of the hypertext paradigm within postcolonial theory.

First of all, hypertext in its most commonly encountered form, the World Wide Web, provides a particularly important way for the empire to write back. As Susan Nash Smith has shown in her work on Azerbaijan, former colonies use the Internet as a means of defining and communicating a newly recreated identity. In essence, the smallest country with access to the Internet can speak for itself in ways impracticable if not virtually impossible in the world of print. Take the example of Zimbabwe. When I went to Harare in August 1997 with Gunnar Liestøl and Andrew Morrison of the University of Oslo to help set up a local website and discuss educational applications of the Web, I discovered that not only did Zimbabwe have a rich postcolonial literature, quite different from that of, say, Nigeria, but it also had its literary critics. Nothing particularly surprising here, perhaps, except that like so much scholarship and criticism produced by citizens of former colonies, it remained unknown to European and American postcolonialists because it never entered the distribution channels for printed books and periodicals outside Africa. In other words, Zimbabwe had its own literary critics who could write about their country's literature, but there was little chance of postcolonial scholars in the West ever reading their essays.

Since I had already begun to create a section about Zimbabwe in my *Postcolonial Literature and Culture Web*, I obtained permission from Rino Zhuwarara, chair of the department of English at University of Zimbabwe, to include his substantial "Introduction to Zimbabwean Fiction in English," which I divided into ten sections for its appearance on the web, and at the same time Anthony Chennells contributed his "Rhodesian Discourse, Rhodesian Novels and the Zimbabwean Liberation War." Sometime later, Irene Staunton, then publishing director of Baobab Books, donated both her essays on the literary aftermath of Zimbabwe's seventeen-year civil war and her *Mothers of the*

Revolution, an oral history of women's experience of the conflict. During the next few years, Naume M. Ziyambi contributed sections of her study of women's groups in Zimbabwe while Maurice Taonezvi Vambe added his "Gender and Class Issues in the Postcolonial Zimbabwean Novel." As I was writing this section, Phillip Chidavaenzi, a Zimbabwean journalist, sent in another essay.

The chief value of placing these essays online is simply that Zimbabweans can speak—or rather, write—for themselves rather than having critics from the Europe, the United Kingdom, and the United States write for them. Such self-representation on the Web will not solve the current terrible problems in Zimbabwe, nor will it instantly rival publications produced by major European and American presses. It will, however, lessen the degree to which postcolonial criticism tends to repeat the pattern of colonizers, imposing cultural definitions from Europe upon the local scene. To a small extent, this and similar projects begin to fulfill Geert Lovinck's claim that "future media politics is about empowerment, not about representing the Other. The goal of the democratization of the media is the elimination of all forms of mediated representation and artificial scarcity of channels. There are now the technical possibilities to let people speak for themselves" (*Dark Fiber,* 39).

The political value of placing such materials on a website (whose servers incidentally reside in Singapore and the United States) lies in the fact that these essays become location independent, accessible from anywhere in the world to someone connected to the Internet. Such location independence also means that they evade censorship by local authorities, something that contributors to the site appreciate. The *Postcolonial Literature and Culture Web* receives contributions from Nigerians living in South Africa, South Africans living in Ghana, and a host of contributors from Canada, South Asia, Australia, and so on.

Several times now I have found myself letting contributors redefine the scope of the *Postcolonial Literature and Culture Web,* which I originally intended to include only Anglophone literature, after they wrote to me pointing out that they could find no other sites to include their work. Newly added sections for Canada, Sri Lanka, and the Caribbean obviously fit the original description of the site, but when Louise Vijoen, who teaches in the department of Afrikaans and Dutch, University of Stellenbosch, submitted materials on writing in Afrikaans and others sent in essays on francophone culture, I found room for them on the site, too. Writings on contemporary Nepali literature followed, but the biggest stretch appears in the section on Morocco, where the languages discussed are Arabic, Berber, and French. This addition came about after attendees at a conference in Casablanca, who had heard me talk about the case of Zimbabwean self-representation, asked if I'd put up some of their materials. And so it goes.

In the almost two decades since I began working with the cultural and educational uses of hypertext, I have not infrequently encountered skepticism about its value. The way that the World Wide Web permits those in postcolonial countries to represent themselves, thus partially redressing a major imbalance in postcolonial studies, strikes me as one undoubted success.

Infotech, Empires, and Decolonization

It is perhaps fitting that hypertext and the Web, late-twentieth-century information technologies, offer some solutions to postcolonial dilemmas, since much eighteenth- and nineteenth-century colonialism depended on the imposition of writing and printing on indigenous oral cultures. According to McLuhan, of all the clashes of civilizations that produce "furious release of energy and change, there is none to surpass the meeting of literate and oral cultures." In fact, the arrival of "phonetic literacy is, socially and politically, probably the most radical explosion that can occur in any social structure" (*Understanding Media,* 55). Postcolonial fiction, which often describes this "explosion," takes differing approaches to this change of information regimes.

First, novelists like Yvonne Vera and Charles Mongoshi make the collision of oral and writing cultures a significant part of their narratives. In Vera's *Nehanda,* a novel about the Chimurenga, Zimbabwe's nineteenth-century war of independence, the chief argues that his people's oral culture has a life and truth missing from "the stranger's own peculiar custom"—writing on paper. Sounding much like Plato's Socrates, the chief tells his listeners because "our people know the power of words,"

they desire to have words continuously spoken and kept alive. We do not believe that words can become independent of the speech that bore them, of the humans who controlled and gave birth to them. Can words exchanged today on this clearing surrounded by waving grass become like a child left to be brought up by strangers? Words surrendered to the stranger, like the abandoned child, will become alien—a stranger to our tongues.

The paper is the stranger's own peculiar custom. Among ourselves, speech is not like rock. Words cannot be taken from the people who create them. People are the words. (39–40)

Vera makes the additional point that although the Shona do not possess European alphanumeric writing, they do employ multiple systems of written signs. The narrator explains that the chief "bears proud marks on his fore-head, and on his legs," and although "the stranger" sees only scars, these marks "distinguish him," signifying his individuality and status. In addition, he occasionally "bears other signs that are less permanent, painted for particular rituals and festivities. He can even invent signs that will immediately be understood by his people as his own. Indeed, these signs help to communicate sacred messages among the people" (40).

The gap between Shona and British attitudes toward language produces confusion and comedy when a well-meaning Anglican priest tries to convert Kaguvi, one of the leaders of the Chimurenga, to Christianity. Yvonne Vera brilliantly dramatizes a clash of cultures in which two sincere, believing individuals misconceive each other's positions. Kaguvi, whom Vera presents as the embodiment of oral culture, finds the notion that a printed book could contain divinity intensely problematic, in part because for him, like Socrates, writing separates the words of the speaker from his or her presence. Since he does not come from a print culture, the kind of multiplicity characteristic of a book puzzles him, and as he points out to the Christian, his is a "strange" god who "is inside your book, but he is also in many books." In contrast, to this book-bound divinity, he explains: "My god lives up above. He is a pool of water in the sky. My god is a rain-giver. I approach my god through my ancestors and my mudzimu. I brew beer for my god to praise him, and I dance. My mudzimu is always with me, and I pay tribute to my protective spirit" (105).[13]

In Charles Mungoshi's *Waiting for the Rain,* which is set half a century after *Nehanda,* another wise old man rejects the white man's information technology. Mungoshi, who refuses to sentimentalize either modern or traditional ways, creates complex portraits of the gap between generations as the second Chimurenga, or war of liberation, begins. When the young would-be revolutionaries approach the Old Man in hopes that he will tell them what he recalls from the first uprising against British colonial oppression, he refuses to pass on his knowledge, in part because they badly misread his character, in part because he feels their acceptance of Western ways dooms them from the start. Mungoshi presents the Old Man's refusal as fundamentally related to his rejection of modern information technologies, which he sees as inhuman, though he first presents this rejection as a matter of moral values. Because the young men made the mistake of telling him that they would put what he tells them in a book, the publication of which might make him "rich and famous," he first focuses on that, since he claims that "by the way he said it to me it seems that's all he is interested in. Riches and fame. As if that were everything. As if we haven't seen enough destruction through those two things." Although his rejection at first seems to derive entirely from a kind of moral superiority, what he says next immediately complicates our judgment of him and his interpretations of political and cultural reality, since the Old Man in part seems merely defeatist:

And even if I were to talk to them, what can he and his friends do? Take up arms and fight the white man? They will be defeated before they even fire the first shot. They are already defeated. What kind of fighting is it when you are clutching and praying to your enemy's gods? I don't know what he was talking about but he is certainly playing someone else's drum. Each time I see my wife Japi take in a handful of sugar, I know how complete and final the white man's conquest has been.

According to the Old Man, then, any modern rebellion would fail since the rebels have adopted Western ways. Given that Mungoshi wrote this novel after the war for liberation had been won, the reader has to question the speaker's views. Does he mean that any rebellion would fail, or that any rebellion, even if apparently successful, would only produce a pseudo-victory since traditional culture would have been lost?

In one sense, or at one level, this statement appears to reflect the simple xenophobia of the defeated: anything from outside the group is bad, dangerous, destructive. Thus he finds his gluttonous wife's love of refined sugar, like the widespread habit of drinking tea or listening to radios, equally dangerous because each represents something from outside his culture, though he does not mention that many of his people's basic foodstuffs, such as maize, also come from outside—in this case from the Americas.

In another sense, however, the Old Man raises the basic political issues implicit in different information technologies: he accepts the value only of speech within a small group, and any technologies of cultural memory that permit thoughts to be recorded or transported out of the presence of the speaker seem to him fundamentally wrong. Whether in the form of handwriting, book, or radio, modern information technology removes the need for one person to be in the presence of another for them to communicate. Like Socrates (as Derrida reminds us), the Old Man emphasizes the human costs but not the benefits of such technology.

In addition to making the collision of oral and writing cultures a paradigm of the colonial experience, African, Maori, Samoan, and other novelists self-consciously try to create the effect of an oral culture in their works, often by the use of proverbs, rituals, and formal orations. Chidi Okonkwo points out that "the importance of orality in the first generation of [postcolonial] novels certainly eluded the first generation of readers and critics, whose responses defined European approaches to the novel for half a century" (33). Jayalakshmi V. Rao has catalogued dozens of the proverbs Chinua Achebe cites in *Arrow of God, Things Fall Apart,* and *No Longer at Ease,* and she also shows his use of folktales in his writings. In *Antills of the Savannah,* Ikem's speech to his countrymen who have come to the capitol to inform the government of their plight, like many such addresses in recent African novels, attempts to recreate the glories of an oral culture within the novel, the genre that epitomizes print culture. Duzia's funeral oration for Adda in Ken Saro-Wiwa's "A Death in Town," though far briefer, works in much the same way by demonstrating the force and elegance of the speaker's words.

The central importance of the colonial importation—and imposition—of writing and print on indigenous oral cultures problematizes Ngugi Wa Thiong'o's famous decision to abandon writing novels in English and thenceforth write only in his native language, Gikuyu. Ngugi's declaration prompted a fiery, often bitter debate, containing, as it had to, claims of authenticity and revolutionary commitment. I don't propose to rehearse these debates again, but I would like to draw attention to a set of associated issues, which place the entire subject in an important light. Following McLuhan and Goody, I want to emphasize that the movement from an oral to a print culture brings such radical changes in conceptions of self, authorship, society, and verbal arts that the question of which language is more authentic to a particular indigenous people seems incredibly ill-conceived. Whether writing in Gikuyu or English, Ngugi thinks and expresses himself in a way foreign to precolonial times. This is not all. The modern novel, a literary genre that derives from—and epitomizes—print culture, represents a major European influence on a colonized people. The novel, in other words, is a major colonial imposition on African and other oral cultures. One can appropriate it, and turn it against its originators: think of *Wide Sargasso Sea*'s rewriting of *Jane Eyre* or *Jack Maggs* as a rebuttal to *Great Expectations.* But one simply cannot *write* an authentic work of orality. Of course, there are many political justifications for promoting the language of a nation newly freed from an imperial power, and one does not have to go to Africa for examples of languages that countries have revived or even created as an act of decolonization: the invention of Norwegian after Norway's independence from Denmark in the nineteenth century and Gaelic after Ireland's independence from Britain in the twentieth.

Hypertext as Paradigm
for Postcoloniality

Having observed a few examples of the way existing hypermedia projects do in fact answer some needs of various postcolonial countries, I'd now like to examine the usefulness of hypertext as a paradigm for understanding postcolonial (or decolonized) cultures. It is hardly surprising that hypertext and postcolonial theory, both of which have important parallels to poststructuralist thought, have much in common, but because hypertext lexias take part in a networked structure, resist simple linearity, and show that a complex entity can exist within multiple contexts without losing its identity they have proved particularly useful when discussing postcolonialities. Although my students and I created an early version of *The Postcolonial Web* in 1992, I don't think I realized the full implications of hypertext for postcolonial theory until Jaishree K. Odin, associate professor at the University of Hawaii, Manoa, sent in her essay, "The Performative and Processual: A Study of the Hypertext/Postcolonial Aesthetic." Drawing on the first version of *Hypertext,* Odin pointed out how this information technology modeled key issues of postcolonial theory:

The postcolonial critique of unitary models of subjectivity reveals that all such models are based on binary thinking that creates categories like self and other, male and female, first world and third world where the first term is always the privileged term. Rejecting binary models, postcolonial theorists describe both subjectivity as well as experience decentered and pluralistic. The electronic media can be used as metaphor for describing what is happening to the culture at large as the Culture (represented by the dominant group) is being displaced by minority cultures which demand recognition of their histories as well as cultural productions. Just as in networked computers diverse, sometimes contradictory information, can exist simultaneously in hypertext format, so it is in culturally diverse societies with different, sometime contradictory narratives. The person's location based on race, class, and gender determines what perspective will be taken.

The rejection of oversimplifying, even falsifying binary oppositions has, I would urge, an immediate practical consequence. Postcolonial theory and practice is riddled with less-than-helpful oppositions, pre- and postcolonial being just one of them. As Neal Lazarus pointed out in his pioneering *Resistance in Postcolonial African Fiction* (1990), the reductive rhetoric of anticolonialism led to serious postliberation problems because it did not prepare the newly freed countries for differences of attitude and approach among former allies. "It implied that there was only one struggle to be waged, and it was a negative one: a struggle *against* colonialism, not a struggle *for* anything specific." In a desire "above all to remain free of ideological factionalism," anticolonial rhetoric in Africa relied too heavily on empty abstractions. "To it, there was only today and tomorrow, bondage and freedom. It never paused long enough to give its ideal of 'freedom' a content. Specifically, it implicitly rationalized, exposed the movement to the risk of division. Typically, therefore, the radical anticolonial writers tended to romanticize the resistance movement and to underestimate—even theoretically to suppress—the dimensions within it." Unaware of "groups and individuals working with quite different, and often incompatible, aspirations for the future" (5), many of the revolutionaries set themselves up for failure and disillusionment.

The history of how the terrible situation in Zimbabwe has been reported in Western mass media provides an example of the way a clichéd binary opposition—here black African versus white European—long prevented the world from understanding what was taking place. As Richard Mugabe's government plunged further into corruption and inefficiency, he proclaimed that he was satisfying the demands of Zimbabweans for land that had been appropriated by the British during the days when the country was a British colony called Rhodesia. The international media, which likes its news prepackaged, immediately presented Mugabe's version of events because, after all, isn't the split between black and white, colonized and colonizer, rich farmer and poor, dispossessed farm worker the central fact of this situation? Well, actually it isn't. In fact, it doesn't exist. As a black Zimbabwean contributor to *The Postcolonial Web* (she identifies herself publicly only as "Fliss") pointed out in a series of communications in May 2000, Mugabe brilliantly manipulated the racial bigotries of the international press:

1. "Contrary to popular belief, the white commercial farmers do *not* own 70% of the fertile land in Zimbabwe. 900 farms comprising roughly 2 million hectares were acquired by the government for redistribution and they were indeed distributed—to the cronies and relatives of President Mugabe, in some cases in such ridiculous portions that one minister owns 17 farms."

2. "The farms of mixed and black people have also been invaded in a pattern that suggests the invasions are political rather than racial, eg. the farm workers that are beaten and harassed are those that support the opposition, the white farmers that were killed were prominent supporters of the main opposition party as were all the black people killed so far."

3. [Those who attacked white and black farmers were not, as the press reported, veterans of the war of independence, most of whom were in their six-

ties.] "Except for a very few exceptions, these so-called war veterans average 18 years of age, and unless there's a war that happened a couple of years ago of which the rest of Zimbabwe has never heard . . . The 'war veterans' are transported by government vehicles to each new farm, armed and paid by the government."

4. "This is not the first time that Zanu PF (Robert Mugabe's ruling party) has attempted what might be seen as an 'ethnic cleansing.' In the mid-late '80s, Mugabe unleashed his Cuban-trained 5th brigade on the Matabele peoples in the South of the country, committing genocides on a scale that rivals any in Kosovo or Rwanda. Zimbabwe is made up of several different ethnic and racial groups. The Shona speaking ones are the Karanga, the Mazezuru, and the Manica. There are also the Ndebele (Matabele), Tonga, Batoka, N'dau. Then there are the Asians, Coloureds (mixed people) and Whites. Mugabe is of the Mazezuru as are all his cabinet, all his ministers, actually anyone in his government or party of any importance."

5. "Many of the people who own farms today bought them legitimately after independence at fair market value. Many of them are fourth generation Zimbabweans who embraced the hand of reconciliation that Mugabe offered at independence. Why is he choosing to take the land back now when he's had twenty years with willing co-operation from the British and the white farmers?"

6. "60% of the population is below the age of 40, well-educated and looking for a future, a job, trade and industry, not a subsistence small land-holding."

Obviously, I am not claiming that hypertext-as-paradigm can produce political miracles by simply changing habits of thought dependent on falsifying binaries. Actual hypertext, hypertext as an information technology in the form of the World Wide Web, can at least permit individual voices to be heard. Summoning skepticism and hypertextual habits of mind can, however, lead us to ask, "What connections (links) are missing? What complex network of events percolates up through the linear political narrative we've been offered?"

Geert Lovink, who reports in his experience with Net activism in Albania, Taiwan, and India, rejects usual formulations of

the problem of the local and global. Net activists and artists are confronted with the dilemma between the supposedly friction-free machinic globality and the experience that social networks, in order to be successful, need to be rooted in local structures. Internet culture pops up in places where crystals of (media) freedom have been found before. At the same time the net is constantly subverting the very same local ties it grows out of while creating new forms of "glocality." The choice of global or local is a false one. (*Dark Fiber*, 63)

Users of the Internet, particularly in postcolonial regions, experience the simultaneous advantage and dilemma of existing as a complex palimpsest of identities, locations, and responsibilities. Lovink, who is well aware of problems of accessibility, commercialization of the Internet, and the difficulties of using the Internet to foster democracy, nonetheless describes examples of success.[14]

Forms of Postcolonialist Amnesia

The crude binary oppositions that permeate so much of colonial and postcolonial studies produce three especially scandalous forms of historical amnesia. First, displaying a blatant disregard of modern history, postcolonialists all too often write as though the British empire of the eighteenth and nineteenth centuries is the major, and often the only, form of European colonialism that existed, or at least the only one worthwhile studying. After surveying discussions of subject on the World Wide Web, Eric Dickens complained, "Most of the websites, pages, searches, etc., concentrate on countries where the British Empire had colonies. Even when conferences are held in Finland, it's still British culture, British Black literature, and so forth, which are the focus. And the research itself is often done in Australia or the United States, where a British colonial past dominates, if subliminally. Whatever happened to discussing Portuguese, Spanish, French, Dutch, Belgian, Russian, Swedish, Danish, Roman, etc., etc., colonialism over the centuries?" In other words, instead of conceiving colonialism and its aftermath as a complex network of relations, too many postcolonialists present their field in terms of a simple opposition that omits much of European history.

Second, too much of postcolonial studies not only focuses exclusively on England and its empire but this narrowness appears in a crude eurocentrism. Chidi Okonkwo, a postcolonial scholar originally from Nigeria who writes about African, Oceanic, and Maori texts, points out the importance of recognizing both contemporary and historical "imperial ambitions among non-European peoples." Looking at the present, Okonkwo reminds us that "Indonesia's annexation of East Timor in 1975, with the attendant atrocities by which the colonial occupation has been perpetuated (about a third of the indigenous population), provides a stark reminder that neither the imperialist nor the genocidal impulse is exclusively European" (25). Okonkwo, who shows no interest in justifying European imperialism, argues that ignorant eurocentrism of most colonial and postcolonial scholarship omits central facts in the history of the colonized:

> Though the field of postcolonial studies is focused on Euro-Christian imperialism and colonialism, the spread of Islam into Africa (and Europe) by Arabs of the Arabian peninsula constitutes a colonization enterprise whose beginnings predate that of Europe and whose efforts have proved equally enduring. It is a failure of postcolonial theory, and further proof of its Eurocentrism, that it ignores this major dimension of history. (26)

European, much less the British, projects of colonial expansion did not, in other words, take the form of a linear, isolated narrative. Take this history of South and Southeast Asia as an example of imperial complexities. In the eighteenth century the Burmese, whom Western journalists often quaintly and condescendingly characterize as a gentle people, had major imperial ambitions that resulted in war with Thailand and the consequent destruction of the great Thai capital of Sukothai (which prompted the later founding of Bangkok as a new capital). The Burmese, who wished to expand toward India, collided with the British who were trying to protect their trade routes. The British had no particular interest in colonizing Burma, but its imperial ambitions led to three Anglo-Burmese wars, in each of which the Burmese lost more territory. My point is that one has to know something about the layered, often confusing local history of a nation to write with any authority about its colonial and postcolonial identities.

This essential need for both comparative and historical knowledge appears when studying Soviet Russia and its former colonies, which represent yet another crucial part of the story missing from most postcolonial studies. According to Dickens, Finland and the Baltic states—Estonia, Lithuania, and Latvia—"finally shook off colonial rule in 1917. By the end of World War II, only Finland remained a free country and has flourished ever since. The other three were once again swallowed up until 1991 by the Soviet Union. Surely here again there are fruitful comparisons to be drawn and lessons to be learnt for newly independent states in Africa and Asia." The creation of new national identities for themselves remains a fundamental problem for such decolonized nations.

These post-Soviet attempts to create the sense and identity of a nation obviously bear some interesting parallels to the situation in Africa and Asia, but many of these Eastern bloc postcolonialists emphasize significant differences between the British and Soviet empires. For example, in *Imperial Knowledge: Russian Literature and Colonialism*, Ewa Thompson claims that Soviet-era colonies, such as Belarus and the Ukraine, never were able to redefine the center/margin relation of empire and colony:

> Unlike Western colonies, which have increasingly talked back to their former masters, Russia's colonies have by and large remained mute, sometimes lacking Western-educated national elites and always lacking the encouragement of Western academia that foregrounding issues relevant to them would afford. They continue to be perceived within the paradigms relevant to Russia, the objects of Russian perception rather than subjects responding to their own histories, perceptions, and interests. In that connection, the perception of postcolonialist commentators that history is "the discourse through which the West has asserted its hegemony over the rest of the world" is incorrect.

In fact, Thompson argues, both pre-Soviet and Soviet Russian ideology has so permeated European and American discourse that we readily employ the misleading division into "West and non-West" that disregards Russia's successful "effort to manufacture a history, one that stands in partial opposition to the history created by the West on the one hand, and on the other to the history sustained by the efforts of those whom Russia had colonized. In doing so, Russia has successfully superimposed portions of its own narrative on the Western one."

Hypertext as Paradigm in

Postcolonial Theory

The example of the mass media's misinterpretation of the current situation in Zimbabwe, like the problems inherent in the rhetoric of African liberation, demonstrates how easily we fall into the habit of binary thought. Derrida's deconstruction reminds us that many of the concepts and categories we routinely oppose—powerful/weak, colonizer/colonized, male/female, inside/outside—actually share many qualities that, when recognized, weaken the force of the original contrast. In fact, many such culturally affirmed oppositions are like the standard use of red and green in traffic lights: the contrast has its practical uses, certainly, but in reality these colors exist along a spectrum and not in binary opposition. Binary oppositions are generally rhetorical techniques or thought-forms that have limited practical and political uses, just as long as they are not misunderstood.

The value of hypertext as a paradigm exists in its essential multivocality, decentering, and redefinition of edges, borders, identities. As such, it provides a paradigm, a way of thinking about postcolonial issues, that continually serves to remind us of the complex factors at issue. As Odin convincingly argues, "The perpetual negotiation of difference that the border subject engages in creates a new space that demands its own aesthetic. This new aesthetic, which I term 'hypertext' or 'postcolonial', represents the need to switch from the linear, univocal, closed, authoritative aesthetic involving passive encounters characterizing the performance of the same to that of non-linear, multivocal, open, non-hierarchical aesthetic involving active encounters that are marked by repetition of the same with and in difference." In "The Performative and Processual: A Study of the Hypertext/Postcolonial Aesthetic," the study in which Odin advances her proposal to use hypermedia as an effective means of understanding various aspects of postcolonial situations, she concentrates on analyses of Leslie Silko and Shelley Jackson. Looking at some problematic aspects of liberation rhetoric, as Lazarus has done, as well as the fiction and autobiography of major decolonization writers demonstrates the value of her approach. Antoinette's problems with her double or triple identity in *Wide Sargasso Sea,* Soyinka's more joyous presentation of his complex multiple heritages in *Ake* and *Isara: A Voyage around Essay,* and Kerewin's complex ethnic, sexual, and artistic identity in Hume's *The Bone People* all testify to the need in postcolonial situations for what Odin terms "non-linear, multivocal, open, non-hierarchical aesthetic involving active encounters that are marked by repetition of the same with and in difference."

The writings of Salman Rushdie, Sara Suleri, and many others support Odin's claim that "the intertextual and interactive hypertext aesthetic is most suited for representing postcolonial cultural experience because it embodies our changed conception of language, space, and time. Language and place are here no longer seen as existing in abstract space and time, but involve a dynamic interaction of history, politics, and culture." Rushdie's meditations in *Shame* on roots, rootlessness, migration, and being between exemplify what Odin means. Rushdie's narrator explains that he knows "something of this immigrant business. I am an emigrant from one country (India) and a newcomer in two (England, where I live, and Pakistan, to which my family moved against my will)." According to him, if gravity equates with belonging somewhere, he and other wanderers among various cultures "have come unstuck from more than land. We have floated upwards from history from memory, from Time" (90–91). The best thing about people who have moved between worlds, say, Rushdie, is "their hopefulness," the worst "the emptiness of their luggage." Here many postcolonial novelists would disagree, for they find that they travel with too much baggage rather than too little.

In *Meatless Days,* for example, Suleri blends languages, geographies, and life stories emphasizing the heavy weight of public and private histories. Occasionally, multiple identities defined by their simultaneous existence in too many contexts produce tragedy, as when she introduces the possibility that her father's political views may have played a part in her sister's murder; more often, Suleri produces comedy, or at least a wry glance at her multiple worlds—for example, when discovering that *kapura* are testicles and not sweetbreads, she find her relationship with her Welsh mother, family, various homes, and nationalities become threateningly complex and incoherent. She, like Rushdie, Hume, and so many other postcolonial novelists, creates narratives "composed of cracks, in-between spaces, gaps where linearity and homogeneity are rejected in favor of heterogeneity and discontinuity." Like writers of hyperfiction, such as Jackson and Joyce, postcolonial novelists "use strategies of disruption and discontinuity" in multilinear narratives in which "meaning does not lie in the tracing of one narrative trajectory, but rather in the relationship that various tracings forge with one other." As Sage Wilson, one of the undergraduate contributors to *The Postcolonial Web,* put it, "postcolonial thought refuses to wipe the slate clean." Past traditions, oral culture, English colonial education, syncretic religions, personal identities are all contaminated, mixed, hybrid, and one has to find ways of depicting—and living with—such complexity. Hypertext as paradigm at least offers an effective, understandable means of thinking about this congeries of complex and conflicting issues.

The Politics of Access: Who Can Make Links, Who Decides What Is Linked?

Mixed with the generally democratic, even anarchic tendencies of hypertext is another strain that might threaten to control the most basic characteristics of this information medium. Readers in informational hypertext obviously have far more control over the order in which they read individual passages than do readers of books, and to a large extent the reader's experience also defines the boundaries of the text and even the identity of the author, if one can conveniently speak of such a unitary figure in this kind of dispersed medium.

The use of hypertext systems like the Web involves four kinds of access to text and control over it: reading, linking, writing, and networking. Access to the hypertext begins with the technology required to read and produce hypertext, and this technology has only recently become widely available in the limited form of blogs. Once it becomes widespread enough to serve as a dominant, or at least major, form of publication, issues of the right and power to use such technology will be multiplied.

One can easily envision reading a text for which one has only partial permission, so that portions of it remain forbidden, out of sight, and perhaps entirely unknown. An analogy from print technology would be having access to a published book but not to the full reports by referees, the author's contract, the manuscript before it has undergone copyediting, and so on. Conventionally, we do not consider such materials to be *part* of the book. Electronic linking has the potential, however, radically to redefine the nature of the text, and since this redefinition includes connection of the so-called main text to a host of ancillary ones (that then lose the status of ancillary-ness), issues of power immediately arise. Who controls access to such materials, the author, the publisher, or the reader?

Linking involves the essence of hypertext technology. Already we have seen the invention of web software that provides the capacity to create links to texts over which others have editorial control. This ability to make links to lexias for which one does not possess the right to make verbal or other changes has no analogy in the world of print technology. One effect of this kind of linking is to create an intermediate realm between the writer and the reader, thus further blurring the distinction between these roles.

When discussing the educational uses of hypertext, one immediately encounters the various ways that reshaping the roles of reader and author quickly reshape those of student and teacher, for this information medium enforces several kinds of collaborative learning. Granting students far more control over their reading paths than does book technology obviously empowers students in a range of ways, one of which is to encourage active explorations by readers and another of which is to enable students to contextualize what they read. Pointing to such empowerment, however, leads directly to questions about the politics of hypertext.

Hypertext demands the presence of many blocks of text that can link to one another. Decisions about relevance obviously bear heavy ideological freight, and hypertext's very emphasis on connectivity means that excluding any particular bit of text from the metatext places it comparatively much farther from sight than would be the case in print technology. When every connection requires a particular level of effort, particularly when physical effort is required to procure a copy of an individual work, availability and accessibility become essentially equal, as they are for the skilled reader in a modern library. When, however, some connections require no more effort than does continuing to read the same text, *un*connected texts are experienced as lying much farther away, and availability and accessibility become very different matters.

Complete hypertextuality requires gigantic information networks of the kind now taking form on the Web. This vision of hypertext as a means of democratic empowerment depends ultimately on the individual reader-author's access to enormous networks of information. As Norman Meyrowitz admits, "Down deep, we all think and believe that hypertext is a vision that sometime soon there will be an infrastructure, national and international, that supports a network and community of knowledge linking together myriad types of information for an enormous variety of audiences" (2). The person occupying the roles of reader and author must have access to information, which in practice means access to a network—the Internet. For the writer this access becomes essential, for in the hypertext world access to a network is publication.

Considered as an information and publication medium, hypertext presents in starkest outline the contrast between availability and accessibility. Texts can be available somewhere in an archive, but without cataloguing, support personnel, and opportunities to visit that archive, they remain unseen and unread. Since search tools have made materials within a hypertext environment much easier to obtain, it simultaneously threatens to make any of those not present seem even more distant and more invisible than absent documents in the world of print are felt to be. The political implications of this contrast seem clear enough: gaining access to a network permits a text to exist as a text in this new information world. Lyotard, who argues that knowledge "can fit into the new channels, and become operational, only if learning is translated into quantities of information," predicts that "anything in the constituted body of knowledge that is not translatable in this way will be abandoned and that the direction of new research will be dictated by the possibility of its eventual results being translatable into computer language" (*Postmodern Condition,* 4). Antonio Zampolli, the Italian computational linguist and past president of the Association of Literary and Linguistic Computing, warns about this problem when he suggests an analogy between the Gutenberg revolution and what he terms the *informatization* of languages: "Languages which have not been involved with printing, have become dialects or have disappeared. The same could happen to languages that have not been 'informatized'" (47) transferred to the world of electronic text storage, manipulation, and retrieval. As Lyotard and Zampolli suggest, individual texts and entire languages that do not transfer to a new information medium when it becomes culturally dominant will become marginalized, unimportant, virtually invisible.

Although a treatise on poetry, horticulture, or warfare that existed in half a dozen manuscripts may have continued to exist in the same number of copies several centuries after the introduction of printing, it lost power and status, except as a unique collector's item, and became far harder to use than ever before. Few readers cared to locate, much less make an inconvenient, costly, and possibly dangerous trip to peruse an individual manuscript when relatively far cheaper printed books existed close at hand. As habits and expectations of reading changed during the transition from manuscript to print, the experience of reading texts in manuscript changed in several ways. Although retaining the aura of unique objects, texts in manuscript appeared scarcer, harder to locate, and more difficult to read in comparison with books. Moreover, as readers quickly accustomed themselves to the clarity and uniformity of printed fonts, they also tended to lose or find annoying certain reading skills associated with manuscripts and certain of their characteristics, including copious use of abbreviations that made the copyist's work easier and faster. Similarly, book readers who had begun to take tables of contents, pagination, and indices for granted found locating information in manuscripts particularly difficult. Finally, readers in a culture of print who have enjoyed the convenience of abundant maps, charts, and pictures soon realized that they could not find certain kinds of information in manuscripts at all.

In the past, transitions from one dominant information medium to another have taken so long—millennia with writing and centuries with printing—that the surrounding cultures adapted gradually. Those languages and dialects that did not make the transition remained much the same for a long time but gradually weakened, attenuated, or even died out because they could not do many of the things printed languages and dialects could do. Because during the early stages of both chirographic and typographic cultures much of the resources was devoted to transferring texts from the earlier to the current medium, this masked these transitions somewhat. The first centuries of printing, as McLuhan points out, saw the world flooded with versions of medieval manuscripts in part because the voracious, efficient printing press could reproduce texts faster than authors could write them. This flood of older work had the effect of thus using radically new means to disseminate old-fashioned, conservative, and even reactionary texts.

We can expect that many of the same phenomena of transition will repeat themselves, though often in forms presently unexpected and unpredictable. We can count on hypertext and print existing side by side for some time to come, particularly in elite and scholarly culture, and when the shift to hypertext makes it culturally dominant, it will appear so natural to the general reader-author that only specialists will notice the change or react with much nostalgia for the way things used to be. Whereas certain inventions, such as vacuum cleaners and dishwashers, took almost a century between their initial development and commercial success, recent discoveries and inventions, such as the laser, have required less than a tenth the time to complete the same process. This acceleration of the dispersal of technological change suggests, therefore, that the transition from print to electronic hypertext, if it comes, will therefore take far less time than did earlier transitions.

The history of the print technology and culture also suggests that as the Web becomes even more culturally important than it already is, it will do so by enabling large numbers of people either to do new things or to do old things more easily. Furthermore, such a shift in information paradigms will see another version of what took place in the transition to print culture: an overwhelming percentage of the new texts created, like Renaissance and later how-to-do-it books, will answer the needs of an audience outside the academy and hence will long remain culturally invisible and objects of scorn, particularly among those segments of the cultural elite who claim to know the true needs of "the people." The enormous number of online diaries, political and other parodies, examples of self-published fiction and poetry, and conversion-tales by people with alternative lifestyles reveals that for many such a change has already taken place. The active readers hypertext creates can meet their needs only if they can find the information they want, and to find that information they must have access to the Internet and local text- and databases that require special access. Similarly, authors cannot fully assume the authorial function if they cannot place their texts on a network, something at first impossible on the World Wide Web unless the author had access rights to a server. Blogs with comment functions, as we have observed, allow the Web reader to act as a Web author, too.[15]

Slashdot: The Reader as Writer

and Editor in a Multiuser Weblog

Slashdot, the famous multiuser site that uploaded almost 13,000 blogs in 2003, represents an important experiment in online democracy and large-scale collaboration because it uses its readers to moderate submissions. Ron "CmdrTaco" Malda, one of its founders and editor-in-chief, explains how the editors gradually devised a system to screen readers' contributions. The site grew rapidly, the number of comments increased, and "many users discovered new and annoying ways to abuse the system. The authors had but one option: Delete annoying comments. But as the system grew, we knew that we would never be able to keep up. We were outnumbered." At first Malda invited a few people to help, but the number soon increased to 25, and when they no longer could handle the thousands of posts that arrived every day, "we picked more the only way we could. Using the actions of the original 25 moderators, we picked 400 more. We picked the 400 people who had posted good comments: comments that had been flagged as the cream of Slashdot." When numbers continued to grow, Malda decided to have anyone who logged in to *Slashdot* and "read an average number of times—no obsessive compulsive reloaders, and nobody who just happened to read an article this week"—serve as a moderator.

Moderators receive "points of influence" that expire after three days, or when they use up their allotment, since "each comment they moderate deducts a point. When they run out of points, they are done serving until next time it is their turn." In evaluating comments, *Slashdot* moderators select "an adjective from a drop down list that appears next to comments containing descriptive words like 'Flamebait' or 'Informative.' Bad words will reduce the comment's score by a single point, and good words increase a comment's score by a single point. All comments are scored on an absolute scale from –1 to 5," although they are adjusted according to one's past performance, or "karma."[16] Malda emphasizes to moderators, who are urged to judge impartially, that they should "concentrate more on promoting than demoting," since their goal should be to "sift through the haystack and find needles. And to keep the children who like to spam Slashdot in check." The editors, who still moderate about 3 percent of contributions, try to protect the user-moderators by screening major spammers, since "a single malicious user can post dozens of comments, which would require several users to moderate them down, but a single admin can take care of it in seconds."

Slashdot has several means to prevent abuse by moderators, the most simplest and basic of which is the rule that they cannot participate in a discussion they moderate. The more complex means involves metamoderation, or M2, which the site uses to check the quality of decisions, removing bad moderators and rewarding good ones with additional mod points. Any active user who has been a member of the site for at least several months can serve as a metamoderator as well as a moderator.

Slashdot represents a fascinatingly successful experiment in large-scale online collaboration and reader empowerment. It does not, however, embody cyber-utopian Internet anarchy, for, as Malda's history of the site reveals, he quickly discovered that *Slashdot* needed a moderator to protect it from vandals. Furthermore, as soon as he and the other editors tried to share power with users, they realized that they had "to limit the power of each person to prevent a single rogue from spoiling it for everyone." Although the editors still maintain control of *Slashdot* and have final say on overall policies, they have transformed it into a collaborative venture, which means that in large part, they have turned most of the enterprise over to the users, who act at different times as contributors, readers, moderators, and judges of the moderators.

Comparing this multiuser blog to offline publishers, one realizes that it has retained certain features of the print world, such as editors with separate responsibilities, but at the same time it shares editorial authority with its large body of users, making most of *Slashdot* an example of decentered, distributed power. Interestingly enough, protecting the communal enterprise while sharing power and responsibility proved as crucial to the enterprise as software. Although *Slashdot* obviously employs many links, its hypertextuality appears chiefly in its changing network of reader-user-editors.

Pornography, Gambling, and Law on the Internet—Vulnerability and Invulnerability in E-Space

The Associated Press reported on December 5, 1994, that Robert and Carleen Thomas, who operated a computer bulletin board in California, were convicted in Memphis, Tennessee, on eleven counts of transmitting obscene materials to a members-only computer bulletin board via a telephone line. "The prosecution of the Thomases marked the first time that operators of a computer bulletin board were charged with obscenity in the city where the material was received, rather than where it originated." The Thomases, who live in Milpitas, California, near San Francisco, claimed that the prosecutors shopped around until they found a Bible Belt jurisdiction to increase chances of conviction. "If the 1973 Supreme Court standard is applied to cyberspace," the AP story continues, "juries in the most conservative parts of the country could decide what images and words get onto computer networks, said Stephen Bates, a senior fellow with the Annenberg Washington Program, a communications think-tank." To be sure this case involves digital networked culture and not hypertext itself since the crime with which the Thomases were charged involved a commercial bulletin board rather than the Web. Nonetheless, the same issues are involved.

In the Thomas case the virtual space that permits disseminating information at great speed turned out, according to the presiding judge, also to have extended the legal and hence physical space, grotesquely many have argued, in which one is legally vulnerable. The Internet in effect was understood to have dissolved one kind of legal boundary—that of the more liberal municipal authorities and of the state of California—while simultaneously extending that of Tennessee to override wishes of voters and judiciary in another state.

The Thomases' conviction's legal implications certainly have more importance to the United States with its conflicting legal jurisdictions than to many other countries. Their case also presents some odd features, one of the most obvious being that local Tennessee Internet providers offering the same kind of sexually explicit materials were supposedly not prosecuted either before or after the prosecutor went after the California couple. But the issue of jurisdiction in virtual space reminds us that in cyberspace the basic definitions of rights and responsibilities, law and its limits, are currently up for grab—and, as James Boyle suggests, since law works by analogy to often outmoded conditions, one can expect that crucial precedents will be made by those unaware of differences between physical and virtual space.

Granted that many people find such erotica offensive, the recent hysteria about pornography and exploitation of children on the Internet seems more than a little fishy, particularly given the fact that an astonishing amount of similar, equally degrading material is available via telephone chat-lines. Unless I have missed that article in my local newspaper, I don't recall reading that politicians and local law enforcement officers have proposed to imprison the CEOs of ATT, the Baby Bells, and local phone companies, much less seize phone lines and equipment. One common interpretation of the high moral dudgeon about possibilities of seduction and corruption on the Internet is that it involves asserting control of the vast great financial potential of its resources. In an article in *PC Magazine* (which I encountered in its World Wide Web form), John C. Dvorak convincingly argues that the entire Thomas case has little to do with the ostensible issues of moral standards: "the purpose of this interstate arrest was to set a legal precedent for all interstate activity done over a computer network. Authorities hope the result of this case (along with that of a parallel case against Thomas pending in the Utah courts) will be effective control of interstate banking, interstate sales tax collection for on-line mall activity, and interstate gambling for the purpose of collecting taxes (which authorities would like to ban outright)."

In this and other ways authorities might hope to control financial resources, in essence using virtual space to reshape physical and legal space within the boundaries of the United States. What can they do, however, when illegal activities originate outside the country? Dvorak sees the entire Thomas matter related essentially to the desire by individual states to control—that is, tax—gambling and other financial transactions by Americans on the Internet, but offshore servers have already shown how difficult this might be in an open society.

The easy, convenient access to Internet resources provided by the Web has, as one might expect, quickly produced distant gambling casinos that, however virtual themselves, require real money—as if money were itself not always virtual! In addition to providing advertising for legal gambling in Las Vegas (Vegas.Com!) and books on the subject, the World Wide Web also hosts both discussion groups and several virtual gambling casinos whose servers are located outside the United States. WagerNet, based in Belize, and Sports International, based in Antigua, permit one to place sports wagers—$50 minimum bet for Sports International—and the Caribbean Casino, which is based in Turks and Caicos Islands, offers blackjack and lotteries as well as wagering on sports. These establishments escape local American laws against gambling both because it takes place in virtual space and because the server lies outside a boundary that would permit law enforcement. According to William M. Bulkley, "the Justice Department says cyberspace casinos are illegal. But the companies' offshore venues may protect them. And authorities will have a tough time detecting who's actually betting because many people will be playing the same games for free" (B1). The effect of cyberspace, in other words, here is the opposite of that observed in the Thomas case: whereas in the pornography case, local authorities (with the assistance of the U.S. postal authorities) asserted their control over another jurisdiction, in this case the limits of U.S. sovereignty means that no control is possible. Perhaps the Thomases should move offshore. Many spammers have already done so.

A decade has passed since the appearance of online casinos, and Matt Richtel reports that "a new generation of online services like Betfair has emerged to allow sports bettors to wager not against the house but directly against each other." Such peer-to-peer betting, which has become popular in Europe, Asia, and Australia, thus far takes two forms: Betfair, a British website, permits would-be bettors to contact each other, whereas Betbug, a newer American Internet service, "is remarkably similar to file-sharing programs like Kazaa and Morpheus, which let people exchange music and other media over the Internet. Anyone downloading the Betbug software will be able to propose a wager, then reach out to everyone else on the network to find a taker for the bet." The Internet services make their money by receiving a small percentage of the winning bet. John O'Malia, "an American entrepreneur based in London," claims that he violates no American laws "because he is not acting as the sports bookmaker by setting the odds or participating in the wager."

"What becomes of government in an electronic revolution?" asks James K. Glassman, who asserts that "government's regulatory functions could weaken, or vanish. It's already a cinch on the Internet to get around the rules; censorship, telecommunications restrictions and patent laws are easily evaded. Even tax collection could become nearly impossible when all funds are transferred by electronic impulses that can be disguised." Glassman describes the cyberpunk science-fiction worlds of William Gibson, Bruce Sterling, and Neal Stephenson, in which the new information technologies prevent national governments from controlling the flow of money and information, thereby inevitably destroying them and transferring their power to other entities, such as multinational corporations and organized crime.

The apparently odd collocation of politics and pornography that appears so explicitly in China and Singapore turns out to be a common theme in the intertwined histories of information technology, democratization, and modernity. In fact, as Lynn Hunt has shown, "pornography as a regulatory category was invented as a response to the perceived menace of the democratization of culture . . . It was only when print culture opened the possibility of the masses gaining access to writing and pictures that pornography began to emerge as a separate genre of representation" (12–13). If one defines pornography as "explicit depiction of sexual organs and sexual practices with the aim of arousing sexual feelings," then it almost always appears accompanied by "something else until the middle or end of the eighteenth century. In early modern Europe, that is, between 1500 and 1800, pornography was most often a vehicle for using the shock of sex to criticize religious and political authorities" (10), and it was therefore linked "to freethinking and heresy, to science and natural philosophy, and to attacks on absolutist political authority" (11).

As these examples suggest, the World Wide Web and the Internet bring with them the threat and promise of democratized access to information—all sorts of information, not all of it savory or sane—but the degree to which information technology will change culture, government, and society very much remains an open question. If, as we have observed, the very slightest changes in technology (the size of a screen, the presence or absence of color, forms of linking) often have surprisingly major effects on the way we read, write, and think in e-space, then one cannot predict if governments will finally control the forms of hypertext we shall encounter, or if they will appear in forms that will prove too powerful for present conceptions of space, power, and the laws that shape them.

Access to the Text and the Author's Right (Copyright)

Access to a network implies access to texts "on" that network, and this access raises the issue of who has the right to have access to a text—access to read it as well as to link to it. Problems and possibilities come with the realization that authorship as it is conventionally understood is a convention. Conceptions of authorship relate importantly to whatever information technology currently prevails, and when that technology changes or shares its power with another, the cultural construction of authorship changes, too, for good or for ill.

A related problem concerns the fate of authorial rights. Michael Heim has pointed out that "as the model of the integrated private self of the author fades, the rights of the author as a persistent self-identity also become more evanescent, more difficult to define. If the work of an author no longer carries with it definite physical properties as a unique original, as a book in definite form, then the author's rights too grow more tenuous, more indistinct" (*Electric Language,* 221). If the author, like the text, becomes dispersed or multivocal, how does society fairly assign legal, commercial, and moral rights?

Before we can begin to answer such a question, we have to recognize that our print-based conceptions of authorial property and copyright even now do harm as well as good. They produce economically irrational effects, hindering as well as stimulating invention. Indeed, as James Boyle reminds us in his splendid book about law and the construction of an information society,

copyright is a fence to keep the public out as well as a scaffolding for the billboards displayed in the marketplace of ideas; it can be used to deny biographers the ability to quote from or to paraphrase letters; to silence parody; to control the packaging, context, and presentation of information. To say that copyright promotes the production and circulation of ideas is to state a conclusion and not an argument. At the very least we might wonder if, *in our particular copyright regime,* the gains outweigh the losses. (18–19)

Boyle forcefully argues that the author paradigm, which provides the center of copyright law and our current visions of intellectual property, "produces effects that are not only unjust, but unprofitable in the long term" (xiv), in part because it only rewards certain kinds of creation to the detriment of others.

Using the examples of the way Western scientists and corporations copyright materials based on information derived from communities in the Third World, he demonstrates how laws supposedly intended to promote innovation by rewarding creators recognize only creativity and originality based on romantic authorship.

Centuries of cultivation by Third World farmers produces wheat and rice strains with valuable qualities—in the resistance of disease, say, or in the ability to give good yields at high altitudes. The biologists, agronomists, and genetic engineers of a Western chemical company take samples of these strains and engineer them a little to add a greater resistance to fungus or a thinner husk . . . The chemical company's scientists fit the paradigm of authorship. The farmers are everything authors should not be—their contribution comes from a community rather than an individual, from tradition rather than innovation, from evolution rather than transformation. Guess who gets the copyright? Next year the farmers may need a license to resow the grain from their crops. (126)

In a situation marked by diametrically opposed conceptions of intellectual property, each side believes the other has stolen from it. Whereas countries like the United States and Japan, that base their conceptions of intellectual property on the author paradigm, accuse Third World nations of pirating their ideas, these countries in turn accuse the United States and Japan of stealing something that belongs to an entire community.

This situation appears particularly bizarre when viewed from the vantage point of American history, since, as Vincent Mosco points out, the United States "was the supreme intellectual property pirate of the nineteenth century" (47). It neither respected foreign copyright, such as that on Dickens's novels, nor gave copyright protection to foreigners unless they published first in the United States. In fact, "it was not until 1891, when the U.S. had a thriving publishing industry and literary culture of its own, that it extended copyright protection to foreign work." Mosco then asks the difficult question: "If, as most analysts admit, this was a key to successful national economic development then, why is it wrong for Mexico, India, Brazil, or China to follow this model now? What makes copying CDs in China theft, when copying *Great Expectations* in nineteenth-century America was deemed simply good business practice?" (47).

An even more crucial problem with copyright is that notions of intellectual property based on the author paradigm, which supposedly reward and hence stimulate originality, "can actually *restrict* debate and slow down innovation—by limiting the availability of the public domain to future users and speakers" (155). Those who write about intellectual property often point out that many corporations elect to rely on trade secrets rather than copyright law to protect their inventions, and, anyway, as Boyle urges, "innovators can recover their investment by methods other than intellectual property—packaging, reputation, being first to market, trading on knowledge of the more likely economic effects of the innovation, and so on" (140). If electronic information technology threatens to reconfigure our conceptions of intellectual property, we can take reassurance from several things, among them not only that our fundamentally problematic ideas of copyright often do not achieve what they are supposed to do but also that other means of rewarding innovation already exist.

As we have observed, one problem challenging print-based conceptions of intellectual property in an age of the digital word and image involves our changing understanding of authorship. A second problem concerning intellectual property derives from the nature of virtual textuality, any example of which by definition exists only as an easily copiable and modifiable version—

as a derivative of something else or as what Baudrillard would call a simulacrum. Traditional conceptions of literary property derive importantly from ideas of original creation, and these derive in turn from the existence of multiple copies of a printed text that is both fixed and unique. Electronic text processing changes, to varying degrees, all aspects of the text that had made conceptions of authorial property practicable and even possible. Heim correctly warns that an outmoded conception of "proprietary rights based on the possession of an original creation no longer permits us to adapt ourselves to a world where the technological basis of creative work makes copying easy and inevitable," and that to protect creativity we "must envision a wholly new order of creative ownership" (*Electric Language,* 170). But the problem we face, Boyle warns, is that our "author-vision" of copyright and intellectual property "downplays the importance of fair use and thus encourages an absolutist rather than a functional idea of intellectual property" (139).

As Steven W. Gilbert testified before a congressional committee, technology already both extends conventional conceptions of intellectual property and makes its protection difficult and even inconceivable:

It may soon be technically possible for any student, teacher, or researcher to have immediate electronic access from any location to retrieve and manipulate the full text (including pictures) of any book, sound recording, or computer program ever published—and more. When almost any kind of "information" in almost any medium can now be represented and processed with digital electronics, the range of things that can be considered "intellectual property" is mind-boggling. Perhaps the briefest statement of the need to redefine terms was made by Harlan Cleveland in the May/June 1989 issue of *Change* magazine: "How can 'intellectual property' be 'protected'? The question contains the seed of its own confusion: it's the wrong verb about the wrong noun." (16)

Attitudes toward the correct and incorrect use of a text written by someone else depend importantly on the medium in which that text appears. "To copy and circulate another man's book," H. J. Chaytor reminds us, "might be regarded as a meritorious action in the age of manuscript; in the age of print, such action results in law suits and damages."[17] From the point of view of the author of a print text, copying, virtual textuality, and hypertext linking must appear wrong. They infringe upon one person's property rights by appropriating and manipulating something over which another person has no proper rights. In contrast, from the point of view of the author of hypertext, for whom collaboration and sharing are of the essence of "writing," restrictions on the availability of text, like prohibitions against copying or linking, appear absurd, indeed immoral, constraints. In fact, without far more access to (originally) print text than is now possible, true networked hypertextuality cannot come into being.

Difficult as it may be to recognize from our position in the midst of transition from print to electronic writing, "it is an asset of the new technology," Gilbert reminds us, "not a defect, that permits users to make and modify copies of information of all kinds—easily, cheaply, and accurately. This is one of the fundamental powers of this technology and it cannot be repressed" (18). Therefore, one of the prime requisites for developing a fully empowering hypertextuality is to improve, not technology, but laws concerning copyright and authorial property. Otherwise, as Meyrowitz warned, copyrights will "replace ambulances as the things that lawyers chase" (24). We do need copyright laws protecting intellectual property, and we shall need them for the foreseeable future. Without copyright, society as a whole suffers, for without such protection authors receive little encouragement to publish their work. Without copyright protection they cannot profit from their work, or they can profit from it only by returning to an aristocratic patronage system. Too rigid copyright and patent law, on the other hand, also harms society by permitting individuals to restrict the flow of information that can benefit large numbers of people.

Hypertext demands new classes or conceptions of copyright that protect the rights of the author while permitting others to link to that author's text. Hypertext, in other words, requires a new balancing of rights belonging to those entities whom we can describe variously as primary versus secondary authors, authors versus reader-authors, or authors versus linkers. Although no one should have the right to modify or appropriate another's text any more than one does now, hypertext reader-authors should be able to link their own texts or those by a third author to a text created by someone else, and they should also be able to copyright their own link sets should they wish to do so. A crucial component in the coming financial and legal reconception of authorship involves developing schemes for equitable royalties or some other form of payment to authors. We need, first of all, to develop some sort of usage fee, perhaps of the kind that ASCAP levies when radio stations transmit recorded music; each time a composition is broadcast the copyright owner earns a minute sum that adds up as many "users" employ the same information—an apposite model, it would seem, for using electronic information technology on electronic networks.

Gilbert warns us that we must work to formulate new conceptions of copyright and fair use, since "under the present legal and economic conventions, easy use of the widest range of information and related services may become available only to individuals affiliated with a few large universities or corporations" (14). Thus, dividing the world into the informationally rich and informationally impoverished, one may add, would produce a kind of techno-feudalism in which those with access to information and information technology would rule the world from electronic fiefdoms. William Gibson, John Shirley, and other practitioners of cyberpunk science fiction have convincingly painted pictures of a grim future, much like that in the movie *Blade Runner,* in which giant multinational corporations have real power and governments play with the scraps left over. Now is the time to protect ourselves from such a future. Like many others concerned with the future of education and electronic information technology, Gilbert therefore urges that we must develop *"new economic mechanisms to democratize the use of information, and economic mechanisms beyond copyright and patent.* It would be a tragedy if the technology that offers the greatest hope for democratizing information became the mechanism for withholding it. We must make information accessible to those who need it . . . Any pattern that resembles information disenfranchisement of the masses will become more obviously socially and politically unacceptable" (17–18).

Most of the discussions of copyright in the electronic age that I have encountered recently fall into two sharply opposing camps. Those people, like Gilbert, who consider issues of authorial property from the vantage point of the hypertext reader or user of electronic text and data emphasize the need for access to them and want to work out some kind of equitable means of assigning rights, payment, and protection to all parties. Their main concern, nonetheless, falls on rights of access. Others, mostly representatives of publishers, often representatives of university presses, fiercely resist any questioning of conventional notions of authorship, intellectual property, and copyright as if their livelihoods depended on such resistance, as indeed they well might. They argue that they only wish to protect authors and that without the system of refereed works that controls almost all access to publication by university presses, standards would plummet, scholarship would grind to a halt, and authors would not benefit financially as they do now. These arguments have great power, but it must be noted that commercial presses, which do not always use referees, have published particularly important scholarly contributions and that even the most prestigious presses invite thesis advisors to read the work of their own students or have scholars evaluate the manuscripts of their close friends. Nonetheless, publishers do make an important point when they claim that they fulfill an important role by vetting and then distributing books, and one would expect them to retain such roles even when their authors begin to publish their texts on networks.

Although almost all defenses of present versions of copyright I have encountered clearly use the rights of the author or society in large part as a screen to defend commercial interests, one issue, that of the author's moral rights, is rarely discussed, certainly not by publishers. As John Sutherland explains in "Author's Rights and Transatlantic Differences," Anglo-American law treats copyright solely in terms of property. "Continental Europe by contrast enshrines moral right by statute. In France and West Germany the author has the right to withdraw his or her work after it has been (legally) published—something that would be impossible in Britain or the United States without the consent of the publisher . . . In [France and West Germany], publishers who acquire rights to the literary work do not 'own it,' as do their Anglo-American counterparts. They merely acquire the right to 'exploit' it."[18]

The occasion for Sutherland's article raises important questions about rights of the hypertext as well as the print author. In 1985 an American historian, Francis R. Nicosia, published *The Third Reich and the Palestine Question* with the University of Texas Press, which subsequently sold translation rights to Duffel-Verlag, a Neo-Nazi publisher whose director "is (according to Nicosia) identified by the West German Interior Ministry as the publisher of the *Deutscher Monatschefte,* a publication that, among other matters, has talked about 'a coming Fourth Reich in which there will be no place for anti-Fascists. The path to self-discovery for the German people will be over the ruins of the concentration camp memorials.'" Believing that an association with Duffel-Verlag will damage his personal and professional reputation, the author has complained vigorously about his American publisher's treatment of his book. Traditional Anglo-American law permits the author no recourse in such situations, but Sutherland points out that "on October 31, 1988, Ronald Reagan signed into law America's ratification of the Berne Convention," which grants the author moral rights including that which prevents a publisher from acting in ways "prejudicial to his honour and reputation."

The question arises, would an author whose text appears on a hypertext environment, such as the Web, find that text protected more or less than a comparable print author? At first glance, one might think that Nicosia would find himself with even fewer rights if his work appeared as a hypertext, since anyone, including advocates of a Fourth Reich, could link comments and longer texts to *The Third Reich and the Palestine Question.* In 1991 when I was thinking in terms of full read-write systems like Intermedia, I wrote that such an answer is incorrect for two reasons. First, in its hypertext version Nicosia's monograph would not appear isolated from its context in the way its print versions does. Second (and this is really a restatement of my previous point), a read-write hypertext version would permit Nicosia to append his objections and any other materials he wished to include. Linking, in other words, has the capacity to protect the author and his work in a way impossible with printed volumes. Allowing others to link to one's text therefore does not sacrifice the author's moral rights. The problem with this response is that since we've ended up with the current World Wide Web, my original argument turns out to be of only theoretical value. The one way on the Web that an author could state (publish?) his objections to what someone else has done to his work involves creating a site in which he explains them. Since those interested in the relationship of Nazi Germany to the Middle East who are reading a Web version would likely search the Internet, his response would appear in search results very close to his book.

On the Web, links can also associate one's work with unexpected or unwanted materials. Exploring focus.com, an online museum of experimental digital photography, which includes one of my images, I followed links to the sites of other photographers, looked at galleries of their work, and then clicked on a link for recommended sites, which brought me to a one called "CNPN-Best Erotic Nudes—Nude Gallery—Fine Art Nude Photography," which contains links to dozens of photographers' sites, including that of Kim Weston, grandson of the great master of landscape and the nude. What I found jarring, however, was that above the list of art photographers appeared links to hardcore pornographic sites. In all fairness I have to mention that all the art photographers listed appear against a blue background, whereas all the porno ads appear in black boxes (except for some in sidebars). Checking the supposed art sites, I found that they in fact belonged to an international group of photographers who did the kind of work one finds in galleries and museums. These sites, almost all of which were extremely elegant, included artists' biographies, lists of exhibitions, and statements about their work as well as Web galleries containing selections from it. Although some of the images I encountered could be described as erotic, a large number were experimental, highly abstract, or emphasized landscape settings—hardly the kind of material that would interest anyone seeking pornography. Curious about how such professional work ended up juxtaposed to hardcore pornography, I e-mailed more than a dozen photographers. In response, one replied that he did not know that CNPN listed him, but most of them responded that when they agreed to be listed on CNPN the site contained no links to pornography. Some were very angry or disappointed and removed return links on their sites to CNPN, others were resigned. Hans Molnar, a German photographer who had joined in earlier days, pointed out that "no one is in control of the internet and can dictate where links end up . . . Do a search with any search engine [and] porn links are also listed amongst real fine art nude photography, so the only way to avoid the whole thing [is] just don't have a web site and remain unknown."

The one photographer who asked to join the list after CNPN added prominent links to hardcore for-pay sites explained that he felt forced to do so in order to publicize his work because apparently more relevant (and more respectable) sites refused to include his work because they classified nudes with pornography! The obvious power of Internet portals to censor the sites they choose shows one side of the Internet. Students of photography can of course still search for "nude photography," but many will be put off when they encounter a document that places advertisements for sexually explicit sites first—or they will follow links to them before those of the art photographers. Here is a case where censorship falls prey to the law of unintended circumstances.

When considering the implications of Internet linking for these art photographers, I pointed out that "on the Web, links can also associate one's work with unexpected or unwanted materials." A far more disturbing fact is that on the Web, other people can add links to your documents! When applied to the World Wide Web, the open hypermedia systems described in chapter 1 permit others to add links to any document. Of course, only users with access to a website with a server that has the requisite link services can read your documents with links inserted by someone else, but such websites can be giant Internet portals like Yahoo, or they can be websites maintained by political parties, militant groups, NGOs, or individuals anywhere on the political spectrum. Any person or organization using Active Navigation's Portal Maximizer has the power to place links in product advertisements, proposed legislation, political speeches, educational materials, newspaper articles, and scientific and scholarly writing. Imagine political opponents annotating each other's speeches, or Holocaust memorialists and Holocaust deniers annotating each other's sites. Of course, the original Web document remains unaffected— these are virtual documents, remember—but search engines like Google can rate the annotated document higher than the original one if enough people read it and link to it. True, websites using Portal Maximizer or similar software must be open to public access, or they could not influence large numbers of readers, but password-protected sites have their own danger: original Web authors have no way of knowing that their writings have been annotated, and hence they cannot respond. A final "of course": readers do not have to follow the inserted links. Still, the ability to link unmoderated commentary to another author's text markedly reduces the authority of that author, and if only one party has access to a tool like Portal Maximizer, the original author is at a great disadvantage. One can view such technology as potentially democratizing, or, if it is only available to a few, as potentially dangerous, which raises the question,

Is the Hypertextual World of the Internet Anarchy or Big Brother's Realm?

At the present moment, it shows the potential to be both, perhaps even at the same time. In countries like China, Singapore, and Zimbabwe, which have a history of Internet censorship and surveillance, Big Brother already seems present, and as technology develops it might allow ways to either thwart him or produce the means that ensure he cannot be thwarted. At the moment, throughout most of the world information anarchy seems to reign, at least according to those who would rein it in: anyone with Google or other search tools can locate Michael Moore's attacks on George W. Bush, multiple treasure troves of literature, denials of the Holocaust, health information, egotistical ramblings of twenty-five-year-olds convinced that everyone should care about their daily lives, underground fiction of all kinds, the anti-Americanism of *Baghdad is Burning*, detailed computer information of a high professional standard, maps to museums and restaurants, and means of purchasing almost anything one does and doesn't want.

We face two great dangers, as many commentators have long pointed out, the first of which is that the best information, the finest art, and the most valuable new ideas will be swamped by the sheer mass of material—something analogous to the supposedly terrible effects of the explosion of cheap reading enabled by high-speed printing that, Cardinal Newman complained, was destroying *real* culture, or the way the Jet Propulsion Lab (JPL) finds itself swamped by tapes of messages sent back from unmanned missions throughout the solar system that will almost certainly decay before they have been deciphered. The other contrary possibility is that the newest versions of data mining and computer-based surveillance will permit those with control of *the machines* total control over all information and the people who read, write, and exchange it. Some, like the opponents of Gmail, claim the systems are already in place for total surveillance, but the experience of the JPL and the recent failures of U.S. intelligence suggest that such is not the case. I certainly don't have the answers, but, as I write, now that the rain has stopped and the sun has come out in Providence at 6:07 p.m. on July 19, 2004, I have enough hope to believe that the libratory potential of hypermedia will enable good things to happen. I could be wrong.

Notes

Chapter 1. Hypertext

1. An important caveat: here, right at the beginning, let me assure my readers that although I demonstrate that Barthes and Derrida relate in interesting and important ways to computer hypertext, I do not take them—or semiotics, poststructuralism, or, for that matter, structuralism—to be essentially the same.

2. In fact, some of the most exciting student projects and published examples of hypermedia take the form of testing, applying, or critiquing specific points of theory, including notions of the author, text, and multivocality. Cicero Ignacio da Silva's *Plato On-line: Nothing, Science and Technology* (2003–4) exemplifies a particularly carnivalesque, rambunctious experiment with conventional attitudes toward authorship and its relation to conceptions of a work. The Brazilian scholar explains in *Plato On-line*, which has no pagination, that "in order to test my hypothesis that there is no work without a 'signature,' and there is no 'safe' means to authenticate the signature of a text and in a text on the internet," he created "hundreds" of websites for fictional research institutes, scientific journals, and survey centers "hosted by free-of-charge providers (geocities, tripod, among others)" upon which he placed computer-generated texts created by a combination of "PERL and Java Script programming" from "fragments of text from the internet." Each text is signed with "Algorithm [author's name]," such as "Algorithm Giles Deleuze," and the resultant text is "purposefully unstructured and rarely makes any sense." All the texts he keeps on the Internet appear in Portuguese, which Babelfish then translates into English, French, German, and Japanese. Finally, *Plato On-line* makes the element of spoof quite clear when it announces that it is "a serious journal interested only in publishing texts written by electric generators. This magazine does not have the intention to publish anything that makes sense . . . The names of the authors are not true and all the names are not from authors who exist [but from] programmed algorithms." Nonetheless, da Silva has discovered that readers persist in submitting "articles, reviews on articles, and comments on the texts, etc." Moreover, despite the fact that

his computer-generated texts signed with a clearly suspicious-sounding name do not make sense, he has found that readers take them seriously enough to quote them in both blogs and scholarly work, such as graduate theses. The presence of what da Silva calls a signature—a name similar to that of an established author—convinces readers that they are reading a genuine text, even if it does not make grammatical and other sense. (I would add that the appearance of these jumbled texts on sites that supposedly represent serious-sounding, if fictional, institutions also convinces people that authorship and text are genuine.)

3. Although the following pages examine some aspects of the history of hypertext theory, they do not provide a history of earlier pioneering systems, such as NLS, Augment, HES, FRESS, Guide, and Hyperties, and later developments, since valuable basic surveys can be found in Nielsen, *Multimedia and Hypertext* and Hall, Davis, and Hutchings, *Rethinking Hypermedia*, 11–32.

4. A second important caveat: by hypertext I mean only one of at least five possible forms of the digital word. In addition to hypertext, there are four other important kinds of electronic textuality, each of which can exist within hypertext environments, though not itself hypertextual:

1. Graphic representations of text. Using computer graphics to represent text produces images of it that cannot be searched, parsed, or otherwise manipulated linguistically. The resulting images can be animated, made to change in size, accompanied by sound, and so on. This kind of e-text, which is familiar from television advertising, is often created using Macromedia Director and Flash.

2. Simple alphanumeric digital text. This form of electronic text, which functions linguistically, appears in electronic mail, bulletin boards, and word-processing environments.

3. Nonlinear text. In contrast to hypertext, which enables multisequential reading, this form is best thought of as nonlinear. According to Espen Aarseth (whose "Nonlinearity and Literary Theory," in *Hyper/Text/Theory*, ed. Landow, provides the essential discussion of its subject) the various forms of nonlinear textuality include (a) computer games, (b) text-based collaborative environments, such as Multi-User Domains (MUDs) and Multi-User Domains that employ Object-Oriented programming methods (MOOs), and (c) cybertext, or text generated on the fly. See essays by Carreño, Donguy, Lenoble, Vuillemin, and Balpe in *A:\ Littérature: Colloque Nord Poésie et Ordinateur*. See Meyer, Blair, and Hader for a MOO for the World Wide Web.

4. Simulation. Text in simulation environments can range from computationally produced alphanumeric text (and hence have much in common with the nonlinear form) to instances of fully immersive virtual (or artificial) reality. For discussions of the educational use of such simulation environments within electronic books, see my "Twenty Minutes into the Future, or How Are We Moving beyond the Book?" For general discussions of virtual reality, see Benedikt, ed., *Cyberspace*; Heim, *The Metaphysics of Virtual Reality*; Earnshaw, Gigante, and Jones, eds., *Virtual Reality Systems*; and Wexelblat, *Virtual Reality*.

5. A third (and last) caveat: as I pointed out in the introduction to *Hyper/Text/ Theory*, some hypertext environments, which are not chiefly text- or image-based, employ logical and conceptual links as a means of assisting organization, collaborative work, and decision making. Systems like Xerox PARC's Acquanet and IDE thus far have appealed to workers in computer and cognitive science investigating the business applications of information technology. For Acquanet, see the articles by Catherine C. Marshall listed in the bibliography; for IDE see those by Daniel Russell. Clara Mancini's 2003 doctoral dissertation includes a brief summary with screen shots of various systems of semantic hypertext.

6. The developers of Microcosm, currently the most advanced hypertext system yet developed, similarly argue: "*There should be no artificial distinction between author and reader.* Many systems have an authoring mode and a reader mode; such a system is not open from the reader's point of view. We believe that all users should have access to all parts of the system; this does not imply that one user will be able to access or change another's data, but implies that this aspect should be controlled by the granted rights of access to the operating system. Users should be able to create their own links and nodes within their private workspace, then change the access rights so that other users may view or edit them as required" (Hall, Davis, and Hutchings, *Rethinking Hypermedia*, 30).

7. The original text here read, "Intermedia, the hypertext system with which I work," but shortly after the print publication of *Hypertext*, my students and I found ourselves forced to use several other systems after Apple Computers, which had funded a portion of the project, fundamentally changed its version of UNIX, thus halting development—and eventually even the use—of Intermedia. Two fully illustrated articles describe IRIS Intermedia in detail: Yankelovich, Meyrowitz, and Drucker, "Intermedia"; and Bernard J. Haan, Kahn, Riley, Coombs, and Meyrowitz. "IRIS Hypermedia Services." The Intermedia section of my *Cyberspace, Hypertext, and Critical Theory* website, which contains a detailed introduction to the system with many screenshots, can be found at http://www.cyberartsweb.orgt/ht/HTatBrown/Intermedia.html. This URL also provides information about obtaining Paul Kahn's archival video, *Intermedia: A Retrospective*, from the Association of Computing Machinery.

8. One could make the same point about contributors to discussion lists, but since these lists are intended to take the form of group discussions, new contributions don't seem unusual, and one experiences what seems a very different form of collaboration.

9. Writers have offered other classifications of links, often in terms of binary oppositions. Thus in 1988, Paul Kahn compared objective to subjective links, an opposition chiefly relevant to so-called legacy text—text, that is, translated into hypertext from print or other paper presentation. According to Kahn, footnotes and cross references represent objective links, because they are present in the original text structure, whereas subjective links are added by the person translating the document into hypertextual form. Kahn's objective versus subjective links appear closely related to Anna Gunder's analog and digital links ("Aspects of Linkology," 112–13). Gunder also distinguishes between internal and external links: "Links within a work are called *internal* links while links running between works are labeled *external* links" (113).

10. Such media reversals continue today, though for different reasons: Geert Lovink's *Dark Fiber: Tracking Critical Internet Culture*, which MIT Press published in 2002, reproduces essays that appeared on various Internet discussion groups between 1995 and 2001. In this case, the characteristic qualities of networked digital text discussed later in this section have been exchanged for the relative fixity and stability of print.

11. At conferences I've several times found myself defending Bolter and Grushin's valuable idea of remediation from charges that it is too simple or limiting. Markku Eskelinen and Raine Koskimaa, for example, claim that "the concept of remediation carries worrying stabilizing effects with it. Whatever new form, mode or medium there is, there's no time to study it and build a decent scholarship around it, as we supposed to be immediately stuck with remediating it" (9). I don't see how pointing out that various information technologies remediate one another has any limiting effects, and Bolter and Grushin's emphasis that we have to consider the place of any particular form of IT (such as hypertext) within a media ecology, strikes me as an essential place to begin, in large part because it goes a long way toward preventing misunderstandings about supposed total oppositions of earlier and later technologies (such as, say, print and hypertext).

12. *In Writing Space,* Bolter explains some of these costs: "Electronic text is the first text in which the elements of meaning, of structure, and of visual display are fundamentally unstable. Unlike the printing press, or the medieval codex, the computer does not require that any aspect of writing be determined in advance for the whole life of a text. This restlessness is inherent in a technology that records information by collecting for fractions of a second evanescent electrons at tiny junctions of silicon and metal. All information, all data, in the computer world is a kind of controlled movement, and so the natural inclination of computer writing is to change" (31).

13. Terry Eagleton's explanation of the way ideology relates the individual to his or her society bears an uncanny resemblance to the conception of the virtual machine in computing: "It is as though society were not just an impersonal structure to me, but a 'subject' which 'addresses' me personally—which recognizes me, tells me that I am valued, and so makes me by that very act into a free, autonomous subject. I come to feel, not exactly as though the world exists for me alone, but as though it is significantly 'centred' on me, and I in turn am significantly 'centred' on it. Ideology, for Althusser, is the set of beliefs and practices which does this centring" (*Literary Theory,* 172).

14. Marie-Laure Ryan's *Narrative and Virtual Reality,* which provides a valuable discussion of virtuality with specific emphasis on its relation to immersion (25–47), suggests "three distinct senses of *virtual:* an optical one (the virtual as illusion), a scholastic one (the virtual as potentiality), and an informal technological one (the virtual as computer mediated)" (13).

15. Hayles's demand that we recognize the importance of embodiment and materiality in a digital age derives from her recognition of the absurdity of some postmodern claims: "Every epoch," she points out, "has beliefs, widely accepted by contemporaries, that appear fantastic to later generations . . . One contemporary belief likely to stupefy future generations is the postmodern orthodoxy that the body is

primarily, if not entirely, a linguistic and discursive formation . . . Although researchers in the physical and human sciences acknowledged the importance of materiality in different ways, they nevertheless collaborated in creating the postmodern ideology that the body's materiality is secondary to the logical or semiotic structures it encodes" (192). Compare J. David Bolter and Diane Gromola's discussions of "the myths of disembodiment" in *Windows and Mirrors*, 117–23.

16. Mitchell wittily narrates the evolution of computers (rather than monitors or displays) from the vantage point of an architect-designer: "Mainframes were designed as large-scale items of industrial equipment, and at their best—in the hands of Charles Eames, for example—achieved a tough, hard-edged, machine-age clarity of form. They were often put on display in special, glass-enclosed rooms. The bulky computer workstations of the 1970s and 1980s were medium-scaled wheeled furniture—not too different from writing desks, pianos, and treadle sewing machines, but styled for laboratory rather than domestic environments. PCs evolved from clumsy beige boxes to sleekly specialized, various colored and shaped versions for offices, classrooms, and homes. Now that they are fading into history, after a life of approximately twenty years, they look increasingly like surrealist constructions—the chance encounter of a typewriter and a television on a desktop. Portables started out mimicking luggage (right down to the handles and snaps), then appropriated the imagery of books that could open, close, and slip into a briefcase" (*Me++*, 70–71).

17. Mitchell points out that the effect on work-practice of such location-independent information has turned out differently than many predicted: "The emerging, characteristic pattern of twenty-first-century work is not that of telecommuting, as many futurists had once confidently predicted; it is that of the mobile worker who appropriates multiple, diverse sites as workplaces" (153).

18. This brings up the entire subject of computer humor and parody, often directed at Microsoft products. Anyone who's found annoying the Microsoft Office Assistant in earlier versions of Word, which pops up with the intrusive statement that you seem to be writing a letter and asks if you want help, will appreciate Dave Deckert's parody: one encounters what appears to be a screenshot of a document from an earlier version of Microsoft Word (5.1?), in which a user has typed "Dear World, I just can't take it anymore. I've decided"—at which point a cartoon image of dancing paperclip pops up on the screen accompanied by the message "Looks like you're committing suicide," followed below by the text "Office Assistant can help you write a suicide note. First, tell us how you plan to kill yourself." This text appears above two rows of buttons, the top one of which offers the options "Pills," "Jump," "Pastry," and the bottom row has "Tips," "Options," and "Close" (dgd-filt@visar.com, 2000). Another parody, apparently by a British user, mocks both the instability of the Microsoft Windows operating system and its often unexpected hidden settings. On a panel labeled "Hidden Settings (Not to be edited)," one discovers a series of options that purports to explain difficulties users encounter every day. The first line has a box containing a check next to "Crash every 2 Hours," the "2" and "Hours" appearing within option boxes, and the following lines contain in similar format the instruction to crash after 5000 "bytes of un-saved changes." Other factory-set options include those for "Save," which produces "incredibly large files" and Auto Recovery

that "takes Bloody Ages." The final factory-set option involves "Annoy me with the sodding paper clip" either constantly or "<u>w</u>hen I least expect it."

Cartoons published worldwide, which show how much computing has become part of our everyday lives, similarly present users' attitudes toward personal computers. In a brilliant four-panel *Doonesbury* cartoon, Gerry Trudeau conveys the frustrations of people who installed Windows 95. In the first panel, which shows the communal nature of personal computing by so-called early adaptors, Mike approaches two co-workers, one of whom is seated at a PC and is told, "We're loading in the new Windows 95 operating system," and when he asks in the next panel how it's going, the bearded, bespectacled man seated at the computer replies, "Don't know yet. I'm still trying to clear enough memory for it." In the third panel, in which the three men appear in white silhouette against black background, we receive the software installer's message: "Attention User: You call this capacity? Reboot when you're ready to play."—a fine parody of the error messages those trying to install Windows 95 on older machines often received! The final panel effectively dramatizes the way users came to fear both their PCs and the company that created their operating systems as the man seated at the computer exclaims, "Son of a . . . It's dissing my hard drive!" only to be cautioned by the man behind him, "Back off, Hank. Don't want to make it lose face . . ." Yet other cartoons satirize Microsoft's monopolistic practices. In Bill Arend's *Foxtrot* the older of two brothers comes upon his sibling sitting at a computer "reading about a big Windows source code leak," 600 Mb of which are "all over the internet." In the third panel the younger brother points out that people probably have already guessed "some of what's in it," after which the final panel shows the parody code onscreen:

```
        BEGIN
        IF browser_type=
        "Internet_Explorer"
THEN    smooth.sailingELSE
        IF (browser_type=
        "Netscape") AND
        "justice_department NOT looking)
THEN
        REPEAT
        Crash (random)
```

Computer cartoons have many other subjects, including crashes that destroy home and office work, the youth of skilled computer users, overblown claims about the World Wide Web, annoying animated graphics, and suggestions that the devil invented computing—or at least is a heavy user: an Italian cartoon of the 1980s shows a devil seated at a computer terminal in Hades. Some parodies mock the user's expectations more than they satirize software manufacturers' products. In another parody that presents a fantasy version of Microsoft Word, the drop-down menu labeled "Tools" contains the following options: "<u>U</u>ndo stupid changes," "Take Back <u>F</u>lippant Comment," "<u>C</u>reate Brilliant Idea," "<u>E</u>xtend Deadline," "Read <u>B</u>osses' Minds," "<u>T</u>erminate Smart-ass IT Technician," "<u>I</u>ncrease Salary," "<u>R</u>eclaim Wasted

Evenings," "Extend Weekend," and, finally, "Find Perfect Mate." This parody, which says more about Microsoft users than about the company, suggests that the cyberspace myth and the dotcom crash derive in large part from our secret desires that computers make our lives better without much effort on our part.

19. Janet Murray asserts the importance of agency in true interactivity: "Because of the vague and pervasive use of the term *interactivity*, the pleasure of agency in electronic environments is often confused with the mere ability to move a joystick or click on a mouse. But activity alone is not agency . . . As an aesthetic pleasure, as an experience to be savored for its own sake, it is . . . more commonly available in the structured activities we call games" (128–29).

20. Maurie-Laure Ryan offers a critique of Baudrillard from another vantage point (31–25).

21. Scott Blake's *Bar Code Jesus* (1999) plays interestingly with computer-related codes as the basis of a visual reality composed of the images we see on a computer screen. In this piece, Blake manipulates the ubiquitous bar code (as opposed to the far "deeper" machine code) to take us in stages from a recognizable image to the codes that produce it. The viewer first encounters a fairly low-resolution image of the face of Jesus, above which appears a panel that permits the viewer to zoom in seven states or stages, enlarging a portion of the image in its frame each time. Diving into the image with the control panel transforms it from a recognizable face to three successive images that resemble mosaic until, at the fifth level, one arrives at barcodes. The next two zooms resolve the image barcodes until the viewer arrives at one-inch-high vertical lines (bars) and the number associated with each. In an animated version, the zooming in and out occurs at a dizzying pace. The playfulness of the project appears in the fact that these barcodes would not actually produce an image when read by a computer; Blake is just using their visual appearance as building blocks.

22. Chartier, *The Culture of Print*, 139. Chartier bases his remarks in part on Marie-Elizabeth Ducreux, "Reading unto Death: Books and Readers in Eighteenth-Century Bohemia," also in *The Culture of Print*, 191–230.

Chapter 2. Hypertext and Critical Theory

1. I am thinking of Richard Rorty's description in *Philosophy and the Mirror of Nature*, 378, of edifying philosophy as a conversation: "To see keeping a conversation going as a sufficient aim of philosophy, to see wisdom as consisting in the ability to sustain a conversation, is to see human beings as generators of new descriptions rather than beings one hopes to be able to describe accurately. To see the aim of philosophy as truth—namely, the truth about the terms which provide ultimate commensuration for all human inquiries and activities—is to see human beings as objects rather than subjects, as existing en-soi rather than as both pour-soi and en-soi, as both described objects and describing subjects." To a large extent, Rorty can be thought of as the philosopher of hypertextuality.

2. Examples include GodSpeed Instant Bible Search Program from Kingdom Age Software in San Diego, California, and the Dallas Seminary CD-Word Project, which builds upon Guide™, a hypertext system developed by OWL International

(Office Workstations Limited). See Steven J. DeRose, "Biblical Studies and Hypertext," in *Hypermedia and Literary Studies*, ed. Delany and Landow, 185–204.

3. Borges, "The Aleph," in *The Aleph and Other Stories*, 13: "In that single gigantic instant I saw millions of acts both delightful and awful; not one of them amazed me more than the fact that all of them occupied the same point in space, without overlapping or transparency. What my eyes beheld was simultaneous, but what I shall now write down will be successive, because language is successive . . . The Aleph's diameter was probably little more than an inch, but all space was there, actual and undiminished. Each thing (a mirror's face, let us say) was infinite things, since I saw it from every angle of the universe."

4. For a description of early networks that preceded the Internet, see LaQuey, "Networks for Academics." For a description of the proposed National Research and Education Network, see Gore, "Remarks on the NREN"; and Rogers, "Educational Applications of the NREN."

5. Gregory L. Ulmer pointed this fact out to me during our conversations at the October 1989 Literacy Online conference at the University of Alabama in Tuscaloosa.

Chapter 3. Reconfiguring the Text

1. In fact, a primitive form of hypertext appears whenever one places an electronic text on a system that has capacities for full-text retrieval or a built-in reference device, such as a dictionary or thesaurus. For example, I wrote the manuscript of the first version of the book you are reading on an Apple Macintosh II, using a word-processing program called Microsoft Word; my machine also ran On Location, a program that quickly located all occurrences of an individual word or phrase, provided a list of them, and, when requested, opened documents containing them. Although somewhat clumsier than an advanced hypertext system, this software provides the functional analogue to some aspects of hypertext.

2. When I first used *intratextuality* in an article some years ago to refer to such referential and reverberatory relations within a text, or within a metatext conceived as a "work," I mistakenly believed I had coined the term. So did my editor, who was not enthusiastic about the coinage. But we were both wrong: Tzvetan Todorov used it in "How to Read" (1969), which appears in *The Poetics of Prose*, 242.

3. IBM mainframe computers running the CMS operating system call each user's electronic mailbox or message center the "reader."

4. To indicate the presence of one or more links, Intermedia placed a link marker, which took the form of a small horizontal rectangle containing an arrow, at the beginning of a passage. Apple's HyperCard permitted a wide range of graphic symbols ("buttons") to indicate the unidirectional links that characterize this program. *CD Word*, which was based on an amplification of Guide, employed an ingenious combination of cursor shapes to indicate linked material. For example, if one moved the cursor over a word and the cursor changed into a horizontal outline of an arrow, one knew the cursor was on a reference button, and clicking the mouse would produce the linked text. Following this procedure on the title page and clicking the cursor when on *Bibles* produced a list of abbreviations that included versions of the scriptures. Then, moving the cursor over RSV changed it to a crosshair shape, which in-

dicated the presence of a replacement button; clicking the mouse button produced the phrase "Revised Standard Version."

5. Bolter, *Writing Space*, 63–81, provides an excellent survey of visual elements in writing technologies from hieroglyphics to hypertext. The periodical *Visible Language*, which has appeared since 1966, contains discussions of this subject from a wide variety of disciplines, ranging from the history of calligraphy and educational psychology to book design and human/computer interaction.

6. In discussing Barthes's *Elements of Semiology*, Lavers exemplifies the usual attitude toward nonalphanumeric information when she writes that Barthes's notion of narrative "acknowledges the fact that literature is not only 'made of words' but also of representational elements, although the latter can of course only be conveyed in words" (134). That pregnant "of course" exposes conventional assumptions about textuality.

7. Geert Lovink caustically complains: "Interaction design seems to have lost its battle against interface stupidity. The office metaphor of the previous decade has been exchanged for an adaptation of the newspaper front page outlook as the dominant information architecture" ("Cyberculture in the Dotcom Age," in *Dark Fiber*, 334)

8. These pieces greatly resemble the student projects in Macromedia Director carried out at the Rhode School of Design in the mid-1980s in digital typography courses conducted by Krystoff Lenk and Paul Kahn. These projects, which I have discussed elsewhere, take the form of animating the texts of poems by Berthold Brecht and Mary Oliver, so that lines move across the screen, appear and disappear, in ways that perform the poem. Occasionally, sound was added to the text as well.

9. One of the most important pioneering discussions of the importance of fixity in print culture is McLuhan's *Gutenberg Galaxy*. See also Eisenstein, *Printing Press;* and Bolter, *Writing Space*.

10. These paragraphs are directly inspired by Noah Wardop-Fruin's eloquent talk at Brown University's *E-fest* (April 2004), reminding us that Nelson's stretchtext demonstrates he does not limit hypertext to that created by links.

Chapter 4. Reconfiguring the Author

1. For a discussion of to what degree hypermedia in both read-only and read-write forms does or does not empower readers, see chapter 8.

2. Marie-Laure Ryan makes some properly forceful observations about extreme claims that hypertext makes readers into writers: "To the skeptical observer, the accession of the reader to the role of writer . . . is a self-serving metaphor that presents hypertext as a magic elixir: 'Read me, and you will receive the gift of literary creativity.' If taken literally—but who really does so?—the idea would reduce writing to summoning words to the screen though an activity as one, two, three, click . . . Call this writing if you wish; but if working one's way through the maze of an interactive text is suddenly called writing, we will need a new word for retrieving words from one's mind to encode meanings" (9). The context of this astute warning makes clear that Ryan mistakenly includes me among critics who believe in the complete merging of reader and writer. As the complete sentence she quotes makes clear, the phrase she emphasizes with italics—"*of ourselves* as authors"—refers to the way

linking changes the author's conception of his or her power and authority. In fact, the sentence implies a distinction between readers and authors.

3. See the final sections of chapter 8 for a discussion of the political implications of open hypermedia applications for the Web.

4. Lévi-Strauss's observation in a note on the same page of *The Raw and the Cooked* (12) that "the Ojibiwa Indians consider myths as 'conscious beings, with powers of thought and action'" has some interesting parallels to remarks by Pagels on the subject of quasi-animate portions of neural nets: "Networks don't quite so much compute a solution as they settle into it, much as we subjectively experience our own problem solving . . . There could be subsystems within supersystems—a hierarchy of information and command, resembling nothing so much as human society itself. In this image the neuron in the brain is like an individual in society. What we experience as consciousness is the 'social consciousness' of our neuronal network" (126, 224).

5. Lévi-Strauss also employs this model for societies as a whole: "Our society, a particular instance in a much vaster family of societies, depends, like all others, for its coherence and its very existence on a network—grown infinitely unstable and complicated among us—of ties between consanguineal families" (*Scope of Anthropology,* 33).

6. Said in fact prefaces this remark by the evasive phrase, "it is quite possible to argue," and since he nowhere qualifies the statement that follows, I take it as a claim, no matter how nervous or half-hearted.

7. I originally wrote in 1991 that Heim would be correct only "in some bizarrely inefficient dystopic future sense—'future' because today [1991] few people writing with word processors participate very frequently in the lesser versions of such information networks that already exist, and 'bizarrely inefficient' because one would have to assume that the billions and billions of words we would write would all have equal ability to clutter the major resource that such networks will be." The reason for Heim's prescience comes, as we shall observe in chapter 8, from the new technologies of Internet surveillance, web browser cookies, Google-like search tools, and data mining.

8. An example of the way changes in an author's beliefs weaken the value of the author function—the traditional conception of the unitary author—appears in the works of Thomas Carlyle: whereas in *The French Revolution* he clearly accepts the necessity of violence and sympathizes with lower classes, he became increasingly reactionary and racist in his later works. In arguing for the unity of any particular Carlylean text one cannot casually refer to "Carlyle" unless one specifies to which Carlyle one refers.

9. According to the scientists that Galegher, Egido, and Kraut studied, people in these fields work collaboratively not only to share material and intellectual resources but also because "working with another person was simply more fun than working alone. They also believed that working together increased the quality of the research product, because of the synthesis of ideas it afforded, the feedback they received from each other, and the new skills they learned. In addition to these two major motives, a number of our respondents collaborate primarily to maintain a preestab-

lished relationship. In a relationship threatened by physical separation, the collaboration provided a reason for keeping in touch. Finally some researchers collaborated for self-presentational or political reasons, because they believed that working with a particular person or being in a collaborative relationship per se was valuable for their careers. Of course, these motives are not mutually exclusive" (152).

10. For a classical statement of the historicizing elements in humanistic study, see Erwin Panofsky, "The History of Art as a Humanistic Discipline," *Meaning in the Visual Arts,* 1–25.

11. The large number of individuals credited with authorship of scientific papers—sometimes more than one hundred—produces problems, too, as does the practice of so-called honorary authorship according to which the head of a laboratory or other person of prestige receives credit for research whose course he or she may not have followed and about which he or she may know very little. In this latter case problems arise when the names of such scientists of reputation serve to authenticate poor quality or even falsified research. See Walter W. Stewart and Ned Feder, "The Integrity of Scientific Literature," *Nature* 15 (1987): 207–14; cited by Ede and Lunsford.

12. According to Joanne Kaufman's article in the July 21, 2004 *Wall Street Journal,* despite the recent success of more than a dozen "double-bylined novels . . . concerns about the bottom line continue fueling resistance to double bylines," particularly in novels. Mary O'Shaunessy recalls that publishers told the sisters "that they couldn't think of any best sellers in recent history that had two authors' names. They originally wanted to use Pam's name because she was a Harvard law school graduate and, with the book a legal thriller, it seemed like it would be an easier sale in terms of the reader. We had to fight to get any recognition for me." Kaufman quotes several publishers with reasons for putting single-author bylines on collaboratively written books, including "a feeling that novels should be written by one person and could only come from one mind and one point of view." This attitude does seem to be changing. More important, an increasing number of authors write books together.

Chapter 5. Reconfiguring Writing

1. Coover, who is known for his postmodern experimental fiction, argues that the linear narrative of the traditional novel is an obsolete, politically offensive genre:

> For all its passing charm, the traditional novel, which took center stage at the same time the industrial mercantile democracies arose—Hegel called it "the epic of the middle class world"—is perceived by its would-be executioners as being the virulent carrier of patriarchal, colonial, canonical, proprietary, hierarchical, and authoritarian values of a past which is no longer with us.
>
> Much of the novel's alleged power is imbedded in *the line,* that compulsory author-directed movement from the beginning of a sentence to its period, from the top of the page to the bottom, from the first page to the last. Of course, through print's long history, there have been countless counter-strategies to the line's power from marginalia and footnoting to the creative innovations of novelists like Sterne, Joyce, Queneau, Cortàzar, Calvino, and Pavic, and not excluding the form's father Cervantes himself, but true freedom from the tyranny of the line is perceived as only really possible now at last with the advent of *hyper-*

text, where the line in fact does not exist unless one invents and implants it. ("End of Books," 1, 11)

I hardly agree with the overheated charge that the traditional novel served only as "the virulent carrier of patriarchal, colonial, canonical, proprietary, hierarchical, and authoritarian values," in part because the narratives of many traditional novels, particularly those with multiple plots, are not accurately described as linear and in part because much postcolonial fiction appears in the form of the traditional novel—so much so that African novelists have complained of the difficulties of publishing more experimental fiction in the West, since publishers, they claim, expect a traditional realistic novel.

2. Unfortunately, when one returned to one of them, older versions of Netscape deleted the intervening document titles, thereby turning what been Ariadne's thread into Hansel and Gretel's breadcrumbs. Furthermore, Netscape and similar browsers not only did not retain records of the complete reading path when one backtracked, they also deleted it entirely both after each session and when users closed the viewer window, even though one hadn't quit the application.

3. Intermedia provided two forms of preview information. First, its web view announced destinations of all links from the current lexia; activating a link marker with a single mouse click—clicking *twice* followed the link—darkened the icons for all the lexias linked to it. Intermedia also permitted authors to attach descriptions to each anchor, and these descriptions appeared in menus automatically generated when one followed a link leading to two or more lexias. In contrast to Intermedia, Storyspace allows authors to attach descriptions, not to anchors but to links themselves, though the reader perceives the result as much the same. As useful as these features were and are, hypermedia authors still have to assist readers by employing various techniques that constitute a rhetoric of departure.

4. Clara Mancini's Ph.D. thesis, "Towards Cinematic Hypertext," devotes several chapters to surveying various attempts to define coherence by psycholinguistics and gestalt psychology. Although she does not mention the fact, almost all the proposed "discourse coherence relations," such as "exemplifies," "supports," and "disproves," precisely match the forms of typed links proposed in pioneering technical papers about hypertext systems. Taking a different approach, Marie-Laure Ryan emphasizes total hypertext structures, which she suggests can take eight different forms, including the graph network, tree, maze, flow chart, braided plot, hidden story, and vector with side branches (246–58).

5. In my earlier work, beginning in 1987, I attempted to sketch out the beginnings of a rhetoric of hypertext and hypermedia, and one way of answering the question, "Is this hypertext any good?" involves looking at the degree to which a particular hypertext observes some of these minimal stylistic rules. This discussion, however, tries to broaden the question, looking for other sources of aesthetic pleasure and success.

6. Is this the result of following a link? If one means by "following a link" that when one carries out this action (clicking) new text appears, then by definition one has followed a link, but in fact it is not clear that one has activated a link or another computational procedure. Both the HTML and Storyspace versions of *(box(ing)*

actually involve links, so that, as in early Hypercard projects, clicking a link actually replaces one document with another, though the reader receives the illusion that the document remains the same and a new word or phrase appears within it. One cannot tell whether or not *Vniverse* works the same way or generates text on the fly, but from the vantage point of the viewer a replacement link or what we may term an *action link* appear identical.

7. Lyons adds: "Thus, the parentheses and interactive interface follow mutually compatible rules to establish what I hope are complementary contributions from writ language on the screen and script code behind the scenes . . . My aim here was simply to make good use of computers to get this ridiculous poem more legible, even as the interactive capability makes a greater range of (potentially confounding) meanings more accessible. You can think of it as magnetic poetry with rules."

8. Strickland's concern with reader empowerment appears in the detailed introduction she has appended to the project.

Chapter 6. Reconfiguring Narrative

1. Dorothy Lee argues that the language of Trobriand Islanders reveals that they "do not describe their activity lineally; they do no dynamic relating of acts; they do not use even so innocuous a connective as and" (157). According to Lee, they do not use causal connections in their descriptions of reality, and "where valued activity is concerned, the Trobrianders do not act on an assumption of lineality at any level. There is organization or rather coherence in their acts because Trobriand activity is patterned activity. One act within this pattern gives rise to a preordained cluster of acts"—much as, Lee explains, when knitting a sweater the "ribbing at the bottom does not cause the making of the neckline" (158). Similarly, "a Trobriander does not speak of roads either as connecting two points, or as running from point to point. His paths are self-contained, named as independent units; they are not to and from, they are at. And he himself is at; he has no equivalent for our to or from" (159). Appropriately, therefore, when an inhabitant of the Trobriand Islands "relates happenings, there is no developmental arrangement, no building up of emotional tone. His stories have no plot, no lineal development, no climax" (160), and this absence of what we mean by narrativity relates directly to the fact that "to the Trobriander, climax in history is abominable, a denial of all good, since it would imply not only the presence of change, but also that change increases the good; but to him value lies in sameness, in repeated pattern, in the incorporation of all time within the same point" (161).

Lee, incidentally, does not claim that the people of the Trobriand Islands cannot perceive linearity, just that it possesses solely a negative value in their culture and it is made difficult to use by their customs and language. If one accepts the accuracy of her translations of Trobriand language and her interpretations of Trobriand culture, one can see that what Lee calls nonlineal thought based on the idea of clustering differs significantly from both linear and multilinear thought. Placed on the spectrum constituted by Trobriand culture at one extreme and Western print culture at the other, hypertextuality appears only a moderate distance from other Western cultural patterns. Lee's description of Trobriand structuration by cluster, however, does possibly offer means of creating forms of hypertextual order.

2. Lyotard also proposes that "the decline of narrative can be seen as an effect of the blossoming of techniques and technologies since the Second World War, which has shifted emphasis from the ends of action to its means; it can also be seen as an effect of the redeployment of advanced liberal capitalism after its retreat under the protection of Keynesianism during the period 1930–60, a renewal that has eliminated the communist alternative and valorized the individual enjoyment of good and services" (*Postmodern Condition*, 37–38). His use of "can be seen as" suggests that Lyotard makes less than a full commitment to these explanations.

3. Hypertext is not the first information technology to make closure difficult. In *Writing Space*, Bolter reminds us that "the papyrus scroll was poor at suggesting a sense of closure" (85).

4. I have not substantially added to the following discussion of Joyce's pioneering hyperfiction, since it has been the subject of numerous detailed discussions since I first wrote about it. See, in particular, the chapters by J. Yellowlees Douglas and Terence Harpold in Landow, *Hyper/Text/Theory* as well as Clement, "Afternoon, a Story"; and Coover, "And Now, Boot Up the Reviews," 10.

5. The term *prosopopoeia*, Miller explains, describes "the ascription to entities that are not really alive first of a name, then of a face, and finally, in a return to language, or a voice. The entity I have personified is given the power to respond to the name I invoke, to speak in answer to my speech. Another way to put this would be to say that though my prosopopoeia is a fact of language, a member of the family of tropes, this tends to be hidden because the trope is posited a priori" (*Versions of Pygmalion*, 5).

6. The phrase is from Culler, *Structuralist Poetics*, 207. For Propp, see Vladimir Propp, "Fairy Tale Transformations" (1928), in *Readings in Russian Poetics*, 94–114; *Morphology of the Folktale* (1958); and Propp sections in Groden and Kreisirth, *Guide to Literary Theory*. See also Scholes, *Structuralism in Literature*, 59–141.

7. Janet Murray provides another instance of the way in which people construct connections and coherence from juxtaposition: "In the 1920s the Russian film pioneer Lev Kelshov demonstrated that audiences will take the same footage of an actor's face as signifying appetite, grief, or affection, depending on whether it is juxtaposed with images of a bowl of soup, a dead woman, or a little girl playing with a teddy bear" (160).

8. Goldberg continues: "In *Simulacra and Simulation*, Baudrillard who claims that 'of all the prostheses that mark the history of the body, the double is doubtless the oldest,' discusses science's desire to create life artificially:

> Cloning radically abolishes the Mother, but also the Father, the intertwining of their genes, the imbrication of their differences, but above all, the joint act that is procreation. The cloner does not beget himself: he sprouts from each of his segments. One can speculate on the wealth of each of these vegetal branchings that in effect resolve all oedipal sexuality in the service of 'nonhuman' sex, of sex through immediate contiguity and reduction—it is still the case that it is no longer a question of the fantasy of auto-genesis. The Father and the Mother have disappeared, not in the service of an aleatory liberty of the subject, but in the service of a matrix called code. No more mother, no more father: a matrix. And it is

the matrix, that of the genetic code, that now infinitely 'gives birth' based on a functional mode purged of all aleatory sexuality.

"This statement has many implications for both hypertext and critical theory, particularly about the relationship between the author and her work. The author does not beget herself: she sprouts from each of her segments" ("Comments on *Patchwork Girl*").

9. Williams continues: "Perhaps one may see this tension between order and disorder most clearly in life. *Patchwork Girl*'s functioning mirrors a cell's life. The cytoplasm of links serves as a permeable medium through which disparate parts pass signs. Its global disorder accommodates the local structure of organelles, which may have been conceived autonomously, but together rely on one another's differentiated function to achieve their fullest existence. Cells that incorporated subunits with diverse textures—wrinkled mitochondria, knotted DNA, smooth and rough endoplasmic reticulum—had sufficient complexity as biological collages to form entities such as readers of texts."

10. In his lexia Lars Hubrich argues that in *Patchwork Girl* scars become more than emblems of disfigurement, since we encounter "the story of a long struggle, of an emancipation that ends not in a mourning about the lost battles but in new strength, as the monster explains:

> Scar tissue does more than flaunt its strength by chronicling the assaults it has withstood. Scar tissue is new growth. And it is tougher than skin innocent of the blade.

"In fact, the scars become a new, living organ, opening up a new sensorium that goes straight into the chest of the monster. The scars are hot, responding to other people's input. And they have the ability to share their experience, to inscribe themselves on someone else's skin.

"The scars hold together the individual parts, each one having its own history, and gain their strength from the parts' experiences. But they do not point back, they rather are signs of an active, progressive look into a future that has learned from history.

"I have a navel like any other person. Does Shelley's monster have one? Of course, it has to. Not that it gets mentioned, though, as far as I have read *Patchwork Girl*. It would be rather odd for a monster like the one in the story to have a navel. Its origins lie somewhere else, not at one single point.

"And then we realize what those scars really are: birthmarks. Birthmarks of a new history, arisen from endless struggles. Donna Haraway would smile" ("Stitched identity").

11. In his lexia entitled "A Spotlight on the Haze: Notions of Origin in *Patchwork Girl*," Brian Perkins claims, however, that "hypertext is not so much a harbinger of the new possibilities, but a spotlight on the old machinations. It makes manifest the problems involved in defining the author as producer and the reader as consumer, problems which are not specific to hypertext, but which encompass all of language and signification. The transmission of meaning has forever been a blurry and complicated phenomenon. Hypertexts like *Patchwork Girl* are not novel because the

reader is decisive in determining their meaning, they are novel because they more clearly demonstrate the process which has always been at work."

12. Greco continues: "Any claim that hypertext is a privileged preserve of female or even feminist writing is suspicious for other reasons as well. Who is to say how and why hypertext might in some essential way fulfill a dream of an equal or even superior voice and representation for a group whose voices, interests, and hopes are themselves diverse and difficult to define? Those who make this claim commit themselves to a patronizing ideology of dominance masquerading as support and concern; for it is the privilege of the powerful to appropriate domains of discourse on behalf of others. Moreover, discovering alternatives to 'rational linearity' is not the same as resisting and transforming the structures whose power and authority give rise to the need for alternatives in the first place" (88).

13. There is one way that hypertext has proved clearly relevant to role-playing games, though it tells us more about the use of the World Wide Web than it does about computer games: some of those who participate in continuing non-computer-based role-playing games create websites for both the gameworld and individual characters. For example, one participant of a game set in the nineteenth century has a site in which his character, a Victorian physician, displays the contents of his medical bag.

14. Aarseth makes a much harsher attack on game studies based on literary and cinematic theory:

> The sheer number of students trained in film and literary studies will ensure that the slanted and crude misapplication of "narrative" theory to games will continue and probably overwhelm game scholarship for a long time to come. As long as vast numbers of journals and supervisors from traditional narrative studies continue to sanction dissertations and papers that take narrativity for granted and confuse the story-game hybrids with games in general, good, critical scholarship on games will be outnumbered by incompetence. ("Genre Trouble," in *First Person*, 54)

Henry Jenkins, one of those scholars who comes from film studies, responds in kind:

> Much of the writing in the ludologist tradition is unduly polemical: they are so busy trying to pull game designers out of their "cinema envy" or define a field where no hypertext theorist dares to venture that they are prematurely dismissing the use value of narrative for understanding their desired object of study. For my money, a series of conceptual blind spots prevent them from developing a full understanding of the interplay between narrative and games. ("Game Design as Narrative Architecture," in *First Person*, 120)

15. See my *Elegant Jeremiahs*, 82–115, for discussions of brief narratives with blatantly symbolic meaning in nineteenth- and twentieth-century Anglo-American prose. One chapter, "The Sage as Master of Experience" (132–53), examines passages in the writings of John Ruskin, D. H. Lawrence, Tom Wolfe, and Norman Mailer in which these writers of nonfiction use narrative to create protocinematic forms of description.

16. Kristoffer Gansing, who wants to include games in his theories of interactive cinema, agrees with Aarseth that "we should not be afraid to study gaming for gaming's own sake," but still asserts "there are computer games where narrative has a foregrounded, *explicit* role (adventure games) . . . [and] there are many games where narrative could be described as being *implicit* in game structure (strategy games). If the explicit role is somehow contradictory to the nature of gaming, I leave others to decide, opting instead to focus on narrative simply because it is integral to the idea of an interactive film" (53–54).

17. The article by Henderson, Pruett, Galper, and Copes describing the project states that the simulation does not permit the trainee to kill a patient, for a supervising physician steps in and takes over when that might happen. During the demonstration of the project at a Sloan Foundation–sponsored conference at Dartmouth College in October 1988, two years after the publication of article describing it, I believe the speaker stated that the patient could die; I may be misremembering this point.

18. In the second part of the *Iowa Review* interview, Wardrip-Fruin explains the role of each member of the team and the evolution of the project.

The Brown cave takes the form of an open cube, each of whose surfaces measures 8×8 feet, and is the result of successful application in 1997 to the National Science Foundation for a project entitled "Acquisition of a Cave and Shared Memory Supercomputer." The project, which had thirteen principal investigators from the departments of chemistry, applied math, physics, computer science, and geology, was funded by a $1 million National Science Foundation grant with significant additional cost-sharing from Brown University.

Chapter 7. Reconfiguring Literary Education

1. We have been observing ways that hypertext embodies literary theory, and we should also notice that it also instantiates related pedagogical theory. The hypertextual read-author, for instance, matches R. A. Shoaf's claim that "every reader, in fact, from the beginning student to the seasoned professional, is also a writer, or more accurately a rewriter—and must be aware of that" (80).

2. In 2002 George Lorenzo pointed out in an article in *University Business* that eArmyU "has set out to deliver online distance education to 80,000 soldier-students . . . The program now includes 23 schools and 85 online degree programs . . . All registrations are handled through eArmyU's portal. On-base counselors provide support" (37). I suspect that very few universities realize that one of the biggest experiments with higher education is taking place.

3. "The modularity in question emerges when the Americans take something the Europeans considered as a whole, namely undergraduate education, and break it up into small, self-contained and implicitly recombinable units commonly called course credits or credit hours . . . The implications of the new system show up most clearly in the new artifact to which they give rise: the student transcript . . . The transcript, by tracing one person's passage through the curriculum, is an additive record bounded by the number of credits required for graduation. Equivalence of parts dictates that a course is a course is a course, though locally defined restraints on com-

binability (majors, distribution requirements, and the like) may sometime lead a student to accumulate more credits than the minimum required for graduation" (Blair, 11, 20). A full hypertext version of the present book would, at this point, link to the entire text of Blair's book (most likely through a section or chapter that, in turn, would link to the entire text) and also to the enormous body of internal reports produced in recent decades by individual American colleges and universities discussing the results of such modular approaches.

4. That part of the Intermedia development plan funded by the Annenberg /CPB Project included an intensive three-year evaluation carried out by a team of ethnographers, who taped, attended, and analyzed all class meetings and who frequently surveyed and interviewed students for the two years before the introduction of the hypertext component and for the year following. Many of my observations on conventional education and the educational effects of hypertext on it derive from their data and from conversations with Professor Heywood. See Beeman and colleagues, *Intermedia*.

5. *Hypertext 2.0*, 235–45, narrates in detail the evolution of a small Intermedia collaborative learning project on the poetry of Wole Soyinka to *The Postcolonial and Literature and Culture Web*, and readers interested in the ways such a project developed should consult the earlier work. This website now has about 15,000 documents and images, many of them by contributors from Australia, Canada, India, Japan, Nigeria, Singapore, South Africa, Zimbabwe, and other countries. In the past few years, instructors from various American universities, including Northwestern and DePauw, have had their students submit essays.

6. Hugh Kenner, "The Making of the Modernist Canon," in *Canons*, 371. Writing in terms of the broadest canon, that constituted by the concept of literature and the literary, Eagleton observes: "What you have defined as a 'literary' work will always be closely bound up with what you consider 'appropriate' critical techniques: a 'literary' work will mean, more or less, one which can be usefully illuminated by such means of enquiry" (*Literary Theory*, 80).

7. In 1968, for example, Random House, which purchased seventy-four pages of advertisements to Harper's twenty-nine, "had nearly three times as many books mentioned in the feature 'New and Recommended' as Doubleday or Harper, both of which published as many books as the Random House group" (381). Ohmann also points out "it may be more than coincidental" that in the same year in the *New York Review of Books*, founded by a Random House vice president, "almost one-fourth of the books granted full reviews . . . were published by Random House (again, including Knopf and Pantheon)—more than the combined total of books from Viking, Grove, Holt, Harper, Houghton Mifflin, Oxford, Doubleday, MacMillan, and Harvard so honored; or that in the same year one-fourth of the reviewers had books in print with Random House and that a third of those were reviewing other Random House books, mainly favorably; or that over a five-year period more than half the regular reviewers (ten or more appearances) were Random House authors" (383).

8. According to Hugh Kenner, "Since Chaucer, the domain of English literature had been a country, England. Early in the 20th century its domain commenced to be a language, English" (366).

9. More than a dozen years after hearing this statement, I came upon Vincent Mosco's statement that new technologies only become truly powerful once they become unnoticeable: "The real power of new technologies does not appear during their mythic period, when they are hailed for their ability to bring world peace, renew communities, or end scarcity, history, geography, or politics; rather their social impact is greatest when technologies become banal—when they literally (as is the case of electricity) or figuratively withdraw into the woodwork . . . Indeed, it was not until we stopped looking at electricity as a discrete wonder and began to see it as a contributor to all other forces in society that it became an extraordinary force. Electricity achieved its real power when it left mythology and entered banality" (19–20). Many years ago someone told me that computing would never reach its potential until people entering one's office or home ceased remarking, "I see you have a computer." PCs have truly achieved ordinariness.

10. For a fuller discussion of the University Scholars Program's hypertext paradigm in the context of institutional history and goals, see my essay "The Paradigm is More Important than the Purchase."

Chapter 8. The Politics of Hypertext

1. Nicholas Negroponte, the founder of MIT's Media Lab, is one of Mosco's main targets. According to him, Negroponte "provides one of the more extreme versions of this radical break with history viewpoint. In *Being Digital* (1995) he argues for the benefits of digits (what computer communication produces and distributes) over atoms (us and the material world) and contends that the new digital technologies are creating a fundamentally new world that we must accommodate. In matter-of-fact prose, he offers a prophet's call to say goodbye to the world of atoms, with its coarse and confining materiality, and welcomes the digital world, which its infinitely malleable electrons, able to transcend spatial, temporal, and material constraints" (36). Mosco is more than a little unfair to Negroponte, many of whose predictions have proved correct and whose observations have proved sound—even if they did help stimulate the dotcom mania. Negroponte's discussion of economic factors related to the print demonstrates that he often sounds like a cheerleader for the digital: "A book has a high-contrast display, is lightweight, easy to 'thumb' through, and not very expensive. But getting it to you includes shipping and inventory. In the case of textbooks, 45 percent of the cost is inventory, shipping, and returns. Worse, a book can go out of print. Digital books never go out of print. They are always there" (13). The last two sentences are just silly, since the only reason "Digital books never go out of print" is that they were never literally in print! But publishers, such as my own, do permit e-texts like my *Hypertext-in-Hypertext* to sell out and become unavailable. In all fairness to Negroponte, one must admit that although he does not say so, he probably means that in some future digitized, fully networked world "books never go out of print," but the experience of the World Wide Web hardly makes this seem likely. In the course of managing three large websites, I've observed that sites to which authors invited me to link frequently disappear or change their URLs. Nonetheless, despite this and similar exaggerations, *Being Digital* makes many astute judgments that Mosco fails to acknowledge.

2. Martha McCaughey and Michael D. Ayers's collection of essays, *Cyberactivism: Online Activism in Theory and Practice,* contains discussions of the political uses of the Internet by Amnesty International, NOW, and the Zapatistas, and protests against the World Bank as well as theoretical approaches, such as a Habermasian analysis of the relations of democracy and the Internet. The editors point out that "the Ku Klux Klan (KKK) and other radically conservative organizations have also colonized cyberspace in hopes of achieving their goals" (3).

3. A few years after I offered my speculations about the future of newspapers, Negroponte prophesied a somewhat different vision. Pointing out that both broadcast television and newspapers are produced "with all the intelligence" (19) at the transmitting part of the communicative relationship, he proposes that to change news media for the better we must create "computers to filter, sort, prioritize, and manage multimedia on our behalf—computers that read newspapers and look at television for us, and act as editors when we ask them to do so" (20; see also 84). The results of this filtering would produce a daily news source custom-tailored to each reader's interests, something in fact very close what one receives from the *New York Times* online after one has created a user profile identifying subjects of highest priority. Negroponte's emphasis on filtering, preselection, produces a very different kind of news media than one based on user-directed hypertext. In his vision, users receive only the news that they want to read; in mine, users can also obtain more information when they need it. There is another important difference: whereas Negroponte's filter-centered vision concentrates on current news, a hypertext-centered new media creates more of a communal memory because one can follow links to historical contexts.

4. Eagleton, *Criticism and Ideology,* 44–63. Although Eagleton never cites McLuhan or other students of the history of information technology, he several times compares manuscript and print cultures within the context of Marxist theory; see 47–48, 51–52.

5. Ryan, 60. Ryan also offers an oddly limited description of technology when he writes: "Technology is the human mind working up the natural world into machines. And, as I have argued, it is motivated by the desire of a class of subjects—capitalists—to maintain power over another class of subjects—workers" (92). The problems with this statement include, first, the fact that Ryan confuses "capitalists" with "owners of production" even though he makes clear elsewhere that what he calls the Leninist tradition also relies on heavy technology; and second, such a bizarrely narrow definition apparently restricts technology to heavy machinery, thereby omitting both everything before the Industrial Revolution and everything in the electronic and atomic age other than old-fashioned rust-belt manufacturing. The context makes it difficult to determine whether Ryan's dislike of technology or capitalism leads him to such an obsolete definition.

6. Elizabeth L. Eisenstein makes a particularly astute point when discussing arguments about the role of print technology in radical social change during the Reformation: "Given the convergence of interests among printers and Protestants, given the way that the new media implemented older evangelical goals, it seems pointless to argue whether material or spiritual, socio-economic or religious 'factors' were important in transforming Western Christianity. Not only do these dichotomies

seem to be based on spurious categories, but they also make it difficult to perceive the distinctive amalgam which resulted from collaboration between diverse pressure groups" (406). One does not have to espouse pluralism to recognize that Marxist analyses could easily incorporate evidence provided by Eisenstein.

7. Nelson, *Computer Lib*, 1/4. Nelson also points out: "Tomorrow's hypertext networks have immense political ramifications, and there are many struggles to come. Many vested interests may turn out to be opposed to freedom . . . For rolled into such designs and prospects is the whole future of humanity and, indeed, the future of the past and the future of the future—meaning the kinds of future that become forbidden, or possible" (3/19).

8. In *The Gutenberg Galaxy*, 216, McLuhan quotes Harold Innis, *The Bias of Communication* (Toronto: University of Toronto Press), 29: "The effect of the discovery of printing was evident in the savage religious wars of the sixteenth and seventeenth centuries. Application of power to communication industries hastened the consolidation of vernaculars, the rise of nationalism, revolution, and new outbreaks of savagery in the twentieth century."

9. In print this thrust appears with particular clarity in the radical new discovery that the best way to preserve information lies in disseminating large numbers of copies of a text containing it rather than keeping it secret; see Eisenstein, *Printing Press*, 116.

10. Professor Ulmer made these comments in the course of the 1988 University of Alabama conference *Literacy Online*.

11. He continues on the same page: "The edifying philosophers are thus agreeing with Lessing's choice of the infinite striving for truth over 'all of Truth.' For the edifying philosopher the very idea of being presented with 'all of Truth' is absurd, because the Platonic notion of Truth itself is absurd."

12. Popper, *The Open Society and Its Enemies*, argues that Plato developed his conceptions of humanity, society, and philosophy in reaction to the political disorder of his time. Plato's "theory of Forms or Ideas," according to Popper, has three main functions within his thought: (1) as a methodological device that "makes possible pure scientific knowledge"; (2) as a "clue" to a theory of change, decay, and history; and (3) as the basis of a historicist "social engineering" that can arrest social change (30–31). Popper argues that Plato bases his ideal state on Sparta, "a slave state, and accordingly Plato's best state is based on the most rigid class distinctions. It is a caste state. The problem of avoiding class war is solved, not by abolishing classes, but by giving the ruling class a superiority which cannot be challenged" (46). Popper, who attacks Plato for providing the ultimate ideological basis of fascism, claims that in *The Republic* Plato "used the term 'just' as a synonym for 'that which is in the interest of the best state'. And what is in the interest of this best state? To arrest all change, by the maintenance of a rigid class division and class rule. If I am right in this interpretation, then we should have to say that Plato's demand for justice leaves his political programme at the level of totalitarianism" (89).

13. Working hard to find some point of agreement, the priest adds that his god also "is in the sky," but he then makes a theological claim that appears completely bizarre and inappropriate from a Shona point of view when he tells the man

he wishes to convert that "my God is the true God. He is the way to eternal happiness" (105). Two aspects of Christian belief here puzzle his listener—first, that happiness could be eternal and, second, that hard work is bad and that any form of happiness might involve freedom from what he takes to be a crucial, pleasurable human activity.

14. Lovink quotes a member of a South Asian media collective, who takes an optimistic view of the problem: "I would never use a term like 'digital divide.' We have a print divide in India, an education divide, a railway divide, an airplanes divide. [But] the new economy of India is definitely not conceived as a divide" (210). On commercialization of the Internet, see "Introduction: Twilight of the Digerati," 3, 11–12, and "Information Warfare: From Propaganda Critique to Culture Jamming," 309, 330, both in *Dark Fiber*. For the Amsterdam experiments in using the internet to empower citizens, see "The Digital City—Metaphor and Community," 42–67, in *Dark Fiber*.

15. Earlier versions of *Hypertext* followed the preceding discussion with a short story, "Ms. Austen's Submission," whose heroine encounters the darker implications of a future hypertext author's attempt to gain access to the Net. Anyone wanting to read about the world of future e-publishing as a dystopia, should consult *Hypertext* or *Hypertext 2.0*.

16. Thus, "logged-in users start at 1 (although this can vary from 0 to 2 based on their karma) and anonymous users start at 0." Malda explains: "Slashdot tracks your 'karma.' If you have Positive, Good, or Excellent karma, this means you have posted more good comments than bad, and are eligible to moderate. This weeds out spam accounts. The end result is a pool of eligible users that represent (hopefully) average, positive Slashdot contributors. Occasionally (well, every 30 minutes actually), the system checks the number of comments that have been posted, and gives a proportionate number of eligible users 'tokens' [or moderation points]. When any user acquires a certain number of tokens, he or she becomes a moderator. This means that you'll need to be eligible for many of these slices in order to actually gain access. It all works to make sure that everyone takes turns, and nobody can abuse the system, and that only 'regular' readers become moderators (as opposed to some random newbie)."

17. Chaytor, *From Script to Print*, 1. Cited by McLuhan, *Gutenberg Galaxy*, 87, and credited on the previous page as "a book to which the present one owes a good deal of its reason for being written."

18. Sutherland, "Author's Rights," 554. Sutherland quotes E. Plowman and L. C. Hamilton's explanation in *Copyright* (1980) that in France and Germany moral rights include "the rights to determine the manner of dissemination, to ensure recognition of authorship, to prohibit distortion of the work, to ensure access to the original or copies of the work, and to revoke a license by reason of changed convictions against payment of damages." This and all subsequent quotations from this article in the main text come from page 554.

Bibliography

Printed Materials

A:\ Littérature: Colloque Nord Poésie et Ordinateur. Lille: Université de Lille and Villeneuve D'Ascq: MOTS-VOIR, 1994.

Aarseth, Espen. *Cybertext: Perspectives on Ergodic Literature*. Baltimore: Johns Hopkins University Press, 1997.

———. *Texts of Change: Towards a Poetics of Nonlinearity*. Bergen: University of Bergen, 1991.

Accame, Lorenzo. *La Decostruzione e il Testo*. Florence: G. C. Sansoni, 1976.

Akscyn, Robert M., Donald L. McCracken, and Elise Yoder. "KMS: A Distributed Hypermedia System for Managing Knowledge Organizations." *Communications of the ACM* 31 (1988): 820–35.

Althusser, Louis. *For Marx*. Translated by Ben Brewster. London: Verso, 1979.

Amerika, Mark. "Notes from the Digital Overground." *American Book Review* (December–January 1995–96): 1, 12.

Anderson, Jean. "STELLA: Software for Teaching English Language and Literature." In *Hypermedia at Work: Practice and Theory in Higher Education*. Edited by W. Strang, V. B. Simpson, and D. Slater. Canterbury: University of Kent, 1995. 89–98.

Bakhtin, Mikhail. *Problems of Dostoevsky's Poetics*. Edited and Translated by Caryl Emerson. Minneapolis: University of Minnesota Press, 1984.

Barth, John. "The State of the Art." *Wilson Quarterly* 36 (1996): 37–45.

Barrett, Edward, ed. *Sociomedia: Multimedia, Hypermedia, and the Social Construction of Knowledge*. Cambridge, Mass.: MIT Press, 1992.

———. *Text, ConText, and Hypertext: Writing with and for the Computer*. Cambridge, Mass.: MIT Press, 1988.

Barthes, Roland. "Authors and Writers." In *A Barthes Reader*. Edited by Susan Sontag. New York: Hill and Wang, 1982. 185–93.

———. *The Eiffel Tower and Other Mythologies*. Translated by Richard Howard. New York: Hill and Wang, 1979.

———. *Sade, Fourier, Loyola*. Translated by Richard Miller. New York: Hill and Wang, 1976.

———. *S/Z*. Paris: Éditions du Seuil, 1970. *S/Z*. Translated by Richard Miller. New York: Hill and Wang, 1974.

———. *Mythologies*. Translated by Annette Lavers. New York: Hill and Wang, 1972.

———. *Elements of Semiology*. Translated by Annette Lavers and Colin Smith. London: Jonathan Cape, 1967.

———. *Writing Degree Zero*. Translated by Annette Lavers and Colin Smith. London: Jonathan Cape, 1967.

Bass, Randall. "The Syllabus Builder: A Hypertext Resource for Teachers of Literature." *Journal of Computing in Higher Education* 4 (1993): 3–26.

Bates, Stephen. "The First Amendment in Cyberspace." *Wall Street Journal* (June 1, 1994): A15.

Baudrillard, Jean. *Fatal Strategies*. Translated by Philip Beitchman and W.G.J. Niesluchowski. New York: Semiotext(e)/Pluto, 1990.

———. *The Ecstasy of Communication*. Translated by Bernard and Caroline Schutze. Edited by Sylvère Lotringer. New York: Semiotext(e), 1988.

———. *Simulations*. New York: Semiotext(e), 1983.

Beeman, William O., Kenneth T. Anderson, Gail Bader, James Larkin, Anne P. McClard, Patrick McQuillian, and Mark Shields. *Intermedia: A Case Study of Innovation in Higher Education*. Providence, R.I.: Office of Program Analysis/ Institute for Research in Information and Scholarship, 1988.

Benedikt, Michael, ed. *Cyberspace: First Steps*. Cambridge, Mass.: MIT Press, 1991.

Benjamin, Walter. *Illuminations*. Edited by Hannah Arendt. Translated by Harry Zohn. New York: Schocken, 1969.

Benstock, Shari. *Textualizing the Feminine: On the Limits of Genre*. Norman: University of Oklahoma Press, 1991.

Berger, Peter L., and Thomas Luckmann. *The Social Construction of Reality: A Treatise in the Sociology of Knowledge*. Garden City, N.Y.: Doubleday, 1966.

Berners-Lee, Tim, Robert Calliau, Ari Luotonen, Henrik Frystyk Nielksen, and Arthur Secret. "The World-Wide Web." *Communications of the ACM* 37 (August 1994): 76–82.

Bernstein, Mark. "More Than Legible: On Links that Readers Don't Want to Follow." In *ACM 2000 Hypertext: Proceedings of the Eleventh ACM Conference on Hypertext and Hypermedia*. New York: ACM, 2000. 216–17.

———. "On Writing Hypertext: Tools for Information Farming." In *Proceedings of the Association for Computing in the Humanities, 1994*.

———. "Enactment in Information Farming." In *Hypertext '93 Proceedings*. Seattle: Association for Computing Machinery, 1993.

———. "Storyspace and the Process of Writing." In *Hypertext/Hypermedia Handbook*. Edited by E. Berk and J. Devlin. New York: McGraw-Hill, 1991.

Bernstein, Mark, J. David Bolter, Michael Joyce, and Elli Mylonas. "Architectures for Volatile Hypertexts." In *Hypertext '91 Proceedings*. San Antonio: Association for Computing Machinery, 1991.

BIBLIOGRAPHY

Bernstein, Mark, Michael Joyce, and David Levine. "Contours of Constructive Hypertexts." In *Proceedings of the 1992 European Conference on Hypertext*. Milano: Association for Computing Machinery, 1992.

Bhabha, Homi K. "DissemiNation: Time, Narrative, and the Margins of the Modern Nation." In *Nation and Narration*. Edited by Homi K. Bhabha. London: Routledge, 1990. 291–322.

Bikson, Tora K., and J. D. Eveland. "The Interplay of Work Group Structures and Computer Support." In *Intellectual Team Work: Social and Technological Foundations of Cooperative Work*. Edited by Jolene Galegher, Robert Kraut, and Mark Egido. Hillsdale, N.J.: Erlbaum, 1990. 245–90.

Blackwell, Lewis, and David Carson. *The End of Print: The Graphic Design of David Carson*. San Francisco: Chronicle, 1996.

Blair, John G. *Modular America: Cross-Cultural Perspectives on the Emergence of an American Way of Life*. New York: Greenwood, 1988.

Bloom, Harold, Paul de Man, Jacques Derrida, Geoffrey H. Hartman, and J. Hillis Miller. *Deconstruction and Criticism*. London: Routledge & Kegan Paul, 1979.

Bolter, J. David. *Writing Space: The Computer in the History of Literacy*. Hillsdale, N.J.: Erlbaum, 1990.

Bolter, J. David, and Diane Gromola. *Windows and Mirrors: Interaction Design, Digital Art, and the Myth of Transparency*. Cambridge, Mass.: MIT Press, 2003.

———. "Beyond Word Processing: The Computer as a New Writing Space." *Language & Communication* 9 (1989): 129–42.

———. *Turing's Man: Western Culture in the Computer Age*. Chapel Hill, N.C.: University of North Carolina Press, 1984.

Bolter, J. David, and Richard Grushin. *Remediation*. Cambridge, Mass.: MIT Press, 2001.

Borges, Jorge Luis. *The Aleph and Other Stories, 1933–1969*. Translated by Norman Thomas di Giovanni. New York: Bantam, 1971.

———. *Other Inquisitions, 1937–1952*. Translated by Ruth L. C. Simms. New York: Washington Square, 1966.

Bornstein, George, and Ralph G. Williams, eds. *Palimpsest: Editorial Theory in the Humanities*. Ann Arbor: University of Michigan Press, 1993.

Boyle, James. *Shamans, Software, and Spleens: Law and the Construction of Information Society*. Cambridge, Mass.: Harvard University Press, 1996.

Brown, Peter J. "Creating Educational Hyperdocuments: Can It Be Economic?" In *Hypermedia at Work: Practice and Theory in Higher Education*. Edited by W. Strang, V. B. Simpson, and D. Slater. Canterbury: University of Kent, 1995. 9–20.

Bruns, Gerald L. "Canon and Power in the Hebrew Scriptures." In *Canons*. Edited by Robert von Hallberg. Chicago: University of Chicago Press, 1984.

Bulkley, William M. "New On-Line Casinos May Thwart U.S. Laws." *Wall Street Journal* (May 10, 1995): B1, B8.

Bush, Vannevar. "Memex Revisited." In *Science is Not Enough*. New York: William Morrow, 1967. 75–101.

———. *Endless Horizons*. Washington, D.C.: Public Affairs, 1946.

BIBLIOGRAPHY

———. "As We May Think." *Atlantic Monthly* 176 (July 1945): 101–8.

Calvino, Italo. *If on a Winter's Night a Traveler.* Translated by William Weaver. San Diego: Harcourt Brace Jovanovitch, 1981.

Carlyle, Thomas. "Signs of the Times." In *Collected Works.* London: Chapman and Hall, 1858. 98–118.

Casetti, Francesco. *Communicative Negotiation in Cinema and Television.* Milan: V&P Strumenti, 2002.

Catano, James. "Poetry and Computers: Experimenting with Communal Text." *Computers and the Humanities* 13 (1979): 269–75.

Chartier, Roger. "Meaningful Forms." Translated by Patrick Curry. *Liber* 1 (1989): 8–9.

———. *The Culture of Print: Power and the Uses of Print in Early Modern Europe.* Translated by Lydia G. Cochrane. Princeton: Princeton University Press, 1987.

———. *The Cultural Uses of Print in Early Modern France.* Translated by Lydia G. Cochrane. Princeton: Princeton University Press, 1987.

Chatman, Seymour. *Story and Discourse: Narrative Structure in Fiction and Film.* Ithaca, N.Y.: Cornell University Press, 1978.

Chaytor, H. J. *From Script to Print.* Cambridge: Heffer and Sons, 1945.

Cixous, Hélène. *Readings: The Poetics of Blanchot, Joyce, Kafka, Kleist, Lispector, and Tsvetayeva.* Minneapolis: University of Minnesota Press, 1981.

Cixous, Hélène, and Catherine Clement. *The Newly Born Woman.* Translated by Betsy Wing. Minneapolis: University of Minnesota Press, 1986.

Clement, Jean. "Afternoon, a Story: From Narration to Poetry in Hypertextual Books." In *A:\ Littérature:: Colloque Nord Poésie et Ordinateur.* Lille: Université de Lille and Villeneuve D'Ascq: MOTS-VOIR, 1994.

Collaud, G., J. Monnard, and J. Pasquier-Boltuck. *Untangling Webs: A User's Guide to the Woven Electronic Book System.* Fribourg, Switzerland: University of Fribourg (IAUF), 1989.

Conklin, E. Jeffrey. "Hypertext: An Introduction and Survey." *IEEE Computer* 20 (1987): 17–41.

Coombs, James H. "Hypertext, Full Text, and Automatic Linking." SIGIR 90 (Technical Report). Providence, R.I.: Institute for Research in Information and Scholarship, 1990.

Coover, Robert. "And Hypertext Is Only the Beginning. Watch Out!" *New York Times Book Review* (August 29, 1993): 8–9.

———. "And Now, Boot Up the Reviews." *New York Times Book Review* (August 29, 1993): 10–12.

———. "Hyperfiction: Novels for the Computer." *New York Times Book Review* (August 29, 1993): 1, 8–10.

———. "The End of Books." *New York Times Book Review* (June 21, 1992): 1, 11, 24–25.

———. "Endings: Work Notes." Manuscript, 1990.

———. "He Thinks the Way We Dream." *New York Times Book Review* (November 20, 1988): 15.

———. *Pricksongs and Descants.* New York: New American Library, 1969.

Cortázar, Julio. *Hopscotch.* Translated by Gregory Rabassa. New York: Random House, 1966.

BIBLIOGRAPHY

Cotton, Bob, and Richard Oliver. *Understanding Hypermedia: From Multimedia to Virtual Reality.* London: Phaidon, 1993.

Crane, Gregory. "Redefining the Book: Some Preliminary Problems." *Academic Computing* 2 (February 1988): 6–11, 36–41.

Culler, Jonathan. *Framing the Sign: Criticism and Its Institutions.* Norman: University of Oklahoma Press, 1988.

———. *On Deconstruction: Theory and Criticism after Structuralism.* Ithaca, N.Y.: Cornell University Press, 1982.

———. *The Pursuit of Signs: Semiotics, Literature, Deconstruction.* Ithaca, N.Y.: Cornell University Press, 1981.

———. *Structuralist Poetics: Structuralism, Linguistics and the Study of Literature.* Ithaca, N.Y.: Cornell University Press, 1975.

Cyberactivism: Online Activism in Theory and Practice. Edited by Martha McCaughey and Michael D. Ayers. New York: Routledge, 2003.

Daniele, Daniela. "Travelogues in a Broken Landscape: Robert Smithson's Mixed-Media Tribute to William Carlos Williams." *Rivista di Studi Anglo-Americani* 8 (1994): 95–104.

Dasenbrock, Reed Way. "What to Teach When the Canon Closes Down: Toward a New Essentialism. In *Reorientations: Critical Theories and Pedagogies.* Edited by Bruce Hendricksen and Thaïs Morgan. Urbana: University of Illinois Press, 1990. 63–76.

deCerteau, Michel. *The Practice of Everyday Life.* Translated by Steven Rendall. Berkeley: University of California Press, 1984.

Deegan, Marilyn, Nicola Timbrell, and Lorraine Warren. *Hypermedia in the Humanities.* Oxford: Universities of Oxford and Hull, 1993.

Delany, Paul, and George P. Landow, eds. *Hypermedia and Literary Studies.* Cambridge, Mass.: MIT Press, 1991.

Deleuze, Gilles, and Félix Guattari. *A Thousand Plateaus: Capitalism and Schizophrenia.* Translated by Brian Massumi. Minneapolis: University of Minnesota Press, 1987.

Del Monaco, Emanuella, and Alessandro Pamini. *Ernest Lubitsch: L'Arte della Variazone nel Cinema.* Rome: Ente dello Spectacola, 1995.

DeRose, Steven J. *CD Word Tutorial: Learning CD Word for Bible Study.* Dallas: CD Word Library, 1990.

———. "Expanding the Notion of Links." *Hypertext '89 Proceedings.* New York: Association of Computing Machinery, 1989. 249–57.

DeRose, Stephen J., David G. Durand, Elli Mylonas, and Allen H. Renear. "What Is Text Really?" *Journal of Computing in Higher Education* 1 (1990): 3–26.

De Roure, David C., Nigel G. Walker, and Leslie A. Carr. "Investigating Link Service Infrastructures." *ACM 2000 Hypertext.* New York: ACM, 2000. 67–76.

Derrida, Jacques. *La Dissemination.* Paris: Éditions du Seuil, 1972. *Dissemination.* Translated by Barbara Johnson. Chicago: University of Chicago Press, 1981.

———. "Living On." In *Deconstruction and Criticism.* Edited by Harold Vloom et al. London: Routledge & Kegan Paul, 1979. 75–176.

———. *Writing and Difference.* Translated by Alan Bass. Chicago: University of Chicago Press, 1978.

BIBLIOGRAPHY

―――. *De la Grammatolgie*. Paris: Les Éditions de Minuit, 1967. *Of Grammatology*. Translated by Gayatri Chakravorty Spivak. Baltimore: Johns Hopkins University Press, 1976.

―――. "Structure Sign and Play in the Discourse of the Human Sciences." In *The Structuralist Controversy: The Languages of Criticism and the Sciences of Man*. Edited by Richard A. Macksey and Eugenio Donato. Baltimore: Johns Hopkins University Press, 1972.

"Designing Hypermedia Applications" issue of *Communications of the ACM* 38 (August 1995).

Dickey, William. "Poem Descending a Staircase: Hypertext and the Simultaneity of Experience." In *Hypermedia and Literary Studies*. Edited by Paul Delany and George P. Landow. Cambridge, Mass.: MIT Press, 1991. 143–52.

Digital Dialectic: New Essays on New Media, The. Edited by Peter Lunenfeld. Cambridge, Mass.: MIT Press, 1999.

"Digital Libraries" issue of *Communications of the ACM* 38 (April 1995).

Digital Media Revisited: Theoretical and Conceptual Innovations in Digital Domains. Edited by Gunnar Liestøl, Andrew Morrison, and Terje Rasmussen. Cambridge, Mass.: MIT Press, 2003.

Duchastel, Philippe C. "Discussion: Formal and Informal Learning with Hypermedia." In *Designing Hypertext/Hypermedia for Learning*. Edited by David H. Jonassen and Heinz Mandl. Heidelberg: Springer-Verlag, 1990. 135–46.

Ducreux, Marie-Elizabeth. "Reading unto Death: Books and Readers in Eighteenth-Century Bohemia." In *The Culture of Print: Power and the Uses of Print in Early Modern Europe*. Edited by Roger Chartier. Translated by Lydia G. Cochrane. Princeton: Princeton University Press, 1989. 191–229.

Dyson, Esther. "If You Don't Love It, Leave It." *New York Times Sunday Magazine* (July 16, 1995): 26–27.

Eagleton, Terry. *Literary Theory: An Introduction*. Minneapolis: University of Minnesota Press, 1983.

―――. *Criticism and Ideology: A Study in Marxist Theory*. London: NLB, 1976.

Earnshaw, R. A., M. A. Gigante, and H. Jones, eds. *Virtual Reality Systems*. London: Academic, 1993.

Eco, Umberto. *The Open Work*. Translated by Anna Cancogni. Cambridge, Mass.: Harvard University Press, 1989.

―――. *A Theory of Semiotics*. Bloomington: Indiana University Press, 1979.

Ede, Lisa, and Andrea Lunsford. *Singular Texts/Plural Authors: Perspectives on Collaborative Writing*. Carbondale: Southern Illinois University Press, 1990.

Eisenstein, Elizabeth L. *The Printing Press as an Agent of Change: Communications and Cultural Transformations in Early-Modern Europe*. Cambridge: Cambridge University Press, 1980.

Eisenstein, Sergei. *The Film Sense*. Translated by Jay Ledyda. New York: Harvest, n.d.

eLearning: Didattica e innovazione in universita. Edited by Patrizia Ghislandi. Trento: Universita degli Studi di Trento, [2004?].

Eskelinen, Markku, and Raine Koskimaa. "Introduction: Towards a Functional

BIBLIOGRAPHY

Theory of Media." In *CyberText Yearbook 2001.* Jyväskylä: Research Centre for Contemporary Culture, University of Jyväskylä, 2002. 7–12.

Fiormonte, Domenico. *Scrittura e filologia nell'era digitale.* Turin: Bollati Boringhieri, 2003.

First Person: New Media as Story, Performance, and Game. Edited by Noah Wardip-Fruin and Pat Harigan. Cambridge, Mass.: MIT Press, 2004.

Flaxman, Rhoda L. *Victorian Word Painting and Narrative: Toward the Blending of Genres.* Ann Arbor, Mich.: UMI Research Press, 1987.

Foster, Hal, ed. *The Anti-Aesthetic: Essays on Postmodern Culture.* Port Townsend, Wash.: Bay, 1983.

Foucault, Michel. "What is an Author?" In *Language, Counter-Memory, Practice: Selected Essays and Interviews.* Translated by Donald F. Bouchard and Sherry Simon. Ithaca, N.Y.: Cornell University Press, 1977. 113–38.

———. *The Archeology of Knowledge and the Discourse on Language.* Translated by A. M. Sheridan Smith. New York: Harper & Row, 1976.

———. *The Order of Things: An Archeology of the Human Sciences.* New York: Vintage, 1973.

French, Howard. "Big Brother and the WWW—China in Action." *The New York Times* (June 27, 2004). Online version at http://www.nytimes.com/2004/06/27/international/asia/27chin.html.

Frow, John. *Marxism and Literary History.* Cambridge, Mass.: Harvard University Press, 1986.

Galegher, Jolene, Carmen Egido, and Robert Kraut, eds. *Intellectual Teamwork.* Hillsdale, N.J.: Erlbaum, 1990.

Gansing, Kristoffer." The Myth of Interactivity or the Interactive Myth?: Interactive Film as an Imaginary Genre." In *Melbourne DAC.* Edited by Adrian Miles. Melbourne: RMIT University, 2003. 51–58.

Geertz, Clifford. *Works and Lives: The Anthropologist as Author.* Stanford, Calif.: Stanford University Press, 1988.

Genette, Gérard. *Figures of Literary Discourse.* Translated by Alan Sheridan. New York: Columbia University Press, 1982.

———. *Narrative Discourse: An Essay in Method.* Translated by Jane E. Lewin. Ithaca, N.Y.: Cornell University Press, 1980.

Gibson, William. *Mona Lisa Overdrive.* New York: Bantam, 1988.

———. *Burning Chrome.* New York: Ace, 1987.

———. *Neuromancer.* New York: Ace, 1984.

Giedion, Sigfried. *Mechanization Takes Command: A Contribution to Anonymous History.* New York: Norton, 1969.

Gilbert, Steven W. "Information Technology, Intellectual Property, and Education." *EDUCOM Review* 25 (Spring 1990): 14–20.

Glassman, James K. "What Becomes of Government in an Electronic Revolution?" *Providence Journal-Bulletin* (September 3, 1995): D13.

Golovchinsky, Gene, and Catherine C. Marshall. "Hypertext Interaction Revisited." In *ACM 2000 Hypertext: Proceedings of the Eleventh ACM Conference on Hypertext and Hypermedia.* New York: ACM, 2000. 171–79.

Gore, Albert. "Remarks on the NREN." *EDUCOM Review* 25 (Summer 1990): 12–16.

Gray, Chris Hables, Heidi J. Figueroa-Sarriera, and Steven Mentor. *The Cyborg Handbook*. London: Routledge, 1995.

Greco, Diane. "Hypertext with Consequences: Recovering a Politics of Hypertext." In *Hypertext '96*. New York: Association for Computing Machinery, 1996. 85–92.

Grigely, Joseph. *Textualterity: Art, Theory, and Textual Criticism*. Ann Arbor: University of Michigan Press, 1995.

Groden, Michael, and Martin Kreiswirth, eds. *The Johns Hopkins Guide to Literary Theory and Criticism*. Baltimore: Johns Hopkins University Press, 1994.

Grudin, Robert. *Book: A Novel*. New York: Random House, 1992.

Gunder, Anna. "Aspects of Linkology: A Method for Description of Links and Linking." In *CyberText Yearbook 2001*. Jyväskylä: Research Centre for Contemporary Culture, University of Jyväskylä, 2002. 111–39.

———. "Forming the Text, Performing the Work—Aspects of Media, Navigation, and Linking." In *Human IT*. Boras, Sweden: Centre for Information Technology Studies as a Human Science (ITH), 2001. 81–206.

Guyer, Carolyn. "Buzz-Daze Jazz and the Quotidian Stream." Unpublished MS of a paper delivered at MLA Panel, "Hypertext, Hypermedia: Defining a Fictional Form." December 1992.

———. Journal kept during writing of *Quibbling*. Unpublished MS.

———. "Something about *Quibbling*." *Leonardo* (October 1992): 258.

Haan, Bernard J., Paul Kahn, Victor A. Riley, James H. Coombs, and Norman K. Meyrowitz. "IRIS Hypermedia Services." *Communications of the ACM* 35 (1992): 36–51.

Hall, Wendy. "Making Hypermedia Work in Education." In *Hypermedia at Work: Practice and Theory in Higher Education*. Edited by W. Strang, V. B. Simpson, and D. Slater. Canterbury: University of Kent, 1995. 1–19.

Hall, Wendy, Hugh Davis, and Gerard Hutchings. *Rethinking Hypermedia: The Microcosm Approach*. Boston: Kluwer, 1996.

Hamilton, David P. "Japanese Embrace a Man Too Eccentric for Silicon Valley: After Years of Failure in the U.S., Ted Nelson Is Continuing His Quest for Xanadu." *Wall Street Journal* (April 26, 1996): A1, A10.

Harpold, Terence. "Threnody: Psychoanalytic Digressions on the Subject of Hypertexts." In *Hypermedia and Literary Studies*. Edited by Paul Delany and George P. Landow. Cambridge, Mass.: MIT Press, 1991. 171–84.

Hayles, N. Katherine. *How We Became Posthuman: Virtual Bodies in Cybernetics, Literature, and Informatics*. Chicago: University of Chicago Press, 1999.

Heim, Michael. *The Metaphysics of Virtual Reality*. New York: Oxford University Press, 1993.

———. *Electric Language: A Philosophical Study of Word Processing*. New Haven: Yale University Press, 1987.

Henderson, J. V., R. K. Pruett, A. R. Galper, and W. S. Copes. "Interactive Videodisc to Teach Combat Trauma Life Support." *Journal of Medical Systems* 10, no. 3 (June 1986): 271–76.

BIBLIOGRAPHY

Hernandez, Carlos Moreno. *Literatura e Hipertexto: De la cultura manuiscrita a la cultura electonica.* Madrid: Universidad Nacional de Educaion a Distancia, 1998.

Hertz, J. J., ed. *The Pentateuch and Haftorahs.* 2nd ed. London: Soncino, 1962.

Howard, Alan. "Hypermedia and the Future of Ethnography." *Cultural Anthropology* 3 (1988): 304–15.

Hunt, Lynn, ed. *The Invention of Pornography: Obscenity and the Origins of Modernity, 1500–1800.* New York: Zone, 1993.

Hutzler, Charles. "China Finds New Ways to Restrict Access to the Internet." *Wall Street Journal* (September 2004): B1–B2.

Huyssen, Andreas. *After the Great Divide: Modernism, Mass Culture, and Postmodernism.* Bloomington: Indiana University Press, 1987.

Institute for Research in Information and Scholarship. *The Dickens Web: User's and Installation Guide.* Providence, R.I.: Institute for Research in Information and Scholarship, 1990.

IRIS Intermedia System Administrator's Guide: Release 3.0. Providence, R.I.: Institute for Research in Information and Scholarship, 1989.

IRIS Intermedia User's Guide: Release 3.0. Providence, R.I.: Institute for Research in Information and Scholarship, 1989.

Ivins, William M. *Prints and Visual Communication.* New York: DaCapo, 1969.

Jameson, Fredric. *The Political Unconscious: Narrative as a Socially Symbolic Act.* Ithaca, N.Y.: Cornell University Press, 1981.

———. *Marxism and Form: Twentieth-Century Dialectical Theories of Literature.* Princeton: Princeton University Press, 1971.

Janson, H. W., with Dora Jane Janson. *History of Art: Survey of Major Visual Arts from the Dawn of History to the Present Day.* New York: Abrams, 1962.

Johnson, Barbara. *A World of Difference.* Baltimore: Johns Hopkins University Press, 1987.

Jonassen, David H. "Hypertext Principles for Text and Courseware Design." *Educational Psychologist* 21 (1986): 269–92.

———. "Information Mapping: A Description, Rationale and Comparison with Programmed Instruction." *Visible Language* 15 (1981): 55–66.

Jonassen, David H., and R. Scott Grablinger. "Problems and Issues in Designing Hypertext/Hypermedia for Learning." In *Designing Hypertext/Hypermedia for Learning.* Edited by David H. Jonassen and Heinz Mandl. Heidelberg: Springer-Verlag, 1990. 3–26.

Jonassen, David H., and Heinz Mandl, eds. *Designing Hypertext/Hypermedia for Learning,* Heidelberg: Springer-Verlag, 1990.

Jones, Loretta L., Jennifer L. Karloski, and Stanley G. Smith. "A General Chemistry Learning Center: Using the Interactive Videodisc." *Academic Computing* 2 (September 1987): 36–37, 54.

Joyce, Michael. "My Body the Library." *American Book Review* (December–January 1995–96): 6, 31.

———. *Of Two Minds: Hypertext Pedagogy and Poetics.* Ann Arbor: University of Michigan Press, 1995.

BIBLIOGRAPHY

———. "Storyspace as a Hypertext System for Writers of Varying Ability." In *Hypertext '91*. New York: Association of Computing Machinery, 1991. 381–88.

Kahn, Joseph, Kathy Chen, and Marcus W. Brauchli. "Chinese Firewall: China Seeks to Build Version of the Internet that Can Be Censored." *Wall Street Journal* (January 31, 1996): A1, A4.

Kahn, Paul D. "Linking Together Books: Experiments in Adapting Published Material into Intermedia Documents." *Hypermedia* 1 (1989): 111–45.

———. "Isocrates: Greek Literature on CD Rom." In *CD ROM: The New Papyrus: The Current and Future State of the Art*. Edited by Steve Lambert and Suzanne Ropiequet. Redmond, Wash.: Microsoft, 1986.

Kahn, Paul D., Julie Launhardt, Krzysztof Lenk, and Ronnie Peters. "Design Issues of Hypermedia Publications: Issues and Solutions." *EP 90: International Conference on Electronic Publishing, Document Manipulations, and Typography*. Edited by Richard Furuta. Cambridge: Cambridge University Press, 1990. 107–24.

Kahn, Paul D., and Krzysztof Lenk. *Mapping Websites: Digital Media Design*. Craus-Près-Celigny, Switzerland: RotoVision, 2001.

———. "Typography for the Computer Screen: Applying the Lessons of Print to Electronic Documents." *Seybold Report on Desktop Publishing* 7 (July 5, 1993): 3–16.

Kahn, Paul D., and Norman Meyrowitz. "Guide, Hypercard, and Intermedia: A Comparison of Hypertext/Hypermedia Systems." Technical Report No. 88-7. Providence, R.I.: Institute for Research in Information and Scholarship, 1987.

Kapoor, Mitchell. "Democracy and New Information Highway." *Boston Review* (October–November 1993): 19–21.

Kaufman, Joanne. "Publishers are of Two Minds Regarding Double Bylines." *Wall Street Journal* (July 21, 2004): D12.

Kendall, Robert. "Toward an Organic Hypertext." In *ACM 2000 Hypertext: Proceedings of the Eleventh ACM Conference on Hypertext and Hypermedia*. 161–70.

———. "Hypertextual Dynamics in a *Life Set for Two*." In *Hypertext '96*. New York: Association for Computing Machinery, 1996. 74–84.

Kenner, Hugh. "The Making of the Modernist Canon." In *Canons*. Edited by Robert von Hallberg. Chicago: University of Chicago Press, 1984. 363–75.

Kermode, Frank. *The Sense of an Ending: Studies in the Theory of Fiction*. New York: Oxford University Press, 1967.

Kernan, Alvin. *Printing Technology, Letters and Samuel Johnson*. Princeton: Princeton University Press, 1987.

Kerr, Stephen T. "Instructional Text: The Transition from Page to Screen." *Visible Language* 20 (1986): 368–92.

King, Kenneth M. "Evolution of the Concept of Computer Literacy." *EDUCOM Bulletin* 20 (1986): 18–21.

Kittler, Friedrich A. *Discourse Networks 1800 / 1900*. Translated by Michael Metteer and Chris Cullins. Stanford, Calif.: Stanford University Press, 1992.

Kosko, Bart. *Fuzzy Thinking: The New Science of Fuzzy Logic*. London: Flamingo, 1994.

Kuhn, Thomas S. *The Structure of Scientific Revolutions*. 2nd ed. Chicago: University of Chicago Press, 1970.

BIBLIOGRAPHY

Lacan, Jacques. *The Language of the Self: The Function of Language in Psychoanalysis.* Translated by Anthony Wilden. Baltimore: Johns Hopkins University Press, 1968.

Landow, George P. "The Paradigm is More Important than the Purchase: Educational Innovation and Hypertext." In *Digital Media Revisited: Theoretical and Conceptual Innovation in Digital Domains.* Edited by Gunnar Liestøl, Andrew Morrison, and Terje Rasmussen. Cambridge, Mass.: MIT Press, 2003. 35–64.

———. "Newman and the Idea of an Electronic University." In *The Idea of a University.* Edited by Frank Turner. New Haven: Yale University Press, 1996. 339–61.

———. "Twenty Minutes into the Future, or How Are We Moving beyond the Book?" In *The Future of the Book.* Edited by Geoffrey Nunberg. Berkeley: University of California Press, 1996. 209–37.

———., ed. *Hyper/Text/Theory.* Baltimore: Johns Hopkins University Press, 1994.

———. "What's a Critic to Do? Critical Theory in the Age of Hypertext." In *Hyper/Text/Theory.* Edited by George P. Landow. Baltimore: Johns Hopkins University Press, 1994. 1–50.

———. "Electronic Conferences and Samszdat Textuality: The Example of Technoculture." In *The Digital Word: Text-Based Computing.* Edited by George P. Landow and Paul Delaney. Cambridge, Mass.: MIT Press, 1993. 237–49.

———. "Hypertext, Metatext, and the Electronic Canon." In *Literacy Online: The Promise (and Peril) of Reading and Writing with Computers.* Edited by Myron Tuman. Pittsburgh: University of Pittsburg Press., 1991. 67–94.

———. "Connected Images: Hypermedia and the Future of Art Historical Studies." In *Scholarship and Technology in the Humanities.* Edited by May Katzen. London: British Library Research/Bowker Saur, 1990. 77–94.

———. "The Rhetoric of Hypermedia: Some Rules for Authors." *Journal of Computing in Higher Education* 1 (1989): 39–64.

———. *Victorian Types, Victorian Shadows: Biblical Typology and Victorian Literature, Art, and Thought.* Boston: Routledge & Kegan Paul, 1980.

Landow, George P., and Paul Delany, eds. *The Digital Word: Text-Based Computing in the Humanities.* Cambridge, Mass.: MIT Press, 1993.

Landow, George P., and Paul Kahn. "The Pleasures of Possibility: What is Disorientation in Hypertext." *Journal of Computing in Higher Education* 4 (1993): 57–78.

———. "Where's The Hypertext? The Dickens Web as a System-Independent Hypertext." In *ECHT '92.* New York: ACM, 1992.

Lanham, Richard A. *The Electronic Word: Democracy, Technology, and the Arts.* Chicago: University of Chicago Press, 1993.

LaQuey, Tracy. "Networks for Academics." *Academic Computing* 4 (November 1989): 32–34, 39, 65.

Larson, James A. "A Visual Approach to Browsing in a Database Environment." *IEEE Computer* (1986): 62–71.

Lavers, Annette. *Roland Barthes: Structuralism and After.* Cambridge, Mass.: Harvard University Press, 1982.

Lazarus, Neil. *Resistance in Postcolonial African Fiction.* New Haven: Yale University Press, 1990.

Lee, Dorothy. "Lineal and Nonlineal Codifications of Reality." In *Symbolic Anthro-*

pology: A Reader in the Study of Symbols and Meanings. Edited by Janet L. Dolgin, David S. Kemnitzer, and David M. Schneider. New York: Columbia University Press, 1977. 151–64.

Leggett, John J., John L. Schnase, and Charles J. Kacmar. "Hypertext for Learning." In *Designing Hypertext/Hypermedia for Learning.* Edited by David H. Jonassen and Heinz Mandle. Heidelberg: Springer-Verlag, 1990.27–38.

Leitch, Vincent B. *Deconstructive Criticism: An Advanced Introduction.* New York: Columbia University Press, 1983.

Lévi-Strauss, Claude. *The Raw and the Cooked: Introduction to a Science of Mythology: I.* Translated by John and Doreen Weightman. New York: Harper & Row, 1969.

———. *The Scope of Anthropology.* Translated by Sherry Ortner Paul and Robert A. Paul. London: Jonathan Cape, 1967.

Liestøl. Gunnar. *Essays in Rhetorics of Hypermedia Design.* Oslo: Department of Media and Communication, University of Oslo, 1999.

———. "Aesthetic and Rhetorical Aspects of Linking Video in Hypermedia." In *ECHT '94.* New York: Association of Computing Machinery, 1994.

Linguaggio dei nuovi media, Il. Edited by Luca Toschi. Milan: Apogeo, 2001.

Linking the Continents of Knowledge: A Hypermedia Corpus for Discovery and Collaborative Work in the Sciences, Arts, and Humanities. Providence, R.I.: Institute for Research in Information and Scholarship, 1988.

Lorenzo, George. "Operation Education." *University Business* 4 (December 2001– January 2002): 34–42.

Lovink, Geert. *Dark Fiber: Tracking Critical Internet Culture.* Cambridge, Mass.: MIT Press, 2002.

Lyotard, Jean-François. *The Inhuman.* Translated by Geoffrey Bennington and Rachel Bowlby. Stanford, Calif.: Stanford University Press, 1991.

———. *The Postmodern Condition: A Report on Knowledge.* Translated by Geoff Bennington and Brian Massumi. Minneapolis: University of Minnesota Press, 1984.

Machery, Pierre. *A Theory of Literary Production.* Translated by Geoffrey Wall. London: Routledge & Kegan Paul, 1978.

McArthur, Tom. *Worlds of Reference: Lexicography, Learning and Language from the Clay Tablet to the Computer.* Cambridge: Cambridge University Press, 1986.

McCartney, Scott. "For Teens, Chatting on Internet Offers Comfort of Anonymity." *Wall Street Journal* (December 8, 1994): B1, B4.

———. *A Critique of Modern Textual Criticism.* Chicago: University of Chicago Press, 1983.

McClintlock, Robert. "On Computing and the Curriculum." *SIGCUE Outlook* (Spring–Summer 1986): 25–41.

McCloud, Scott. *Understanding Comics: The Invisible Art.* New York: HarperPerennial, [2003?].

McCorduck, Pamela. *Machines Who Think: A Personal Inquiry into the History and Prospects of Artificial Intelligence.* New York: W. H. Freeman, 1979.

McDermott, Dan. "Singapore Unveils Sweeping Measures to Control Words, Images on Internet." *Wall Street Journal* (March 6, 1996): A1.

McGann, Jerome J. "The Complete Writings and Pictures of Dante Gabriel Rossetti: A Hypermedia Research Archive." Unpublished MS.

———. *The Textual Condition*. Princeton: Princeton University Press, 1991.

McGrath, Joseph E. "Time Matters in Groups." In *Intellectual Teamwork*. Edited by Jolene Galegher, Carmen Egido, and Robert Kraut. Hillsdale, N.J.: Erlbaum, 1990. 23–62.

McHale, Brian. *Postmodernist Fiction*. New York: Methuen, 1987.

McKnight, John Richardson, and Andrew Dillon. *The Authoring of Hypertext Documents*. Loughborough, England: Loughborough University of Technology/HUSAT Research Center, 1988.

McLuhan, Marshall. *Understanding Media: The Extensions of Man*. London: Routledge, 2001.

———. *The Gutenberg Galaxy: The Making of Typographic Man*. Toronto: University of Toronto Press, 1962.

McNeill, Laurie. "Teaching an Old Genre New Tricks: The Diary on the Internet." *Biography* 26, no. 1 (2003): 24–47.

McQuillan, Patrick. "Computers and Pedagogy: The Invisible Presence." *Journal of Curriculum Studies* 26 (1994): 631–53.

Mancini, Clara. "Towards Cinematic Hypertext: A Theoretical and Empirical Study." Ph.D. diss. Open University, 2003.

Marchioni, Gary. "Evaluating Hypermedia-Based Learning." In *Designing Hypertext/Hypermedia for Learning*. Edited by David H. Jonassen and Heinz Mandl. Heidelberg: Springer-Verlag, 1990. 355–73.

Marshall, Catherine C., Frank G. Halasz, Russell A. Rogers, and William A. Janssen Jr. "Acquanet: A Hypertext Tool to Hold Your Knowledge in Place." In *Hypertext '91*. New York: Association of Computing Machinery. 1991. 261–75.

Marshall, Catherine C., and Frank M. Shipman III. "Spatial Hypertext and the Practice of Information Triage." In *Hypertext '97*. New York: ACM, 1997. 124–33.

Marshall, Catherine C., and Russell A. Rogers. "Two Years before the Mist: Experiences with Acquanet." In *ECHT '92*. New York: Association of Computing Machinery, 1992. 53–62.

Massumi, Brian. *A User's Guide to Capitalism and Schizophrenia: Deviations from Deleuze and Guattari*. Cambridge, Mass.: MIT Press, 1992.

Matejka, Ladislav, and Krystyna Pomorska, eds. *Readings in Russian Poetics: Formalist and Structuralist Views*. Cambridge, Mass.: MIT Press, 1971.

Matejka, Ladislav, and Irwin R. Titunik. *Semiotics of Art: Prague School Contributions*. Cambridge, Mass.: MIT Press, 1976.

Mayes, Terry, Mike Kibby, and Tony Anderson. "Learning about Learning for Hypertext." In *Designing Hypertext/Hypermedia for Learning*. Edited by David H. Jonassen and Heinz Mandl. Heidelberg: Springer-Verlag, 1990. 227–50.

Meyer, Tom, David Blair, and Suzanne Hader. "*WAXweb*: A MOO-based Collaborative Hypermedia System for WWW." *Computer Networks and ISDN Systems* 28 (1995): 77–84.

Meyrowitz, Norman. "Hypertext—Does It Reduce Cholesterol, Too?" In *Vannevar

BIBLIOGRAPHY

Bush and the Mind's Machine: From Memex to Hypertext. Edited by James M. Nyce and Paul D. Kahn. San Diego: Academic, 1991. 287–318.

Miles, Adrian. "Softvideography: Digital Video as Postliterate Practice." In *Digital Tools and Cultural Contexts: Assessing the Implications of New Media.* Edited by Brian Hawk, James A. Inman, and Ollie Oviedo. Prepublication PDF copy.

Miller, J. Hillis. *Illustration.* Cambridge, Mass.: Harvard University Press, 1992.

———. "Literary Theory, Telecommunications, and the Making of History." In *Scholarship and Technology in the Humanities.* Edited by May Katzen. London: British Library Research/Bowker Saur, 1991. 11–20.

———. *Versions of Pygmalion.* Cambridge, Mass.: Harvard University Press, 1990.

———. *Fiction and Repetition: Seven English Novels.* Cambridge, Mass.: Harvard University Press, 1982.

Minsky, Marvin. *The Society of Mind.* New York: Simon and Schuster, 1986.

Mitchell, William J. *Me++: The Cyborg Self and the Networked City.* Cambridge, Mass.: MIT Press, 2003.

———. *City of Bits: Space, Place, and the Infobahn.* Cambridge, Mass.: MIT Press, 1995.

Mitchell, W.J.T., ed. *On Narrative.* Chicago: University of Chicago Press, 1980.

Moi, Toril. *Sexual/Textual Politics: Feminist Literary Theory.* London: Methuen, 1985.

Morgan, Thaïs E. "Is There an Intertext in This Text?: Literary and Interdisciplinary Approaches to Intertextuality." *American Journal of Semiotics* 3 (1985): 1–40.

Morrell, Kenneth. "Teaching with *Hypercard.* An Evaluation of the Computer-based Section in Literature and Arts C-14: The Concept of the Hero in Hellenic Civilization." Perseus Project Working Paper 3. Cambridge, Mass.: Department of Classics, Harvard University, 1988.

Mosco, Vincent. *The Digital Sublime: Myth, Power, and Cyberspace.* Cambridge, Mass.: MIT Press, 2004.

Moulthrop, Stuart. "Rhizome and Resistance: Hypertext and the Dreams of a New Culture." In *Hyper/Text/Theory.* Edited by George P. Landow. Baltimore: Johns Hopkins University Press, 1994. 299–322.

———. "Beyond the Electronic Book: A Critique of Hypertext Rhetoric." In *Hypertext '91.* New York: Association of Computing Machinery, 1991. 291–98.

———. "Reading from the Map: Metonymy and Metaphor in the Fiction of 'Forking Paths'" In *Hypermedia and Literary Studies.* Edited by Paul Delany and George P. Landow. Cambridge, Mass.: MIT Press, 1991. 119–32.

———. "You Say You Want a Revolution? Hypertext and the Laws of Media." *Postmodern Culture* 1 (May 1991).

———. "Containing the Multitudes: The Problem of Closure in Interactive Fiction." *Association for Computers in the Humanities Newsletter* 10 (Summer 1988): 29–46.

Mowitt, John. *Text: The Genealogy of an Antidisciplinary Object.* Durham, N.C.: Duke University Press, 1992.

———. "Hypertext and 'the Hyperreal.'" In *Hypertext '89 Proceedings.* New York: Association of Computing Machinery, 1989. 259–68.

Multimedia: From Wagner to Virtual Reality. Edited by Randall Packer and Ken Jordan. New York: W. W. Norton, 2001.

BIBLIOGRAPHY

Mungoshi, Charles. *Waiting for the Rain*. London: Heinemann Educational Books, 1975. Reprinted Harare: Zimbabwe Publishing House, c. 1996.

Murray, Janet. *Hamlet on the Holodeck: The Future of Narrative in Cyberspace*. New York: The Free Press, 1997.

Murray, Oswyn. "The Word Is Mightier Than the Sword." *Times Literary Supplement* (June 16–22, 1989): 655.

Mylonas, Elli. "Design by Exploration: An Academic Hypertext." In *Hypermedia at Work: Practice and Theory in Higher Education*. Edited by W. Strang, V. B. Simpson, and D. Slater. Canterbury: University of Kent, 1995. 39–54.

———. "The Perseus Project: Ancient Greece in Texts, Maps and Images." In *Electronic Books—Multimedia Reference Works*. Bergen: Norwegian Computing Centre, 1991. 173–88.

Nation and Narration. Edited by Homi K. Bhabha. London: Routledge, 1990.

Negroponte, Nicholas. *Being Digital*. New York: Knopf, 1995.

Nelson, Theodor Holm. Selection from *Computer Lib/Dream Machines*. In *The New Media Reader*. Edited by Noah Warip-Fruin and Nick Montfort. Cambridge, Mass.: MIT Press, 2003. 301–38.

———. "Hypermedia Unified by Transclusion." *Communications of the ACM* 38 (August 1995): 31–32.

———. *Computer Lib/Dream Machines*. Seattle, Wash.: Microsoft, 1987.

———. *Literary Machines*. Swarthmore, Pa.: Self-published, 1981.

Newman, John Henry. *The Idea of a University*. Edited by Frank Turner. New Haven: Yale University Press, 1996.

Nielsen, Jakob. *Designing Web Usability: The Practice of Simplicity*. Indianapolis: New Riders, 2000.

———. *Multimedia and Hypertext: The Internet and Beyond*. Boston: Academic, 1995.

———. "The Art of Navigating through Hypertext." *Communications of the ACM* 33 (1990): 296–310.

Novak, Joseph D., and D. Bob Gowin. *Learning How to Learn*. New York: Cambridge University Press, 1984.

Nürnberg, Peter J., John J. Leggett, and Erich R. Schneider. "As We Should Have Thought." In *Hypertext '97*. New York: ACM, 1997. 96–101.

Nyce, James M., and Paul Kahn. "Innovation, Pragmatism, and Technological Continuity: Vannevar Bush's Memex." *Journal of the American Society for Information Science* 40 (1989): 214–20.

———, eds. *From Memex to Hypertext: Vannevar Bush and the Mind's Machine*. Boston: Academic, 1991.

Ohmann, Richard. "The Shaping of a Canon: U.S. Fiction, 1960–1975." In *Canons*. Edited by Robert von Hallberg. Chicago: University of Chicago Press, 1984. 377–401.

Okonkwo, Chidi. *Decolonization Agonistics in Postcolonial Fiction*. New York: St. Martins, 1999.

Ong, Walter J. *Orality and Literacy: The Technologizing of the Word*. London: Methuen, 1982.

BIBLIOGRAPHY

————. *Rhetoric, Romance, and Technology: Studies in the Interaction of Expression and Culture*. Ithaca, N.Y.: Cornell University Press, 1971.

Pagels, Heinz R. *The Dreams of Reason: The Computer and the Rise of the Sciences of Complexity*. New York: Bantam, 1989.

Paul, Christian. "Reading/Writing Hyperfictions: The Psychodrama of Interactivity." *Leonardo* 28, no. 4 (1995): 265–72.

Paulson, William R. *The Noise of Culture: Literary Texts in a World of Information*. Ithaca, N.Y.: Cornell University Press, 1988.

Pavic, Milorad. *Dictionary of the Khazars: A Lexicon Novel in 100,000 Words*. Translated by Christina Pribicevic-Zoric. New York: Knopf, 1988.

Peckham, Morse. *Man's Rage for Chaos: Biology, Behavior, and the Arts*. New York: Schocken, 1967.

Penley, Constance. "Brownian Motion: Women, Technology, and Tactics." In *Technoculture*. Edited by Constance Penley and Andrew Ross. Minneapolis: University of Minnesota Press, 1991. 135–61.

Perkins, D. N. "The Fingertip Effect: How Information-Processing Technology Shapes Thinking." *Educational Researcher* (August–September 1985): 11–17.

"Pleasures of the (Hyper)Text, The." *New Yorker* (June 27 and July 4, 1994): 43–44.

Post-colonial Studies Reader, The. Edited by Bill Ashcroft, Gareth Griffiths, and Helen Tiffin. London: Routledge, 1995.

Propp, Vladimir. "Fairy Tale Transformations." In *Readings in Russian Poetics: Formalist and Structuralist Views*. Edited by Ladislav Matejka and Krystyna Pomorska. Cambridge, Mass: MIT Press, 1971. 94–116.

————. *Morphology of the Folk Tale*. Translated by Laurence Scott. 2nd rev. ed. Austin: University of Texas Press, 1968.

Provenzo, Eugene F. *Beyond the Gutenberg Galaxy: Microcomputers and the Emergence of Post-Typographic Culture*. New York: Teachers College Press, 1986.

Reading Digital Culture. Edited by David Trend. Oxford: Blackwell, 2001.

Richtel, Matt, and Heather Timmons. "Gambling Sites Offering Ways to Let Any User Be the Bookie." *New York Times* (July 6, 2004). Electronic edition. http://www.nytimes.com/2004/07/06/technology/06gamble.html.

Ricoeur, Paul. *Time and Narrative*. Translated by Kathleen McLaughlin and David Pellauer. 2 vols. Chicago: University of Chicago Press, 1984.

Rigden, Joan E. "Homebound and Lonely, Older People Use Computers to Get 'Out.'" *Wall Street Journal* (December 8, 1994): B1, B14.

Rizk, Antoine, and Dale Sutcliffe. "Distributed Link Service in the Aquarelle Project." In *ACM 2000 Hypertext*. New York: ACM, 2000. 208–9.

Rogers, Susan M. "Educational Applications of the NREN." *EDUCOM Review* 25 (Summer 1990): 25–29.

Ronell, Avital. *The Telephone Book: Technology, Schizophrenia, Electric Speech*. Lincoln: University of Nebraska Press, 1989.

Rorty, Richard. *Philosophy and the Mirror of Nature*. Princeton: Princeton University Press, 1979.

BIBLIOGRAPHY

Rosen, Jeffrey. "Cheap Speech: Will the Old First Amendment Battles Survive the New Technologies?" *New Yorker* (August 7, 1995): 75–80.

Rosenau, Pauline Marie. *Post-Modernism and the Social Sciences: Insights, Inroads, and Intrusions*. Princeton: Princeton University Press, 1992.

Rosenheim, Andrew. "The Flow That's Becoming a Flood." *Times Literary Supplement* (May 12, 1995): 10–11.

Rossiter, Ned. "Processual Media Theory." In *Melbourne DAC*. Edited by Adrian Miles. Melbourne: RMIT University, 2003. 173–84.

Rushdie, Salman. *Shame*. New York: Aventura/Vintage, 1984.

Russell, Daniel M. "Creating Instruction with IDE: Tools for Instructional Designers." *Intelligent Learning Media* 1 (1990): 1.

Russell, Daniel M., Daniel S. Jordan, Anne-Marie S. Jensen, and Russell A. Rogers. "Facilitating the Development of Representations in Hypertext with IDE." In *Hypertext '89 Proceedings*. New York: Association of Computing Machinery, 1989. 93–104.

Russell, Daniel M., and Peter Piroli. "The Instructional Design Environment: Technology to Support Design Problem Solving." *Instructional Science* 19 (1990): 121–44.

Ryan, Marie-Laure. *Narrative as Virtual Reality*. Baltimore: Johns Hopkins University Press, 2001.

Ryan, Michael. *Marxism and Deconstruction: A Critical Articulation*. Baltimore: Johns Hopkins University Press, 1982.

Saenger, Paul. "Books of Hours and the Reading Habits of the Middle Ages." In *The Culture of Print: Power and the Uses of Print in Early Modern Europe*. Edited by Roger Chartier. Translated by Lydia Cochrane. Princeton: Princeton University Press, 1989. 141–90.

Said, Edward W. *Beginnings: Intention and Method*. New York: Columbia University Press, 1985.

Sandberg, Jared. "Fringe Groups Can Say Almost Anything and Not Worry About Getting Punched." *Wall Street Journal* (December 8, 1994): B1, B4.

———. "Regulators Try to Tame the Untameable On-Line World." *Wall Street Journal* (July 5, 1995): B1, B3.

Saro-Wiwa, Ken. *A Forest of Flowers*. Fort Harcourt, Nigeria: Saros International, 1986; Burnt Mill: Addison Wesley Longman, 1995.

Sawhney, Nitin, David Balcom, and Ian Smith. "HyperCafe: Narrative and Aesthetic Properties of HyperVideo." In *Hypertext '96*. New York: Association for Computing Machinery, 1996. 1–10.

Schmitz, Ulrich. "Automatic Generation of Texts without Using Cognitive Models: Television News." In *The New Medium: ALLC-ACH 90 Book of Abstracts*. Siegen, Germany: University of Siegen, Association for Literary and Linguistic Computing, and the Association for Computers and the Humanities, 1990. 191–95.

———. *Postmoderne Concierge: Die "Tagesschau." Wortwelt und Weltbild der Fernsehnachrichten*. Opladen, Germany: Westdeutscher Verlag, 1990.

Schneiderman, Ben, and Greg Kearsley. *Hypertext Hands-On! An Introduction to a New Way of Organizing and Accessing Information.* Reading, Mass.: Addison-Wesley, 1989.

Scholes, Robert. *Textual Power: Literary Theory and the Teaching of English.* New Haven: Yale University Press, 1985.

———. *Structuralism in Literature: An Introduction.* New Haven: Yale University Press, 1974.

Scholes, Robert, and Robert Kellogg. *The Nature of Narrative.* New York: Oxford University Press, 1966.

Schwartz, Hillel. *The Culture of the Copy: Striking Likeness, Unreasonable Facsimiles.* New York: Zone, 1996.

Shoaf, R. A. "Literary Theory, Medieval Studies, and the Crisis of Difference." In *Reorientations: Critical Theories and Pedagogies.* Edited by Bruce Henricksen and Thaïs Morgan. Urbana: University of Illinois Press, 1990. 77–94.

Silicon Literacies: Communication, Innovation and Education in the Electronic Age. London: Routledge, 2002.

Smith, Barbara Herrnstein. "Narrative Versions, Narrative Theories." In *On Narrative.* Edited by W.J.T. Mitchell. Chicago: University of Chicago Press, 1980. 209–32.

———. *Poetic Closure: A Study of How Poems End.* Chicago: University of Chicago Press, 1968.

Smith, Tony C., and Ian H. Witten. "A Planning Mechanism for Generating Story Text." In *The New Medium: ALLC-ACH 90 Book of Abstracts.* Siegen, Germany: University of Siegen, Association for Literary and Linguistic Computing, and the Association for Computers and the Humanities, 1990. 201–4.

Spiro, Rand J., Richard L. Coulson, Paul J. Felktovich, and Daniel K. Anderson. "Cognitive Flexibility Theory: Advanced Knowledge Acquisition in Ill-structured Domains." In *Program of the Tenth Annual Conference of the Cognitive Science Society.* Hillsdale, N.J.: Erlbaum, 1988.

Spiro, Rand J., Walter P. Vispoel, John G. Schmitz, Ala Samarapungavan, and A. E. Boerger. "Knowledge Acquisition for Application: Cognitive Flexibility and Transfer in Complex Content Domains." In *Executive Control Processes in Reading.* Edited by B. K. Britton and S. McGlynn. Hillsdale, N.J.: Erlbaum, 1987. 177–99.

Steigler, Marc. "Hypermedia and Singularity." *Analog Science Fiction* (1989): 52–71.

Steinberg, S. H. *Five Hundred Years of Printing.* 2nd ed. Baltimore: Penguin, 1961.

Suckale, Robert. *Studien zu Stilbildung und Stilwandel de Madonnenstatuen der Ile-de-France zwischen 1240 und 1300.* Munich: University of Munich, 1971.

Sutherland, John. "Author's Rights and Transatlantic Differences." *Times Literary Supplement* (May 25–31, 1990): 554.

Svedjedal, Johan. "Notes on the Concept of 'Hypertext.'" In *Human IT.* Boras, Sweden: Centre for Information Technology Studies as a Human Science (ITH), 1999. 149–66.

Taylor, Mark C., and Esa Saarinen. *Imagologies: Media Philosophy.* London: Routledge, 1994.

Thomas, Brook. "Bringing about Critical Awareness through History in General Education Literature Courses." In *Reorientations: Critical Theories and Pedagogies.*

Edited by Bruce Henricksen and Thaïs Morgan. Urbana: University of Illinois Press, 1990. 219–47.

Thorpe, James. *Principles of Textual Criticism*. San Marino, Calif.: Huntington Library, 1972.

Timpe, Eugene F. "Memory and Literary Structures." *Journal of Mind and Behavior* 2 (1981): 293–307.

Todorov, Tzvetan. *The Poetics of Prose*. Translated by Richard Howard. Ithaca, N.Y.: Cornell University Press, 1977.

Tosca, Susana Pajares. "A Pragmatics of Links." In *ACM 2000 Hypertext: Proceedings of the Eleventh ACM Conference on Hypertext and Hypermedia*. 77–91.

Ulmer, Gregory L. *Heuretics: The Logic of Invention*. Baltimore: Johns Hopkins University Press, 1994.

———. "Textshop for an Experimental Humanities." In *Reorientations: Critical Theories and Pedagogies*. Edited by Bruce Henricksen and Thaïs Morgan. Urbana: University of Illinois Press, 1990.

———. *Teletheory: Grammatology in the Age of Video*. London: Routledge, 1989.

———. *Applied Grammatology: Post(e)-Pedagogy from Jacques Derrida to Joseph Beuys*. Baltimore: Johns Hopkins University Press, 1985.

———. "The Object of Post-Criticism." In *The Anti-Aesthetic: Essays on Postmodern Culture*. Edited by Hal Forster. Post Townsend, Wash.: Bay, 1983. 83–110.

Utting, Kenneth, and Nicole Yankelovich. "Context and Orientation in Hypermedia Networks." *ACM Transactions on Information Systems* 7 (1989): 58–84.

van Dam, Andries. "Keynote Address Hypertext '87>" *Communications of the Association of Computing Machinery* 31 (July 1988): 887–95.

Vera, Yvonne. *Nehanda*. Harare: Baobab Books, 1993.

von Hallberg, Robert, ed. *Canons*. Chicago: University of Chicago Press, 1984.

Waldrop, M. Mitchell. *Complexity: The Emerging Science at the Edge of Order and Chaos*. New York: Simon and Schuster, 1992.

Walsh, Peter. "Are Words Dead?: Designer David Carson's Revenge on the Stony Roman Alphabet Looks Softer in the Context of Calligraphy." *Providence Phoenix* (March 1, 1996): 2, 3–4.

Walter, John L. "On Phonography: How Recording Techniques Change Music and Musicians." *Times Literary Supplement* (April 26, 1996): 10.

Wardrip Fruin, Noah, and Brion Moss. "The Impermanence Agent: Project and Context." In *CyberText Yearbook 2001*. Jyväskylä: Research Centre for Contemporary Culture, University of Jyväskylä, 2002. 13–58.

Weissman, Ronald F. E. "From the Personal Computer to the Scholar's Workstation." *Academic Computing* 3 (October 1988): 10–14, 30–34, 36, 38–41.

Wexelblat, Alex. *Virtual Reality: Applications and Explorations*. Boston: Academic Publishers Professional, 1993.

Whalley, Peter. "Models of Hypertext Structure and Models of Learning." In *Designing Hypertext/Hypermedia for Learning*. Edited by David H. Jonassen and Heinz Mandl. Heidelberg: Springer-Verlag, 1990. 61–70.

White, Hayden. "The Value of Narrativity in the Representation of Reality." In *On Narrative*. Edited by W.J.T. Mitchell. Chicago: University of Chicago Press, 1980. 1–24.

Wilson, Kathleen S. "Palenque: An Interactive Multimedia Optical Disc Prototype for Children." Working Paper no. 2. New York: Bank Street College of Education/Center for Children and Technology, 1986.

Wittig, Rob. *Invisible Rendezvous: Connection and Collaboration in the New Landscape of Electronic Writing*. Hanover: Wesleyan University Press, 1995.

Wittgenstein, Ludwig. *Philosophical Investigations*. 3rd ed. Translated by G.E.M. Anscombe. New York: Macmillan, 1968.

Wolf, Gary. "The Curse of Xanadu." *Wired* 3 (June 1995): 137–52, 194–202.

Wu, Gordon. "Soft Soap." *Times Literary Supplement* (July 20–26, 1990): 777.

Yankelovich, Nicole. "From Electronic Books to Electronic Libraries: Revisiting 'Reading and Writing the Electronic Book.'" In *Hypermedia and Literary Studies*. Edited by Paul Delany and George P. Landow. Cambridge, Mass.: MIT Press, 1991. 133–41.

Yankelovich, Nicole, Norman Meyrowitz, and Stephen Drucker. "Intermedia: The Concept and the Construction of a Seamless Information Environment." *IEEE Computer* 21 (1988): 81–96.

Yankelovich, Nicole, Norman Meyrowitz, and Andries van Dam. "Reading and Writing the Electronic Book." *IEEE Computer* 18 (October 1985): 15–30.

Zachary, G. Pascal. "Digital Age Spawns "Neo-Luddite" Movement." *Wall Street Journal* (April 12, 1996): B1, B3.

Zampolli, Antonio. "Technology and Linguistics Research." In *Scholarship and Technology in the Humanities*. Edited by May Katzen. London: British Library/Bowker Saur, 1991. 21–51.

Zeleny, Jeff. "Poetic License on the Internet: Odes to Spam Renew/Literary Zest On-Line/Haiku Craze Is Back." *Wall Street Journal* (July 22, 1996): B1–B2.

Zimmer, Carl. "Floppy Fiction." *Discover* (November 1989): 34–36.

Electronic Materials and Videos

Note: Unless otherwise stated, all web materials were accessed in July 2004.

Aarseth, Espen. A hypertext version of Raymond Queneau's *Cent Mille Milliards de Poèmes*. (1961) Environment: Hypercard. n.d.

Adelman, Ian, and Paul Jahn. *Memex Animation*. Environment: Director. Providence: Dynamic Diagrams, 1995. Also available from http://dynamicdiagrams.com/services_ipp_memex.html.

Anderson, David, Robert Cavalier, and Preston K. Covey. *A Right to Die? The Dax Cowart Case*. CD-ROM. London: Routledge, 1996.

Anderson, Laurie, with Huang Hsien-Chien. *Puppet Motel*. CD-ROM. New York: Voyager, 1995.

Blake, Scott. *Bar Code/Jesus 01*. Bar Code Art. http://www.barcodeart.com/art/portrait/bar_code_jesus_02_01.html.

Bly. Bill. We Descend. http://wordcircuits.com/gallery/descend/Title.htm.

Bolter, J. David. *Writing Space: A Hypertext*. Environment: Storyspace. Hillsdale: Erlbaum, 1990.

Bonilla, Diego. *A Space of Time: A Non-Linear Multimedia Story*. Macintosh CD-ROM. Self-published, 2003.

BIBLIOGRAPHY

Bradbury, Nick. "FeedDemon and well-formed Atom feeds." http://nick.typepad.com/blog/2004/01/feeddemon_and_w.html.

Brown, P. J., and Heather Brown. "Integrating Reading and Writing of Documents." *Journal of Digital Information*, Volume 5 Issue 1. Article No. 237, 2004-02-02 .

Brusilovsky, Peter, and Riccardo Rizzo. "Map-Based Horizontal Navigation in Educational Hypertext." *Journal of Digital Information* 3 (2002-07-31): Article No. 156. http://jodi.ecs.soton.ac.uk/Articles/v03/i01/Brusilovsky/.

Buchanan, Dorothy, and Stacie Hibino. *A Voice in the Silence.* Environment: Toolbook. Ann Arbor: Project FLAME, 1996. An unpublished hypermedia project centering on the seventeenth-century Mexican poetess Sor Juana Inese de la Cruz; demonstrated at Hypertext '96.

Carr, Leslie, David De Roure, and Gary Hill. "Ongoing Development of an Open Link Service for the World-Wide Web." http://eprints.ecs.soton.ac.uk/archive/00000768/02/html/.

Cecchi, Alberto. *Ipertesto tratto da "Il castello dei destini incrociati," di Italo Calvino.* Environment: Storyspace. Unpublished disk. c. 1995.

Cook, Steven. *In(f)lections: Writing as Virus: Hypertext as Meme.* http//www.cyberartsweb.org/cpace/infotech/cook/centre.html.

CD Word. The Interactive Bible Library. Environment: Specially amplified Guide. Dallas: CD Word Library, Inc., 1990.

Craven, Jackie. *In the Changing Room.* http://wordcircuits.com/gallery/changing/Intro.htm.

Critical Mass: Graduate Projects. Environment: Projector. Pasadena: Art Center College of Design, 1995.

Daniele, Daniela. *Travelogues in a Broken Landscape: Robert Smithson's Visual/Textual Works.* Environment: Windows 3.1 Help System. Unpublished disks, 1993.

Dickens, Eric . "Widening the Postcolonial Debate." *Postcolonial Web.* http://www.postcolonialweb.org/poldiscourse/edickens1.html.

The Dickens Web. Developer: George P. Landow. Editors: Julie Launhardt and Paul D. Kahn. (1) Environment: Intermedia 3.5. Providence, R.I.: Institute for Research in Information and Scholarship, 1990. (2) Environment: Storyspace. Watertown: Eastgate Systems, 1992. (3) Interleaf Worldview. Unpublished.

di Iorio, Angelo, and Fabio Vitali. "Writing the Web." *Journal of Digital Information,* Volume 5 Issue 1. Article No. 251, 2004-05-27.

Dotze Sentits: Poesia catalona d'avui. Interactive design by Pere Freixa and J. Ignasi Ribe. CD-ROM . Barcelona: Universitat Pompeu Fabra, 1996.

E-blogger. "What is ATOM?" http://help.blogger.com/bin/answer.py?answer=697&topic=36.

Englebart, Douglas C. The Demo (of the online system NLS) on December 9, 1968, in the Augmentation Research Center, Stanford Research Institute, Menlo Park, CA. http://sloan.stanford.edu/mousesite/1968Demo.html.

Exploring the Moon. Developers: Katie Livingston, Jayne Aubele, and James Head. Editors: Paul D. Kahn and Julie Launhardt. Environment: Intermedia 3.5. Providence, R.I.: Institute for Research in Information and Scholarship, 1988.

Fisher, Caitlin. *These Waves of Girls.* 2004. http://www.yorku.ca/caitlin/waves/.

BIBLIOGRAPHY

Fishman, Barry J. *The Works of Graham Swift: A Hypertext Honors Thesis*. Environment: Intermedia. Brown University, 1989.

Flitman, Ian. *Hackney Girl*. 2003. http://www.blipstation.com/.

Forss, Pearl. *Authorship*. http://www.cyberartsweb.org/cpace/theory/.authorship/pearl/index.htm

Goetze, Erik. *Virtual Parks*. http://www.virtualparks.org/ [Quicktime VR panoramas of landscapes throughout the world].

Gmail. "About Gmail." http://gmail.google.com/gmail/help/about.html.

"Gmail is too creepy." http://gmail-is-too-creepy.com/.

Goodman, Hays. "Active Navigation generates links on-the-fly." *Newspapers and Technology: The International Journal of Newspaper Technology* (March 2002). http://www.newsandtech.com/issues/2002/03-02/ot/03-02_activenav.htm.

Gray, Edward, Kirstin Kantner, Chris Spain, and Noah Wardrip-Fruin. *Gray Matters*. (1) originally in Pad ++, (2) Web version (apparently no longer functional) http://artnetweb.com/theoricon/graymatters/.

Greco, Diane. *Cyborg: Engineering the Body Electric*. Environment: Storyspace. Watertown: Eastgate Systems, 1995.

Grossato, Giovanni. *La Scuola Grande di S. Rocco a Venezia*. Windows CD-ROM. Venice: AshMultimedia, 1997.

Guyer, Carolyn. *Quibbling*. Environment: Storyspace. Watertown: Eastgate Systems, 1993.

Gyford, Phil. "Why I turned Pepys' diary into a weblog." *BBC News* (January 2, 2003) http://news.bbc.co.uk/2/hi/uk_news/2621581.stm.

Hafner, Katie. "Old Search Engine, the Library, Tries to Fit Into a Google World." *The New York Times*. June 21, 2004. [Print edition: Section A , Page 1, Column 5.]

Harigan, Pat, Noah Wardrip-Fruin, Janet Murray, and Espen Aarseth. *First Person. Electronic Book Review*. 2004. http://www.electronicbookreview.com/v3/threads/threadtoc.jsp?thread=firstperson.

Howard, Peter. *Xylo*. http://www.wordcircuits.com/gallery/xylo/index.html.

Ikai, Taro, and George P. Landow. "Electronic Zen." *Cyberspace, Hypertext, and Critical Theory Web*. http://www.cyberartsweb.org/cpace/ht/ezen/title.html.

In Memoriam Web, The. Edited by Jon Lanestedt and George P. Landow. Environment: Storyspace. Watertown: Eastgate Systems, 1992.

Intermedia: A Retrospective. Written and narrated by Paul Kahn. 53-minute video. New York: Association for Computing Machinery, 1992.

Intermedia: From Linking to Learning. 27-minute video. Directed by Deborah Dorsey. Cambridge Studios and the Annenberg/Corporation for Public Broadcasting Project, 1986.

Jackson, Shelley. *Patchwork Girl*. Environment: Storyspace. Cambridge: Eastgate Systems, 1995.

Joyce, Michael. *afternoon*. (1) Environment: Storyspace Beta 3.3 Jackson, Mich.: Riverrun Limited, 1987, (2) Environment: Storyspace. Cambridge: Eastgate Systems, 1990.

Kendall, Robert. *A Life Set for Two*. Environment: Visual Basic for Windows. Watertown: Eastgate, 1996.

BIBLIOGRAPHY

Kolb, David. *Socrates in the Labyrinth: Text, Argument, Philosophy.* Environment: Storyspace. Cambridge: Eastgate Systems, 1994.

Lafaille, J. M. *Fragments d'une histoire.* Environment: Windows 3.1 Help System. Unpublished disk, 1994.

Landow, George P., ed. *Cyberspace, Virtual Reality, and Critical Theory Web.* http://www.cyberartsweb.org/index.html.

————. *Hypertext in Hypertext.* Environment: DynaText. Baltimore: Johns Hopkins University Press, 1993.

————. *Intermedia.* http://www.cyberartsweb.org/cpace/ht/HTatBrown/Intermedia.html. Includes screen shots of IRIS Intermedia, bibliography of IRIS publications, and materials on earlier systems, including FRESS.

————., ed. *Postcolonial and Postimperial Literature in English.* http://www.postcolinialweb.org/.

————. "A Review of Dale Porter's *The Thames Embankment.*" *The Victorian Web.* http://www.victorianweb.org/technology/review.html.

————., ed. *The Victorian Web.* http://www. victorianweb.org/.

————., ed. *Writing at the Edge.* Environment: Storyspace. Watertown: Eastgate Systems, 1995.

Landow, Noah M. [personal blog] http://www.macktez.com/users/noah/.

Lanham, Richard A. *The Electronic Word: Democracy, Technology, and the Arts.* Environment: Voyager Expanded Book. Chicago: University of Chicago Press, 1993.

Larsen, Deena. *Marble Springs.* Environment: Storyspace. Watertown: Eastgate Systems, 1994.

————. *Stained Word Window.* 1999. http://wordcircuits.com/gallery/stained/index.html.

Liestøl, Gunnar. *Kon-Tiki Interactive.* CD-ROM. Oslo: Glyndendal Norsk Forlag, 1996.

Lyons, Ian M. *(box(ing)).* Three versions: (1) Visual Basic, (2) Storyspace 2.0, (3) html. 2004.

————. *The Memory Myth Lines.* Director. Macintosh and Windows CD-ROM. 2004.

McDermott, Anne, ed. Samuel Johnson. *A Dictionary of the English Language on CD-ROM.* Environment: DynaText. Cambridge: Cambridge University Press, 1996.

McLaughlin, Tim. *Notes toward Absolute Zero.* Environment: Storyspace. Watertown: Eastgate Systems, 1996.

Marshall, Catherine C., and Frank M. Shipman III. "Spatial Hypertext." http://bush.cs.tamu.edu/~marshall/viki-sidebar.html.

Menezes, Philadelpho, and Wilton Azvedo. *Interpoesia: Poesia Hipmedia Interativa.* CD-ROM. San Paulo: Universitade Presbiteriana Mackenzie, 1998.

Microsoft Art Gallery. Environment by Cognitive Resources, Brighton, England. CD-ROM. N.P: Microsoft, 1993.

Miles, Adrian. "Childmovies." *http://hypertext.rmit.edu.au/vog/vlog/archive/2002/102002.html#3923.*

————. *Videoblog::VOG.* http://hypertext.rmit.edu.au/vog/.

————. *Vlog 2.1.* http://hypertext.rmit.edu.au/vog/vlog/.

Molloy, Judy. *Baithouse.* http://137.132.114.30/~molloy/tain/L01/Element01.htm.

BIBLIOGRAPHY

Molloy, Judy, and Cathy Marshall. *Forward Anywhere*. Environment created specifically for project. Watertown: Eastgate Systems, 1996.

Morrissey, Judd, and Lori Talley. *My Name is Captain, Captain*. Macintosh and Windows CD-ROM. Cambridge: Eastgate Systems, 2002.

Moulthrop, Stuart. *Hegirascope:* A Hypertext Fiction. Version 2. 1997.

———. *Victory Garden*. Environment: Storyspace. Cambridge: Eastgate Systems, 1991.

———. *Forking Paths: An Interaction*. Environment: Storyspace Beta 3.3. Jackson, Mich.: Riverrun Limited, 1987.

Myst. CD-ROM. Novato, Calif.: Broderbund, 1993.

NASA. *Mars Pathfinder: Quicktime VR Panoramas*. http://mpfwww.jpl.nasa.gov/MPF/vrml/ qtvr_monster_lg.html.

Nelson, Theodor Holm. "A Cosmology for a Different Computer Universe: Data Model, Mechanisms, Virtual Machine and Visualization Infrastructure." *Journal of Digital Information*. Article No. 298, 2004-07-16.

New Oxford Annotated Bible with the Apocrypha. New Revised Standard Version. Environment: CompLex. New York: Oxford, 1995.

Obendorf, Hartmut. "The Indirect Authoring Paradigm—Bringing Hypertext into the Web." *Journal of Digital Information*. Article No. 249, 2004-04-07.

Odin, Jaishree K. "The Performative and Processual: A Study of Hypertext/Postcolonial Aesthetic." *Postcolonial Web*. http://www.postcolonialweb.org/poldiscourse/odin/odin1.html.

Pack, Jeffrey. *Growing up Digerate*. http://www.cyberartsweb.org/cpace/infotech/digerate/TITLE.HTML.

Paul, Christiane. *Unreal City: A Hypertext Guide to T. S. Eliot's "The Waste Land."* Environment: Storyspace. Watertown: Eastgate Systems, 1995.

Pepys, Samuel. *The Diary* [as a Weblog]. Edited by Phil Gyford. http://www.pepysdiary.com/.

Perseus 1.0b1: Interactive Sources and Studies on Ancient Greek Civilization. Beta prerelease version. Developers: Gregory Crane and Elli Mylonas. Environment: HyperCard. Washington, D.C.: Annenberg/ Corporation for Public Broadcasting Project, 1990.

Pilgrim, Mark. "What is RSS?" O'Reilly XML.com. http://www.xml.com/pub/a/2002/12/18/dive-into-xml.html.

Privacy Rights Clearing House. "Thirty-One Privacy and Civil Liberties Organizations Urge Google to Suspend Gmail." April 19, 2004. http://www.privacyrights.org/ar/GmailLetter.htm.

Pullinger, Kate, and Talan Memmot. *Branded*. Trace. 2003. http://trace.ntu.ac.uk/frame/branded/index.html.

Rappaport, Joshua. *Hero's Face* in *Writing at the Edge*. Edited by George P. Landow. Storyspace. Cambridge: Eastgate Systems, 1994.

Residents, The. *Freak Show*. CD-ROM. New York: Voyager, 1994.

Roach, Greg. *The Wrong Side of Town (1996)*. http://www.hyperbole.com/lumiere/wrongside.html.

BIBLIOGRAPHY

Robinson, Peter, ed. Chaucer, *The Wife of Bath's Prologue on CD-ROM.* Environment: DynaText. Cambridge: Cambridge University Press, 1996.

Rosenzweig, Roy, Steve Brier, and John Brown. *Who Built America? From the Centennial Celebration of 1876 to the War of 1914.* CD-ROM. New York: Voyager, 1993.

Sanford, Christy Sheffield. *Safara in the Beginning.* http:// gnv.fdt.net/~christys/ safara.html.

Schraefel, M.C., Leslie Carr, David De Roure, and Wendy Hall. "You've Got Hypertext." *Journal of Digital Information.* Article No. 253, 2004-07-16

Seid, Tim. *Interpreting Manuscripts.* Environment: HyperCard. Providence, R.I., 1990.

Strickland, Stephanie. *Vniverse.* 2001 (?). http://www.cddc.vt.edu/journals/newriver/ strickland/vniverse/index.html.

Tamblyn, Christine, Marjorie Franklin, and Paul Tompkins. *She Loves it, She Loves it Not: Women and Technology.* CD-ROM. San Francisco, 1993.

Thomas, Brian. *If Monks Had Macs.* Environment: HyperCard. RiverTEXT.

Thompson, Ewa M. "Russian Identity, Nationalism, Colonialism and Postcolonialism." *Postcolonial Web.*http://www.postcolonialweb.org/poldiscourse/ewt/1.html.

Victorian Web, The. http://www.victorianweb.org. 1992-.

Vizability. Alpha version. Boston: PWS Publishing, 1995.

Wardrip-Fruin, Noah. "Interview [on Cave-writing]." *Iowa Review.* n.d. http:// www.uiowa.edu/~iareview/tirweb/feature/cave/.

Wardrip-Fruin, Noah, Andrew McClain, Shawn Greenlee, Robert Coover, and Josh Carroll. *Screen.* Brown University Cave (virtual reality environment). 2003.

Winkler, Owen. "Blog Software Breakdown." *Asymptomatic.net* http://www .asymptomatic.net/blogbreakdown.htm.

Winter, Robert. *CD Companion to Beethoven Symphony No. 9:* Environment: HyperCard. Santa Monica, Calif.: Voyager, 1989.

A Wrinkle in Time: a collaborative synchronized effort by QTVR producers around the globe. ["December 21st, 1997. Your world, as captured in three dimensions."] http://www.hotspots.hawaii.com/wrinkle.html.

Yun, David. *Subway Story: An exploration of me, myself and I.* 1997. http://www .cyberartsweb.org/cpace/ht/dmyunfinal/frames.html.

Index

436

DATE DUE

IN# 46544		
6/14/16		
		PRINTED IN U.S.A.